Microbial Extracellular Polymeric Substances

Characterization, Structure and Function

Springer

Berlin
Heidelberg
New York
Barcelona
Hong Kong
London
Milan
Paris
Singapore
Tokyo

Jost Wingender · Thomas R. Neu · Hans-Curt Flemming

Microbial Extracellular Polymeric Substances

Characterization, Structure and Function

With 42 Figures

 Springer

Editors:
Dr. Jost Wingender
Dept. of Aquatic Microbiology
University of Duisburg and
IWW Center for Water Research
Geibelstraße 41
D-47057 Duisburg
Germany

Dr. Thomas R. Neu
Dept. of Inland Water Research Magdeburg
UFZ Centre for Environmental Research
Leipzig-Halle
Brückstraße 3 a
D-39114 Magdeburg
Germany

Prof. Dr. Hans-Curt Flemming
Dept. of Aquatic Microbiology
University of Duisburg and
IWW Center for Water Research
Geibelstraße 41
D-47057 Duisburg
Germany

ISBN 3-540-65720-7   Springer-Verlag Berlin Heidelberg New York

Library of Congress Cataloging-in-Publication Data
Microbial extracellular polymeric substances : characterization, structure, and function / Jost Wingender,
   Thomas R. Neu, Hans-Curt Flemming [editors].
   Includes bibliographical references and index.
   ISBN 3-540-65720-7 (alk. paper)
   1. Microbial polymers. 2. Biofilms. I. Wingender, Jost. II. Neu, Thomas R.. III. Flemming, Hans-Curt
QR92.P59.M53   1999   572'.29 – dc21   99-38891

Production Editor: ProduServ GmbH Verlagsservice, Berlin
Typesetting: Fotosatz-Service Köhler GmbH, Würzburg; Cover design: de'blik, Berlin

SPIN: 10552392   2/3020 - 5 4 3 2 1 0 – Printed on acid-free paper

Preface

The variety of microbial aggregates such as biofilms, flocs and sludge are kept together by extracellular polymeric substances (EPS). They represent the construction material which allows the cells to maintain stable microconsortia and to establish synergistic relationships. The EPS play a key role for the understanding of structure, function, properties and development of microbial aggregates. The EPS fill and form the space between the biofilm constituents, e.g., prokaryotic and eukaryotic microorganisms which inhabit the aggregates. It seems that the EPS matrix may serve as a multipurpose functional element of microbial communities, including adhesion, structure, protection, recognition and physiology. Due to the metabolic activity of the cells, gradients develop and create different habitats within small distances, allowing a wide variety of organisms to settle and grow in the aggregate. As many of these organisms produce their specific EPS, it is not surprising that an extremely wide variety of microbial aggregates result. All of them have in common their highly hydrated extracellular matrix which is the place the organisms shape their very own microhabitat in which they live.

The paramount role of the EPS has long been neglected and has been acknowledged only recently. In the first chapter, an attempt is made to give a definition and an overview of what EPS represents, of what they are composed, what their function is, and which role they play in microbial ecology. A large part of the book is devoted to the isolation and characterization of EPS from different systems. The matrix does not seem to be built randomly by some biopolymers but displays functional aspects. This is considered as well as the particular aspect of cell-cell-communication which is of special interest in aggregates where the cells maintain high densities and close proximities. Extracellular enzymes and polysaccharides can interact and this, supports the view that the EPS matrix must be probably recognized as a key factor in a model which considers biofilms and other microbial aggregates as an early form of multicellular organisms.

The purpose of this book is to present information and concepts with regard to current state of knowledge about EPS and, consequently bring the attention of the readers to the material between the cells, which may be of more importance than previously assumed. If it inspires the further understanding of the aggregated way of microbial life, it has fulfilled its purpose.

August 1999

Jost Wingender
Thomas R. Neu
Hans-Curt Flemming

Contents

Polysaccharases in Biofilms – Sources – Action – Consequences!

Extracellular Enzymes Within Microbial Biofilms
and the Role of the Extracellular Polymer Matrix

Interaction Between Extracellular Polysaccharides and Enzymes

Contributors

BEECH, I. B.
(e-mail: Iwona.Beech@port.ac.uk)
University of Portsmouth
Microbiology Research Laboratoy
St. Michaels' Building
White Swan Road
Portsmouth, PO1 2DT
UK

CHRISTENSEN, B. E.
(e-mail: Bjoern.E.Christensen@chem-
bio.ntnu.no)
Norwegian Biopolymer Laboratory
(NOBIPOL)
Department of Biotechnology
Norwegian University of Science
and Technology
N-7034 Trondheim
Norway

DAVIES, D. G.
(e-mail: dgdavies@binghamton.edu)
Department of Biological Sciences
Science III
Binghamton University
Binghamton, N.Y. 13902
USA

DECHO, A. W.
(e-mail: adecho@sph.sc.edu)
Department of Environmental Health
Sciences
School of Public Health
University of South Carolina
Columbia, SC 29208
USA

FLEMMING, H.-C.
(e-mail:
100606.3337@compuserve.com)
Department of Aquatic Microbiology
University of Duisburg and
IWW Center for Water Research
Geibelstrasse 41
D-47057 Duisburg
Germany

GEHRKE, T.
(e-mail: FB6a042@mikrobiologie.uni-
hamburg.de)
Institute of Microbiology
University of Hamburg
Ohnhorststrasse 18
D-22609 Hamburg
Germany

HOFFMANN, M.
(e-mail: adecho@sph.sc.edu)
Department of Environmental Health
Sciences
School of Public Health
University of South Carolina
Columbia, SC 29208
USA

JAEGER, K.-E.
Chair Biology of Microorganisms
Ruhr-University Bochum
D-44780 Bochum
Germany

Jahn, A.
(e-mail: i5phn@civil.auc.dk)
Environmental Engineering
Laboratory
Aalborg University
Sohngaardsholmsvej 57
DK-9000 Aalborg
Denmark

Korber, D.R.
Department of Applied Microbiology
and Food Science
University of Saskatchewan
Saskatoon, SK S7N 5A8
Canada

Lawrence, J.R.
National Hydrology Research Center
11 Innovation Blvd.
Saskatoon, SK, S7N 3H5
Canada

Neu, T.R.
(e-mail: neu@gm.ufz.de)
Department of Inland Water
Research Magdeburg
UFZ Centre for Environmental
Research
Leipzig-Halle
Brückstrasse 3 a
D-39114 Magdeburg
Germany

Nielsen, P.H.
(e-mail: i5phn@civil.auc.dk)
Environmental Engineering
Laboratory
Aalborg University
Sohngaardsholmsvej 57
DK-9000 Aalborg
Denmark

Sand, W.
(e-mail: FB6a042@mikrobiologie.uni-
hamburg.de)
Institute of Microbiology
University of Hamburg
Ohnhorststrasse 18
D-22609 Hamburg
Germany

Sutherland, I.W.
(e-mail: I.W. Sutherland@ed.ac.uk)
Institute of Cell and Molecular
Biology
Edinburgh University
Mayfield Road
Edinburgh EH9 3JH
UK

Tapper, R.C.
(e-mail: Rudi.Tapper@port.ac.uk)
University of Portsmouth
Microbiology Research Laboratory
St. Michaels' Building
White Swan Road
Portsmouth, PO1 2DT
UK

Wingender, J.
(e-mail: hh239wi@uni-duisburg.de)
Department of Aquatic Microbiology
University of Duisburg and
IWW Center for Water Research
Geibelstrasse 41
D-47057 Duisburg
Germany

Wolfaardt, G.M.
(e-mail: gmw@land.sun.ac.za)
Department of Microbiology
University of Stellenbosch
7600 Stellenbosch
South Africa

CHAPTER 1

What are Bacterial Extracellular Polymeric Substances?

Jost Wingender[1] · Thomas R. Neu[2] · Hans-Curt Flemming[1]

[1] Department of Aquatic Microbiology, University of Duisburg and IWW Center for Water Research, Geibelstrasse 41, D-47057 Duisburg, Germany, *E-mail: hh239wi@uni-duisburg.de* (Wingender), *E-mail: 100606.3337@compuserve.com* (Flemming)
[2] Department of Inland Water Research Magdeburg, UFZ Centre for Environmental Research Leipzig-Halle, Brückstrasse 3a, D-39114 Magdeburg, Germany, *E-mail: neu@gm.ufz.de*

Keywords. Extracellular polymeric substances, EPS, Polysaccharide, Alginate, Protein, Biofilm, Immobilization, Aggregates

1
Introduction

The vast majority of microorganisms live and grow in aggregated forms such as biofilms and flocs ("planktonic biofilms"). This mode of existence is lumped in the somewhat inexact but generally accepted expression "biofilm". The common feature of all these phenomena is that the microorganisms are embedded in a matrix of extracellular polymeric substances (EPS). The production of EPS is a general property of microorganisms in natural environments and has been shown to occur both in prokaryotic (Bacteria, Archaea) and in eukaryotic (algae, fungi) microorganisms. Biofilms containing mixed populations of these organisms are ubiquitously distributed in natural soil and aquatic environments, on tissues of plants, animals and man as well as in technical systems such as filters and other porous materials, reservoirs, plumbing systems, pipelines, ship hulls, heat exchangers, separation membranes, etc. (COSTERTON ET AL. 1987; 1995; FLEMMING AND SCHAULE 1996). Biofilms develop adherent to a solid surface (substratum) at solid-water interfaces, but can also be found at water-oil, water-air and solid-air interfaces. Biofilms are accumulations of microorganisms (prokaryotic and eukaryotic unicellular organisms), EPS, multivalent cations, biogenic and inorganic particles as well as colloidal and dissolved compounds. EPS are mainly responsible for the structural and functional integrity of biofilms and are considered as the key components that determine the physicochemical and biological properties of biofilms. EPS form a three-dimensional, gel-like, highly hydrated and often charged biofilm matrix, in which the microorganisms are embedded and more or less immobilized. EPS create a microenvironment for sessile cells which is conditioned by the nature of the EPS matrix. In general, the proportion of EPS in biofilms can vary between roughly 50 and 90% of the total organic matter (CHRISTENSEN AND CHARACKLIS 1990; NIELSEN ET AL. 1997).

2
Definitions of EPS

Microbial EPS are biosynthetic polymers (biopolymers). EPS were defined (GEESEY 1982) as "extracellular polymeric substances of biological origin that participate in the formation of microbial aggregates". Another definition was given in a glossary to the report of the Dahlem Workshop on Structure and Function of Biofilms in Berlin 1988 (CHARACKLIS AND WILDERER 1989). Here, EPS were defined as "organic polymers of microbial origin which in biofilm systems are frequently responsible for binding cells and other particulate materials together (cohesion) and to the substratum (adhesion)".

The abbreviation "EPS" has been used for "extracellular polysaccharides", "exopolysaccharides", "exopolymers" and "extracellular polymeric substances". Polysaccharides have often been assumed to be the most abundant components of EPS in early biofilm research (e.g., COSTERTON ET AL. 1981). That may be the reason why the term "EPS" has frequently been used as an abbreviation for "extracellular polysaccharides" or "exopolysaccharides". However, proteins and nucleic acids (PLATT ET AL. 1985; FRØLUND ET AL. 1996; NIELSEN ET AL. 1997; DIGNAC ET AL. 1998) as well as amphiphilic compounds including (phospho)-lipids (NEU 1996; TAKEDA ET AL. 1998; Sand and Gehrke, this volume) have also been shown to appear in significant amounts or even predominate in EPS preparations from activated sludges, sewer biofilms, trickling filter biofilms, and pure cultures of bacteria. In addition, some researchers described humic substances as components of EPS matrices of soil and water biofilms (BURNS 1989; NIELSEN ET AL. 1997; JAHN AND NIELSEN 1998). In the following, the abbreviation "EPS" is used for "extracellular polymeric substances" as a more general and comprehensive term for different classes of organic macromolecules such as polysaccharides, proteins, nucleic acids, (phospho)lipids, and other polymeric compounds, which have been found to occur in the intercellular spaces of microbial aggregates. At present, other microbial biopolymers such as poly-β-hydroxyalkanoates are not normally regarded as EPS, since they are typically intracellular components of microbial cells. To the authors' knowledge, their extracellular occurrence in biofilms has not yet been reported.

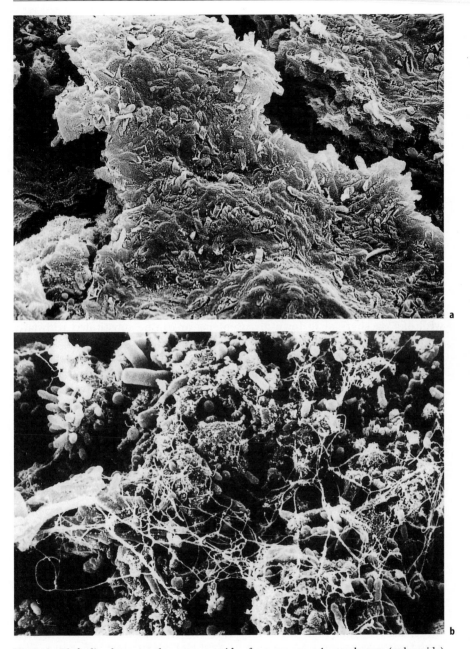

Fig. 1a, b. Biofouling layers on the raw water side of reverse osmosis membranes (polyamide) exposed to different river waters. (a: FLEMMING AND SCHAULE 1989; b: courtesy of G. Schaule)

The structure of EPS varies quite obviously. Figure 1a,b shows the scanning electron micrographs of biofouling layers on various reverse osmosis membranes. The supporting membrane material and the preparation procedures were identical; both membranes were exposed to river water, although of different origin. In Fig. 1a, the cells are embedded in a thick slime matrix, while in Fig. 1b fibrillar structures are predominant. Although it is acknowledged that the dewatering procedure required for scanning electron microscopy imaging produces artifacts, morphological differences are nevertheless obvious and are attributed to the nature of the EPS.

3
Composition, Secretion and Spatial Arrangement of EPS

EPS are organic macromolecules that are formed by polymerization of similar or identical building blocks (Table 1), which may be arranged as repeating units within the polymer molecules such as, e.g., in many polysaccharides. EPS may contain nonpolymeric substituents of low molecular weight which greatly alter their structure and physicochemical properties. Thus, extracellular polysaccharides often carry organic substituents such as acetyl, succinyl, or pyruvyl groups or inorganic substituents such as sulfate. Proteins can be glycosylated with oligosaccharides to form glycoproteins or can be substituted with fatty acids to form lipoproteins.

By definition, EPS are located at or outside the cell surface independent of their origin. The extracellular localization of EPS and their composition may be the result of different processes: active secretion, shedding of cell surface material, cell lysis, and adsorption from the environment.

EPS may be actively secreted by living cells. Various specific pathways of biosynthesis and discrete export machineries involving the translocation of EPS across bacterial membranes to the cell surface or into the surrounding medium have been described for bacterial proteins (for review see, e.g., BINET ET AL. 1997; HUECK 1998; FILLOUX ET AL. 1998) and polysaccharides (for review see, e.g., SUTHERLAND 1990; LEIGH AND COPLIN 1992; ROBERTS 1995, 1996; REHM AND VALLA 1997; BECKER ET AL. 1998; JONAS AND FARAH 1998). Extracellular DNA can be produced by bacteria during growth; it is not known whether DNA is actively secreted or passively released due to increase in cell envelope permeability (for review see LORENZ AND WACKERNAGEL 1994).

Another mechanism of release of extracellular polymers is the spontaneous liberation of integral cellular components such as lipopolysaccharides (LPS) from the outer membrane of Gram-negative bacteria (CADIEUX ET AL. 1983). This may occur through formation of outer membrane-derived vesicles (blebs) which has been described as a common secretion mechanism in Gram-negative bacteria (BEVERIDGE ET AL. 1997; LI ET AL. 1998). Surface blebbing occurs during normal growth and represents a process by which cellular macromolecules including periplasmic compounds and membrane components (nucleic acids, enzymes, LPS, phospholipids) are shed into the extracellular space in the form of membrane vesicles. Release of cellular material by this mechanism may be the result of metabolic turnover processes. Alternatively, membrane vesicles,

Table 1. General composition of bacterial EPS; humic substances are included in the table, since they are sometimes considered as part of the EPS matrix (BURNS 1989; NIELSEN ET AL. 1997; JAHN AND NIELSEN 1998)

EPS	Principal components (subunits, precursors)	Main type of linkage between subunits	Structure of polymer backbone	Substituents (examples)
Polysaccharides	monosaccharides uronic acids amino sugars	glycosidic bonds	linear, branched,	organic: O-acetyl, N-acetyl, succinyl, pyruvyl; inorganic: sulfate, phosphate
Proteins (polypeptides)	amino acids	peptide bonds	linear	oligosaccharides (glycoproteins), fatty acids (lipoproteins)
Nucleic acids	nucleotides	phosphodiester bonds	linear	–
(Phospho)lipids	fatty acids glycerol phosphate ethanolamine serine choline sugars	ester bonds	side-chains	–
Humic substances	phenolic compounds simple sugars amino acids	ether bonds, C-C bonds, peptide bonds	cross-linked	–

into which hydrolytic enzymes are packaged (e.g., peptidoglycan hydrolases), may serve to degrade surrounding cells in the biofilm ("predatory vesicles"; KADURUGAMUWA AND BEVERIDGE 1995; BEVERIDGE ET AL. 1997), liberating nutrients for the vesicle-forming biofilm bacteria.

Death and lysis of cells contribute to the release of cellular high-molecular-weight compounds into the medium and entrapment within the biofilm matrix. Typically, intracellular organic polymers like poly-β-hydroxyalkanoates or glycogen as carbon and energy storage polymers or integral components of cell walls (e.g., peptidoglycan) and membranes (e.g., phospholipids, LPS) may thus become part of the EPS. As a consequence, the biofilm represents a "recycling yard" for intracellular components.

Finally, EPS which are shed from microbial aggregates can be adsorbed in other places. Thus, the sites of synthesis, release, and ultimate localization of EPS components are not necessarily identical.

EPS can be subject to biotic (enzymatic) and/or abiotic degradation, modification and condensation processes. Different enzymes involved in polymer degradation (predominantly hydrolases, less frequently lyases and other enzymes) are abundant in biofilms (Wingender et al., this volume). Degradation of EPS can have various consequences such as the formation of low-molecular-weight cleavage products as potential carbon and energy sources for biofilm bacteria, the sloughing of cells due to depolymerization of structural biofilm polymers (XUN ET AL. 1990; BOYD AND CHAKRABARTY 1994) or the inactivation of extracellular enzyme proteins. Chemical modification of polymers after polymerization occurs in the periplasm of Gram-negative cells (e.g., acetylation of polysaccharides), at the cell surface, or even in the extracellular space (e.g., the conversion of secreted polymannuronate to alginate by mannuronan C-5-epimerase activity in *Azotobacter vinelandii*). Humic substances are examples for polymers, whose chemical structure is supposed to be the result of partial enzymatic degradation of various biopolymers and condensation ("repolymerization") of refractive small organic breakdown products by spontaneous abiotic and enzymatic processes (HEDGES 1988), which may possibly proceed in soil and water biofilms. Thus, the composition and distribution of EPS in microbial aggregates may vary spatially and temporally depending on the prevailing polymerizing, modifying, and degrading activities of the biofilm.

EPS lie outside the cell wall (pseudomurein layer) of archaeobacteria, the cell wall (peptidoglycan layer) of Gram-positive eubacteria and the outer membrane of Gram-negative eubacteria. Since polysaccharides were identified as common constituents of bacterial EPS, the term "glycocalyx" was introduced for the description of polysaccharide-containing structures of bacterial origin lying outside the integral cell surface components of bacteria in analogy to the earlier use of this term for carbohydrate-containing structures on plant and animal cell surfaces (COSTERTON ET AL. 1981, 1992). However, it must be acknowledged that biopolymers other than polysaccharides may occur in substantial amounts or even predominate as EPS components as mentioned above. The term "glycocalyx" encompasses different EPS-containing structures such as capsules, sheaths, and slimes. Capsules ("integral capsules", COSTERTON ET AL. 1992) are normally firmly associated with the cell surface as discrete structures with distinct outlines. Capsular polymers are attached to the cell surface by noncovalent interactions, but may also be covalently bound to phospholipid or lipid-A molecules at the cell surface (ROBERTS 1996). Filamentous bacteria such as certain cyanobacteria or bacteria of the genera *Leptothrix* and *Sphaerotilus* (TAKEDA ET AL. 1998) possess sheaths as linear EPS-containing structures surrounding chains of cells analogous to capsules in nonfilamentous organisms. EPS may only be loosely attached to the cell surface as peripheral capsules (COSTERTON ET AL. 1992) and can be shed into the surrounding environment as a less organized (amorphous) slime. S-layers are also regarded as possible glycocalyx components (COSTERTON ET AL. 1981). S-layers are monomolecular regular assemblies of proteins or glycoproteins, which are associated through noncovalent interactions with the underlying integral cell surface components (SLEYTR 1997). S-layers are ubiquitous as one of the most common cell surface structures in Archaea and Bacteria (SLEYTR 1997; SIDHU AND OLSEN 1997). It was speculated that in *Bacillus* species one of the several functions attributed to the S-layer may be the linking of the capsule to the peptidoglycan layer (SIDHU AND OLSEN 1997). In *Bacillus anthracis* it was observed that the S-layer was formed in the presence of a poly-γ-D-glutamic acid capsule; thus, the two structures were compatible and neither was required for the correct formation of the other (MESNAGE ET AL. 1998). For prokaryotes, "mucilage" is another more general term that includes all EPS-containing structures mentioned above. Microbial mucilages made up of individual macromolecules and colloidal aggregates of them may develop to massive accumulations, which have been studied extensively in aquatic ecosystems (LEPPARD 1995). In all cases, EPS form the outermost surface layers of bacteria as boundary structures mediating contact and exchange processes with their biotic and abiotic environments.

EPS are not essential structures of bacteria in laboratory cultures, since loss of EPS does not impair growth and viability of the cells. Under natural conditions, however, EPS production seems to be an important feature of survival, as most bacteria occur in microbial aggregates such as flocs and biofilms, whose structural and functional integrity is based essentially on the presence of an EPS matrix. Although EPS confer a macroscopically slimy appearance to bacterial masses in many environments and bacteria usually occur in an aggregated state where the cells are surrounded by EPS-containing material, the ubiquity and prime importance of EPS-containing structures has long been overlooked for several reasons. Traditionally, microbiologists used to study and to subculture individual bacterial strains in pure cultures using artificial growth media. Under these in vitro conditions, bacterial isolates did not express EPS-containing structures or even lost their ability to produce EPS. It has been speculated that the energy-requiring process of EPS production does not confer any selective advantage to cells grown in the laboratory in contrast to the situation in competitive multispecies environments in nature and disease, where EPS functions seem to be essential for survival of bacterial populations (COSTERTON ET AL. 1987). Thus, extrapolating data from laboratory studies to EPS-dependent processes in natural and pathogenic environments gave results that did not reflect the actual mode of bacterial existence and activities under real conditions. Furthermore, EPS were originally inadequately visualized by conventional light and electron microscopy (COSTERTON ET AL. 1981). Conventional optical microscopy of microbial aggregates showed that the pores between the cells seemed to be devoid of physical structures. Furthermore, due to its high water content (up to 99%) the glycocalyx collapsed under the dehydrating conditions of sample preparation for electron microscopy. Only stabilization of the glycocalyx by the use of antibodies or lectins and staining of EPS with dyes such as the commonly employed polyanion-specific ruthenium red allowed the visualization of fibrous material lying in the light-microcopically transparent space between the cells and gave an idea of the dimensions of EPS-containing structures around the cells (COSTERTON ET AL. 1992; JACQUES AND GOTTSCHALK 1997). It is assumed that the fibrils are colloidal aggregates of high-molecular-weight EPS molecules (predominantly acidic polysaccharides, but also proteins) representing the dominant physical bridging mechanism between cells and inorganic components within biofilms (LISS ET AL. 1996; LEPPARD 1997). EPS fibrils display a wide variety of patterns in electron micrographs reflecting the different chemical compositions of EPS in mixed biofilm populations (COSTERTON ET AL. 1978; LEPPARD 1995). Examination of different environments by these techniques demonstrated the universal occurrence of EPS-containing structures of those bacteria which colonized, in the form of microcolonies or biofilms, surfaces of natural aquatic environments and technical water systems, the tissue surfaces of plants, animals, and man as well as surfaces of medical devices (COSTERTON ET AL. 1978, 1981, 1987, 1990, 1995; LEPPARD 1995; LISS ET AL. 1996; MORRIS ET AL. 1997). Only recently, confocal laser scanning microscopy (CSLM) in conjunction with fluorescent chemical probes enabled examination of the three-dimensional structure of fully hydrated and intact bacterial biofilms (COSTERTON ET AL. 1992, 1994, 1995; LAWRENCE ET AL. 1991; MORRIS ET AL. 1997). Application of this nondestructive technique changed the traditional view of the homogeneous biofilm as mainly maintained by researchers investigating biofilm reactors, but not shared by light microscopists (Marshall, personal communication; Szewzyk, personal communication). CSLM showed that within biofilms bacteria grew as distinct matrix-enclosed microcolonies that were separated by less dense regions of the biofilm including water channels and pores (COSTERTON ET AL. 1994). The use of fluorescently labeled EPS-specific compounds such as carbohydrate-binding lectins or polyanionically charged dextran allowed study of the extent and spatial arrangements of EPS in natural biofilms and confirmed the heterogenous distribution of EPS within the biofilm matrix (KORBER ET AL. 1995; NEU AND LAWRENCE 1997; WOLFAARDT ET AL. 1998; Neu and Lawrence, this volume).

Different isolation techniques were developed to recover EPS from pure cultures, mixed-species flocs and biofilms from various environments, aiming to determine the quantitative composition of EPS, to elucidate the chemical structure of EPS, to examine the physico-chemical properties of EPS and to study biological functions of EPS (e.g., GEHR AND HENRY 1983; LAZAROWA AND MANEM 1995; JAHN AND NIELSEN 1995; FRØLUND ET AL. 1996; AZEREDO ET AL. 1998; DIGNAC ET AL. 1998; Nielsen and Jahn, this volume). For these studies, EPS were usually extracted and separated from the cells (e.g., by blending, sonication, cation exchange resin treatment, centrifugation, and membrane filtration techniques, alone or in combination), before the composition of the EPS was analyzed, mainly by colorimetric and chromatographic techniques. When single EPS were studied in purified form, they were separated from other molecules, so that the native state of the EPS-containing structures was disrupted and interactions between different EPS were not considered. It has been observed that the frequent isolation process consisting of gravitational separation of EPS from cells of *Pseudomonas diminuta* with subsequent precipitation of polymers from the supernatant by addition of alcohol or acetone led to the disruption of the macromolecules (Schaule and Flemming, unpublished).

Research into EPS composition has been focused on the study of polysaccharides and also to some extent of proteins including enzymes. In microscopic studies, mainly polysaccharide-specific stains were employed to visualize EPS structures. Biochemical analyses of EPS were often restricted to the quantification of total carbohydrate, uronic acids, and proteins, whereas other polymeric constituents of the EPS such as DNA or lipids were only seldom studied.

4
Industrial and Clinical Importance of EPS

Among bacterial EPS, interest has focused particularly on selected bacterial extracellular polysaccharides which are of commercial interest in biotechnology for various industrial and biomedical applications or are of clinical relevance as virulence factors participating in infection processes of plants, animals, and man. The wealth of information about bacterial extracellular polysaccharides with respect to their genetics, biosynthesis, secretion, structure, functions in natural, technical, and pathogenic environments as well as their importance in industrial and medical applications is reflected by numerous review articles (selection of some recent reviews: SUTHERLAND 1982, 1983, 1985, 1994, 1996, 1998; KENNE AND LINDBERG 1983; ISAAC 1985; WHITFIELD 1988; CHRISTENSEN 1989; BERTOCCHI ET AL. 1990; LINDBERG 1990; NEU AND MARSHALL 1990; LEIGH AND COPLIN 1992; ROBERTS 1995, 1996; WHITFIELD AND VALVANO 1993; WEINER ET AL. 1995). Books on microbial polysaccharides providing an introduction to the various aspects of carbohydrate polymers have been published (SUTHERLAND 1977, 1990).

As to their commercial exploitation, bacterial extracellular polysaccharides such as xanthan, gellan, cellulose, hyaluronic acid, and several other β-D-glucans have found various applications, whereas newly discovered polysaccharides and chemical modifications of established polysaccharides offer the potential of novel applications in the future (e.g., BECKER ET AL. 1998; JONAS AND FARAH 1998; SUTHERLAND 1998; DE PHILIPPIS AND VINCENZINI 1998). In infection processes, extracellular polysaccharides may represent virulence factors as capsular and slime polysaccharides, mediating adhesion to tissue surfaces and protecting the invading bacteria from host defense mechanisms (COSTERTON ET AL. 1987; ROBERTS 1996) by allowing the bacteria to live and grow in the microcolony or biofilm mode of growth (COSTERTON ET AL. 1990) and by modulating the host immune response (PETERS ET AL. 1989; PASQUIER ET AL. 1997).

Among the extracellular proteins, enzymes from fungi and bacteria are of special economic interest. Microbial exoenzymes like starch-hydrolyzing enzymes, proteases, cellulases, pectinases, and lipases are produced on an industrial scale for applications in the hydrolysis of macromolecules, but also offer great promise for biosynthetic purposes (PRIEST 1992; JAEGER AND REETZ 1998). Extracellular enzyme and toxin proteins are also involved as virulence factors in the colonization, invasion and destruction of host tissues and in the interference with host immune response mechanisms.

5
Bacterial Alginate – an Example of Bacterial EPS

Bacterial alginates represent an example of a few extracellular polysaccharides, which have been studied in detail under the aspects of their relevance as a general virulence factor in infection processes of plants, animals, and man as well as in terms of their potential commercial exploitation. Alginates belong to the best studied bacterial extracellular polysaccharides and have often been used as model compounds in the study of the physicochemical and biological properties of EPS. Extensive research effort has been focused on alginate from the opportunistic Gram-negative bacterium *Pseudomonas aeruginosa* (GACESA AND RUSSELL 1990). Mucoid strains of this species are characterized by the overproduction of the extracellular polysaccharide alginate (GOVAN 1990). In addition to various extracellular enzymes, alginate has been shown to be an important virulence factor in chronic lung infections of patients with the hereditary disease cystic fibrosis (CF) (GOVAN AND DERETIC 1996). Bacteria colonizing the lung tissue surface are protected in alginate slime-containing microcolonies from the immune defence of the host, resulting in long-term destruction of the lungs by the activated immune system. Alginate-mediated lung infections with mucoid *P. aeruginosa* are the main reason for the premature death of CF patients. This was one reason why substantial efforts have been undertaken since the beginning of the 1980s to elucidate alginate biosynthesis with the aim of finding a therapeutic strategy to prevent the bacterial infection or to eradicate mucoid bacteria from CF patients. Another reason was the potential of microbially produced alginates as industrial polymers. Over the years progress with alginate research has become more rapid and a substantial amount of information about the genetics, biosynthesis, regulation of biosynthesis, physiological roles, and physicochemical properties of alginates, not only from *P. aeruginosa*, but also from other *Pseudomonas* and *Azotobacter* species has accumulated (recent reviews: MAY AND CHAKRABARTY 1994; REHM AND VALLA 1997; GACESA 1998; Davies, this volume).

P. aeruginosa revealed several advantages as a research object: this organism is well characterized with respect to its molecular genetics and physiology, it is ubiquitous in soil and aquatic environments, it can survive and grow in technical water systems, it is able to form biofilms, it secretes a number of various extracellular products (e.g., enzymes, polysaccharides, glycolipids) simultaneously, it is of hygienic relevance, and is involved in infections of plants, animals, and man. Mucoid *P. aeruginosa* as a model organism for EPS-forming biofilm bacteria have been employed in numerous laboratory studies including biofilm research, so that concepts on the functions of EPS are often based on results from alginate research. However, extrapolation of these findings to EPS-mediated processes in natural and engineered environments or under pathogenic conditions are probably of limited value due to the complexity and variety of EPS structures in natural biofilms. Thus, modeling the structure, dynamics, and functions of EPS by using pure cultures of alginate-producing *P. aeruginosa* can only be incomplete. Research into some other capsular polysaccharides like those from *Escherichia coli* (K antigens) and into slime polysaccharides such as xanthan, succinoglycan, cellulose, or more complex capsular polysaccharides from *Rhizobium* species have given some ideas about the genetics and biosynthesis of some other extracellular polysaccharides (WHITFIELD 1988; LEIGH AND COPLIN 1992; ROBERTS 1996; KATZEN ET AL. 1998) which are important in pathogenic processes and as industrial polymers.

6
Functions of EPS

After discovering the universal presence of EPS-containing structures, some general functions have been attributed to EPS such as the formation of a gel-like network keeping the biofilm bacteria together, the mediation of adherence of biofilms to surfaces, the involvement in the establishment of infections, and the protection of bacteria against noxious influences from the environment. Meanwhile, extensive studies have resulted in a number of different and more detailed proposals for possible functions of EPS for biofilm bacteria, some of which are summarized in Table 2 and are discussed in several review articles (e.g., WHITFIELD 1988; CHRISTENSEN 1989; WEINER ET AL. 1995; DE PHILIPPIS AND VINCENZINI 1998; Wolfaardt et al., this volume).

In general, one of the most important functions of extracellular polysaccharides is supposed to be their role as fundamental structural elements of the EPS matrix determining the mechanical stability of biofilms, mediated by non-covalent interactions (FLEMMING 1996; MAYER ET AL. 1999) either directly between the polysaccharide chains or indirectly via multivalent cation bridges. More recent studies suggest that lectin-like proteins also contribute to the formation of the three-dimensional network of the biofilm matrix by cross-linking polysaccharides directly or indirectly through multivalent cations bridges (HIGGINS AND NOVAK 1997). Among activated sludge extracellular polymers, proteins predominated and, on the basis of their relatively high content of negatively charged amino acids, they were supposed to be more involved than sugars in electrostatic bonds with multivalent cations, underlining their key role in the floc structure (DIGNAC ET AL. 1998). In addition, proteins have also been suggested to be involved in hydrophobic bonds within the EPS matrix (DIGNAC ET AL. 1998). However, the main function of extracellular proteins in biofilms is mostly seen in their role as enzymes performing the digestion of exogenous macromolecules and particulate material in the microenvironment of the immobilized cells. Thus, they provide low-molecular-weight nutrients which can readily be taken up and metabolized by the cells. Enzymes within the biofilm matrix may also be involved in the degradation of polysaccharidic EPS causing the release of biofilm bacteria and the spreading of the organisms to new environments (BOYD AND CHAKRABARTY 1994; XUN ET AL. 1990).

A function frequently attributed to EPS is their general protective effect on biofilm organisms against adverse abiotic and biotic influences from the environment (Table 2). As an example, it has frequently been observed that biofilm cells can tolerate significantly higher concentrations of certain biocides including disinfectants and antibiotics than planktonic populations (LECHEVALLIER ET AL. 1988; FOLEY AND GILBERT 1996). This is supposed to be due mainly to physiological changes of biofilm bacteria enhancing their resistance

Table 2. Selection of proposed effects due to microbial EPS

Function	Relevance
Adhesion to surfaces	Initial step in colonization of inert and tissue surface, accumulation of bacteria on nutrient-rich surfaces in oligotrophic environments
Aggregation of bacterial cells, formation of flocs and biofilms	Bridging between cells and inorganic particles trapped from the environment, immobilization of mixed bacterial populations, basis for development of high cell densities, generation of a medium for communication processes, cause for biofouling and biocorrosion events
Cell-cell recognition	Symbiotic relationships with plants or animals, initiation of pathogenic processes
Structural elements of biofilms	Mediation of mechanical stability of biofilms (frequently in conjunction with multivalent cations), determination of the shape of EPS structure (capsule, slime, sheath)
Protective barrier	Resistance to nonspecific and specific host defenses (complement- mediated killing, phagocytosis, antibody response, free radical generation), resistance to certain biocides including disinfectants and antibiotics, protection of cyanobacterial nitrogenase from harmful effects of oxygen
Retention of water	Prevention of desiccation under water-deficient conditions
Sorption of exogenous organic compounds	Scavenging and accumulation of nutrients from the environment, sorption of xenobiotics (detoxification)
Sorption of inorganic ions	Accumulation of toxic metal ions (detoxification), promotion of polysaccharide gel formation, mineral formation
Enzymatic activities	Digestion of exogenous macromolecules for nutrient acquisition, release of biofilm cells by degradation of structural EPS of the biofilm
Interaction of polysaccharides with enzymes	Accumulation/retention and stabilization of secreted enzymes

to biocides, but also to a barrier function of EPS (BROWN AND GILBERT 1993; MORTON ET AL. 1998). It is assumed that the EPS matrix delays or prevents biocides from reaching target microorganims within the biofilm by diffusion limitation and/or chemical interaction with EPS molecules. In the case of chlorine, it has been shown (DE BEER ET AL. 1994) that the chlorine demand by polysaccharide may lead to biofilm areas with low concentrations of chlorine in which the cells can survive. In mucoid *P. aeruginosa*, the protective effect of slime against chlorine was supposed to be based on a chemical reaction of the biocide with alginate as the major slime component resulting in the neutralization of chlorine (WINGENDER ET AL. 1999). However, it is noteworthy that in the case of hydrogen peroxide, mucoid strains of *P. aeruginosa* were even more susceptible than isogenic nonmucoid strains and no chemical reaction between the biocide and alginate was observed (WINGENDER ET AL. 1999). These results demonstrate that the contribution of EPS to the response of bacteria against biocides varies depending on the properties of the biocide applied. Other protective functions of EPS include the contribution to bacterial evasion from various host defense mechanisms (SMITH AND SIMPSON 1990), the protection from desiccation due to enhanced water retention (OPHIR AND GUTNICK 1994), or the prevention of oxygen-mediated inhibition of nitrogen fixation in cyanobacteria (DE PHILIPPIS AND VINCENZINI 1998).

The role of EPS components other than polysaccharides and proteins remains to be established. However, it is expected that EPS such as nucleic acids and lipids significantly influence the rheological properties and thus the stability of biofilms as can be deduced from basic laboratory studies on the properties of polymer mixtures. The colonization of surfaces may also be determined by EPS other than polysaccharides or proteins. Extracellular lipids from *Serratia marcescens* with surface-active properties (serrawettins) have been proposed to help bacteria in surface environments to overcome the strong surface tension of surrounding water, thus facilitating growth on solid surfaces (MATSUYAMA AND NAKAGAWA 1996). In general, extracellular surface-active polymers with lipidic components seem to be involved in the interaction between bacteria and interfaces (NEU 1996).

7
Ecological Aspects of EPS

From an ecological point of view, the EPS matrix provides the possibility that the microorganisms can form stable aggregates of different cells, leading to synergistic microconsortia. This facilitates the sequential degradation of substances not readily biodegradable by single species populations. Many anthropogenic pollutants fall into that category. The spatial arrangement of the microorganisms gives rise to gradients in the concentration of oxygen and other electron acceptors as well as of substrates, products, and pH value (COSTERTON ET AL. 1987; KÜHL AND JØRGENSEN 1992; DE BEER ET AL. 1993; WIMPENNY AND KINNIMENT 1995). Thus, aerobic and anaerobic habitats can arise in close proximity, and, as a consequence, the development of a large variability of species takes place. Genetic material is conserved and more readily taken up than by planktonic cells. As the structure is not rigid, the organisms can move in it which promotes gene exchange (Wuertz et al., in preparation). The EPS matrix sequesters nutrients from the bulk water phase (DECHO 1990), a particularly important mechanism in oligotrophic environments. Thus, biofilms seem to be a favorable form of life and part of a survival strategy of microorganisms (MARSHALL 1996). This matrix influences the sorption of dissolved and particulate substances, including biodegradable compounds (FLEMMING 1995; FLEMMING ET AL. 1996; SPÄTH ET AL. 1998). Binding and accumulation of cations such as Ca^{2+} or Mg^{2+} by EPS may profoundly affect the rheological behavior and diffusivity of biofilms. Uptake of toxic metal ions may contribute to the detoxification of polluted environments.

An important modern concept is the role of EPS in allowing microorganisms to live continuously at high cell densities in stable mixed population biofilm communities. As a consequence, the EPS matrix constitutes a medium for communication processes between constituents cells of biofilms that is only made possible by the close proximity of the bacteria which are held together by the EPS.

Horizontal gene transfer may be facilitated due to the close contact of aggregated cells and the accumulation of DNA in the EPS matrix. Several mechanisms of gene transfer are likely to occur in the natural environment: conjugation, transduction and natural transformation (LORENZ AND WACKERNAGEL 1994). A few studies have demonstrated gene exchange in pure culture biofilms (ANGLES ET AL. 1993; LISLE AND ROSE 1995) and plasmid transfer in river epilithon (BALE ET AL. 1988). Transfer of genetic material may contribute to phenotypic changes of biofilm microorganisms.

Another form of information transfer resides in the phenomenon of quorum sensing which is a signaling mechanism of Gram-negative bacteria in response to population density (SWIFT ET AL. 1996; Decho, this volume). This kind of cell-to-cell communication is mediated by low-molecular-weight diffusible signaling molecules (autoinducers), which are chemically N-acyl-L-homoserine lactones (AHLs). Extracellular accumulation of AHLs above a critical threshold level results in transcriptional activation of a range of different genes with concomitant expression of new phenotypes. AHLs allow bacteria to sense when cell densities in their surroundings reach the minimal level for a coordinate population response to be initiated. AHLs have been discovered in aquatic biofilms (MCLEAN ET AL. 1997) and in biofilms on indwelling urethral catheters (STICKLER ET AL. 1998), demonstrating the occurrence of AHLs in bacterial biofilms in natural and clinical environments. AHLs have been implicated in the shortened recovery process of *Nitrosomonas europaea* biofilm cells from nitrogen starvation due to accumulation of the signaling molecules within the biofilm to levels unobtainable in planktonic populations which showed a prolonged phase of recovery (BATCHELOR ET AL. 1997). With respect to biofilm formation AHLs were considered as mediators of adhesion by switching cells to an attachment phenotype through expression of adhesive polymers and they were supposed to facilitate induction of other genes essential for the maintenance of the biofilm mode of growth (HEYS ET AL. 1997). The involvement of an intercellular AHL in the differentiation of *P. aeruginosa* biofilms has been described

(DAVIES ET AL. 1998; Davies, this volume). Thus, cell-to-cell communication via AHLs seems to be of fundamental importance for biofilm bacteria to adapt dynamically in response to prevailing environmental conditions. Metabolites (halogenated furanones) from the marine macroalga *Delisea pulchra* have been suggested to interfere with AHL regulatory systems (STEINBERG ET AL. 1998), resulting in the inhibition of bacterial colonization of the surface of the seaweeds.

From a practical point of view, EPS as mediators of biofilm formation and stability have to be considered as target structures for remedial actions to remove, prevent, or control undesirable biofilm formation (biofouling) and microbially influenced corrosion of materials (biocorrosion) in industry and medicine.

References

Angles ML, Marshall KC, Goodman AE (1993) Plasmid transfer between marine bacteria in the aqueous phase and biofilms in reactor microcosms. Appl Environ Microbiol 59:843–850

Azeredo J, Oliveira R, Lazarova V (1998) A new method for extraction of exopolymers from activated sludges. Wat Sci Tech 37:367–370

Bale MJ, Fry JC, Day MJ (1988) Transfer and occurrence of large mercury resistance plasmids in river epilithon. Appl Environ Microbiol 54:972–978

Batchelor SE, Cooper M, Chhabra SR, Glover LA, Stewart GSAB, Williams P, Prosser JI (1997) Cell density-regulated recovery of starved biofilm populations of ammonia-oxidizing bacteria. Appl Environ Microbiol 63:2281–2286

Becker A, Katzen F, Puhler A, Ielpi L (1998) Xanthan gum biosynthesis and application: a biochemical/genetic perspective. Appl Microbiol Biotechnol 50:145–152

Bertocchi C, Navarini L, Cesaro A (1990) Polysaccharides from cyanobacteria. Carbohydr Polym 12:127–153

Beveridge TJ, Makin SA, Kadurugamuwa JL, Li Z (1997) Interactions between biofilms and the environment. FEMS Microbiol Rev 20:291–303

Binet R, Létoffé S, Ghigo JM, Delepelaire P, Wandersman C (1997) Protein secretion by Gram-negative bacterial ABC exporters – a review. Gene 192:7–11

Boyd A, Chakrabarty AM (1994) Role of alginate lyase in cell detachment of *Pseudomonas aeruginosa*. Appl Environ Microbiol 60:2355–2359

Brown MRW, Gilbert P (1993) Sensitivity of biofilms to antimicrobial agents. J Appl Bacteriol Symp Suppl 74:87S–97S

Burns RG (1989) Microbial and enzymic activities in soil biofilms. In: Characklis WG, Wilderer PA (eds) Structure and function of biofilms. Wiley, Chichester, pp 333–349

Cadieux JE, Kuzio J, Milazzo FH, Kropinski AM (1983) Spontaneous release of lipopolysaccharide by *Pseudomonas aeruginosa*. J Bacteriol 155:817–825

Characklis WG, Wilderer PA (1989) Glossary. In: Characklis WG, Wilderer PA (eds) Structure and function of biofilms. Wiley, Chichester, pp 369–371

Christensen BE (1989) The role of extracellular polysaccharides in biofilms. J Biotechnol 10:181–202

Christensen BE, Characklis WG (1990) Physical and chemical properties of biofilms. In: Characklis WG, Marshall KC (eds) Biofilms. Wiley, New York, pp 93–130

Costerton JW, Geesey GG, Cheng K-J (1978) How bacteria stick. Sci Am 238:86–95

Costerton JW, Irvin RT, Cheng K-J (1981) The bacterial glycocalyx in nature and disease. Annu Rev Microbiol 35:299–324

Costerton JW, Cheng K-J, Geesey GG, Ladd TI, Nickel JC, Dasgupta M, Marrie TJ (1987) Bacterial biofilms in nature and disease. Annu Rev Microbiol 41:435–464

Costerton JW, Brown MRW, Lam J, Lam K, Cochrane DMG (1990) The microcolony mode of growth in vivo – an ecological perspective. In: Gacesa P, Russell NJ (eds) Pseudomonas infection and alginates. Chapman and Hall, London, pp 76–94

Costerton JW, Lappin-Scott HM, Cheng K-J (1992) Glycocalyx, bacterial. In: Lederberg J (ed) Encyclopedia of microbiology, vol 2. Academic Press, San Diego, pp 311–317

Costerton JW, Lewandowski Z, DeBeer D, Caldwell D, Korber D, James G (1994) Biofilms, the customized microniche. J Bacteriol 176:2137–2142

Costerton JW, Lewandowski Z, Caldwell DE, Korber DR, Lappin-Scott HM (1995) Microbial biofilms. Annu Rev Microbiol 49:711–745

Davies DG, Parsek MR, Pearson JP, Iglewski BH, Costerton JW, Greenberg EP (1998) The involvement of cell-to-cell signals in the development of a bacterial biofilm. Science 280:295–298

De Beer D, van den Heuvel JC, Ottengraf SPP (1993) Microelectrode measurements of the activity distribution in nitrifying bacterial aggregates. Appl Environ Microbiol 59:573–579

De Beer D, Srinivasan R, Stewart PS (1994) Direct measurement of chlorine penetration into biofilms during disinfection. Appl Environ Microbiol 60:4339–4344

Decho AW (1990) Microbial exopolymer secretions in ocean environments: their role(s) in food webs and marine processes. Oceanogr Mar Biol Annu Rev 28:73–153

De Philippis R, Vincenzini M (1998) Exocellular polysaccharides from cyanobacteria and their possible applications. FEMS Microbiol Rev 22:151–175

Dignac M-F, Urbain V, Rybacki D, Bruchet A, Snidaro D, Scribe P (1998) Chemical description of extracellular polymers: implication on activated sludge floc structure. Wat Sci Tech 38:45–53

Filloux A, Michel G, Bally M (1998) GSP-dependent protein secretion in Gram-negative bacteria: the Xcp system of Pseudomonas aeruginosa. FEMS Microbiol Rev 22:177–198

Flemming H-C (1995) Sorption sites in biofilms. Wat Sci Tech 32:27–33

Flemming H-C (1996) The forces that keep biofilms together. In: Sand W (ed) Biodeterioration and biodegradation. Dechema Monographs 133. VCH, Weinheim, pp 311–316

Flemming H-C, Schaule G (1989) Biofouling auf Umkehrosmose- und Ultrafiltrationsmembranen. Teil II:Analyse und Entfernung des Belages. Vom Wasser 73:287–301

Flemming H-C, Schaule G (1996) Biofouling. In: Heitz E, Flemming H-C, Sand W (eds) Microbially influenced corrosion of materials. Springer, Berlin Heidelberg New York, pp 39–54

Flemming H-C, Schmitt J, Marshall KC (1996) Sorption properties of biofilms. In: Calmano W, Förstner U (eds) Environmental behaviour of sediments. Lewis, Chelsea, Michigan, pp 115–157

Foley I, Gilbert P (1996) Antibiotic resistance of biofilms. Biofouling 10:331–346

Frølund B, Palmgren R, Keiding K, Nielsen PH (1996) Extraction of extracellular polymers from activated sludge using a cation exchange resin. Wat Res 30:1749–1758

Gacesa P (1998) Bacterial alginate biosynthesis – recent progress and future prospects. Microbiology 144:1133–1143

Gacesa P, Russell NJ (eds) (1990) Pseudomonas infection and alginates. Biochemistry, genetics and pathology. Chapman and Hall, London

Geesey GG (1982) Microbial exopolymers: ecological and economic considerations. ASM News 48:9–14

Gehr R, Henry JG (1983) Removal of extracellular material. Wat Res 17:1743–1748

Govan JRW (1990) Characteristics of mucoid Pseudomonas aeruginosa in vitro and in vivo. In: Gacesa P, Russell NJ (eds) Pseudomonas infection and alginates. Biochemistry, genetics and pathology. Chapman and Hall, London, pp 50–75

Govan JRW, Deretic V (1996) Microbial pathogenesis in cystic fibrosis:mucoid Pseudomonas aeruginosa and Burkholderia cepacia. Microbiol Rev 60:539–574

Hedges JI (1988) Polymerization of humic substances in natural environments. In: Frimmel FH, Christman RF (eds) Humic substances and their role in the environment. Wiley, Chichester, pp 45–58

Heys SJD, Gilbert P, Eberhard A, Allison DG (1997) Homoserine lactones and bacterial biofilms. In: Wimpenny J, Handley P, Gilbert P, Lappin-Scott H, Jones M (eds) Biofilms: community interactions and control. BioLine, Cardiff, pp 103–112

Higgins MJ, Novak JT (1997) Characterization of exocellular protein and its role in bioflocculation. J Environ Eng 123:479–485

Hueck CJ (1998) Type III protein secretion systems in bacterial pathogens of animals and plants. Microbiol Mol Biol Rev 62:379–433

Isaac DH (1985) Bacterial polysaccharides. In: Atkins EDT (ed) Polysaccharides. Chemie, Weinheim, pp 141–184

Jacques M, Gottschalk M (1997) Use of monoclonal antibodies to visualize capsular material of bacterial pathogens by conventional electron microscopy. Microsc Microanal 3:234–238

Jaeger K-E, Reetz MT (1998) Microbial lipases form versatile tools for biotechnology. Trends Biotechnol 16:396–403

Jahn A, Nielsen PH (1995) Extraction of extracellular polymeric substances (EPS) from biofilms using a cation exchange resin. Wat Sci Tech 1995:157–164

Jahn A, Nielsen PH (1998) Cell biomass and exopolymer composition in sewer biofims. Wat Sci Tech 37:17–24

Jonas R, Farah LF (1998) Production and application of microbial cellulose. Polym Degrad Stabil 59:101–106

Kadurugamuwa JL, Beveridge TJ (1995) Virulence factors are released from *Pseudomonas aeruginosa* in association with membrane vesicles during normal growth and exposure to gentamic: a novel mechanism of enzyme secretion. J Bacteriol 177:3998–4008

Katzen F, Ferreiro DU, Oddo CG, Ielmini MV, Becker A, Pühler A, Ielpi L (1998) *Xanthomonas campestris* pv. campestris gum mutants: effects on xanthan biosynthesis and plant virulence. J Bacteriol 180:1607–1617

Kenne L, Lindberg B (1983) Bacterial polysaccharides. In: Aspinall GO (ed) The polysaccharides, vol 2. Academic Press, New York, pp 287–363

Korber DR, Lawrence JR, Lappin-Scott HM, Costerton JW (1995) Growth of microorganisms on surfaces. In: Lappin-Scott HM, Costerton JW (eds) Microbial biofilms. Cambridge University Press, Cambridge, pp 15–45

Kühl M, Jørgensen BB (1992) Microsensor measurements of sulfate reduction and sulfide oxidation in compact microbial communities of aerobic biofilms. Appl Environ Microbiol 58:1164–1174

Lawrence JR, Korber DR, Hoyle BD, Costerton JW, Caldwell DE (1991) Optical sectioning of microbial biofilms. J Bacteriol 173:6558–6567

Lazarowa V, Manem J (1995) Biofilm characterization and activity analysis in water and wastewater treatment. Wat Res 29:2227–2245

LeChevallier MW, Cawthon CD, Lee RG (1988) Inactivation of biofilm bacteria. Appl Environ Microbiol 54:2492–2499

Leigh JA, Coplin DL (1992) Exopolysaccharides in plant-bacterial interactions. Annu Rev Microbiol 46:307–346

Leppard GG (1995) The characterization of algal and microbial mucilages and their aggregates in aquatic ecosystems. Sci Total Environ 165:103–131

Leppard GG (1997) Colloidal organic fibrils of acid polysaccharides in surface waters: electron-optical characteristics, activities and chemical estimates of abundance. Colloids Surfaces A: Physicochem Eng Aspects 120:1–15

Li Z, Clarke AJ, Beveridge TJ (1998) Gram-negative bacteria produce membrane vesicles which are capable of killing other bacteria. J Bacteriol 180:5478–5483

Lindberg B (1990) Components of bacterial polysaccharides. Adv Carbohydr Chem 48: 279–318

Lisle JT, Rose JB (1995) Gene exchange in drinking water and biofilms by natural transformation. Wat Sci Tech 31:41–46

Liss SN, Droppo IG, Flannigan DT, Leppard GG (1996) Floc architecture in wastewater and natural riverine systems. Environ Sci Technol 30:680–686

Lorenz MG, Wackernagel W (1994) Bacterial gene transfer by natural genetic transformation in the environment. Microbiol Rev 58:563–602

Marshall KC (1996) Adhesion as a strategy for access to nutrients. In: Fletcher M (ed) Bacterial adhesion: molecular and ecological diversity. Wiley, New York, pp 59–87

Matsuyama T, Nakagawa Y (1996) Surface-active exolipids: analysis of absolute chemical structures and biological functions. J Microbiol Meth 25:165–175

May TB, Chakrabarty AM (1994) *Pseudomonas aeruginosa*: genes and enzymes of alginate synthesis. Trends Microbiol 2:151–157

Mayer C, Moritz R, Kirschner C, Borchard W, Maibaum R, Wingender J, Flemming H-C (1999) The role of intermolecular interactions: studies on model systems for bacterial biofilms (in press)

McLean RJC, Whiteley M, Stickler DJ, Fuqua WC (1997) Evidence of autoinducer activity in naturally occurring biofilms. FEMS Microbiol Lett 154:259–263

Mesnage S, Tosi-Couture E, Gounon P, Mock M, Fouet A (1998) The capsule and S-layer: two independent and yet compatible macromolecular structures in *Bacillus anthracis*. J Bacteriol 180:52–58

Morris CE, Monier J-M, Jacques M-A (1997) Methods for observing microbial biofilms directly on leaf surfaces and recovering them for isolation of culturable microorganisms. Appl Environ Microbiol 63:1570–1576

Morton LHG, Greenway DLA, Gaylarde CC, Surman SB (1998) Consideration of some implications of the resistance of biofilms to biocides. Int Biodet Biodegr 41:247–259

Neu TR (1996) Significance of bacterial surface-active compounds in interaction of bacteria with interfaces. Microbiol Rev 60:151–166

Neu TR, Lawrence JR (1997) Development and structure of microbial biofilms in river water studied by confocal laser scanning microscopy. FEMS Microbiol Ecol 24:11–25

Neu TR, Marshall KC (1990) Bacterial polymers: physicochemical aspects of their interaction at interfaces. J Biomaterials Applications 5:107–133

Nielsen PH, Jahn A, Palmgren R (1997) Conceptual model for production and composition of exopolymers in biofilms. Wat Sci Tech 36:11–19

Ophir T, Gutnick DL (1994) A role for exopolysaccharides in the protection of microorganisms from desiccation. Appl Environ Microbiol 60:740–745

Pasquier C, Marty N, Dournes J-L, Chabanon G, Pipy B (1997) Implication of neutral polysaccharides associated to alginate in inhibition of murine macrophage response to *Pseudomonas aeruginosa*. FEMS Microbiol Lett 147:195–202

Peters G, Gray ED, Johnson GM (1989) Immunomodulating properties of extracellular slime substance. In: Bisno AL, Waldvogel FA (eds) Infections associated with indwelling medical devices. American Society for Microbiology, Washington, DC, pp 61–74

Platt RM, Geesey GG, Davis JD, White DC (1985) Isolation and partial chemical analysis of firmly bound exopolysaccharide from adherent cells of a freshwater sediment bacterium. Can J Microbiol 31:675–680

Priest FG (1992) Enzymes, extracellular. In: Lederberg J (ed) Encyclopedia of microbiology, vol 2. Academic Press, San Diego, pp 81–93

Rehm BHA, Valla S (1997) Bacterial alginates: biosynthesis and applications. Appl Microbiol Biotechnol 48:281–288

Roberts IS (1995) Bacterial polysaccharides in sickness and in health. Microbiology 141:2023–2031

Roberts IS (1996) The biochemistry and genetics of capsular polysaccharide production in bacteria. Annu Rev Microbiol 50:285–315

Sidhu MS, Olsen I (1997) S-layers of *Bacillus* species. Microbiology 143:1039–1052

Sleytr UB (1997) Basic and applied S-layer research: an overview. FEMS Microbiol Rev 20:5–12

Smith SE, Simpson JA (1990) The contribution of *Pseudomonas aeruginosa* alginate to evasion of host defence. In: Gacesa P, Russell NJ (eds) (1990) Pseudomonas infection and alginates. Biochemistry, genetics and pathology. Chapman and Hall, London, pp 135–159

Späth R, Flemming HC, Wuertz S (1998) Sorption properties of biofilms. Wat Sci Tech 37:207–210

Steinberg PD, De Nys R, Kjelleberg S (1998) Chemical inhibition of epibiota by Australian sea-weeds. Biofouling 12:227–244

Stickler DJ, Morris NS, McLean RJC, Fuqua C (1998) Biofilms on indwelling urethral catheters produce quorum-sensing signal molecules in situ and in vitro. Appl Environ Microbiol 64:3486–3490

Sutherland IW (1977) Surface carbohydrates of the procaryotic cell. Academic Press, New York

Sutherland IW (1982) Biosynthesis of microbial polysaccharides. Adv Microb Physiol 23:79–150

Sutherland IW (1983) Microbial exopolysaccharides – their role in microbial adhesion in aqueous systems. CRC Crit Rev Microbiol 10:173–201

Sutherland IW (1985) Biosynthesis and composition of gram-negative bacterial extracellular and wall polysaccharides. Annu Rev Microbiol 39:243–270

Sutherland IW (1990) Biotechnology of microbial exopolysaccharides. Cambridge University Press, Cambridge

Sutherland IW (1994) Structure-function relationships in microbial exopolysaccharides. Biotech Adv 12:393–448

Sutherland IW (1996) Extracellular polysaccharides. In: Rehm H-J, Reed G (eds) Biotechnology, vol 6: products of primary metabolism. Chemie, Weinheim, pp 615–657

Sutherland IW (1998) Novel and established applications of microbial polysaccharides. Trends Biotechnol 16:41–46

Swift S, Throup JP, Williams P, Salmond GPC, Stewart GSAB (1996) Quorum sensing: a population-density component in the determination of bacterial phenotype. Trends Biochem Sci 21:214–219

Takeda M, Nakano F, Nagase T, Iohara K, Koizumi J-I (1998) Isolation and chemical composition of the sheath of *Sphaerotilus natans*. Biosci Biotechnol Biochem 62:1138–1143

Weiner R, Langille S, Quintero E (1995) Structure, function and immunochemistry of bacterial exopolysaccharides. J Ind Microbiol 15:339–346

Whitfield C (1988) Bacterial extracellular polysaccharides. Can J Microbiol 34:415–420

Whitfield C, Valvano MA (1993) Biosynthesis and expression of cell-surface polysaccharides in gram-negative bacteria. Adv Microb Physiol 35:135–246

Wimpenny JWT, Kinniment SL (1995) Biochemical reactions and the establishment of gradients within biofilms. In: Lappin-Scott HM, Costerton JW (eds) Microbial biofilms. Cambridge University Press, Cambridge, pp 99–117

Wingender J, Grobe S, Fiedler S, Flemming H-C (1999) The effect of extracellular polysaccharides on the resistance of *Pseudomonas aeruginosa* to chlorine and hydrogen peroxide. In: Keevil C, Holt D, Dow C, Godfree A (eds) Biofilms in aquatic systems. Royal Society of Chemistry, Cambridge (in press)

Wolfaardt GM, Lawrence JR, Robarts RD, Caldwell DE (1998) In situ characterization of biofilm exopolymers involved in the accumulation of chlorinated organics. Microb Ecol 35:213–223

Xun L, Mah RA, Boone DR (1990) Isolation and characterization of disaggregatase from *Methanosarcina mazei* LYC. Appl Environ Microbiol 56:3693–3698

In Situ Characterization of Extracellular Polymeric Substances (EPS) in Biofilm Systems

Thomas R. Neu[1] · John R. Lawrence[2]

[1] Department of Inland Water Research Magdeburg, UFZ Centre for Environmental Research Leipzig-Halle, Brückstrasse 3a, D-39114 Magdeburg, Germany, *E-mail: neu@gm.ufz.de*
[2] National Hydrology Research Center, 11 Innovation Blvd., Saskatoon, SK, S7N 3H5 Canada

Keywords. Biofilm, Extracellular polymeric substances (EPS), Exopolymer, Polysaccharide, Biofilm matrix, Lectin, Fluorescence, Probes, In situ analysis, Confocal laser scanning microscopy

1
Introduction

1.1
Traditional Approaches for Studying Microbial Polysaccharides

Historically, microbial polysaccharides were studied for three reasons. First, polysaccharides represent a structural feature of the microbial cell; therefore they were investigated for pure and basic research interests. Second, polysaccharides were recognized as antigen determinants of the microbial cell surface; the knowledge of their structure was and still is of great importance in medical microbiology. Third, microbial polysaccharides were recognized as a source of polymers with unique properties. These applied aspects of polysaccharides were a reason to study their structure, properties, and production on the pilot and industrial scales.

All microbial polysaccharides were exclusively isolated by two methods. After centrifugation of a liquid culture, they were either isolated from the cell free culture supernatant by precipitation or they were recovered from the remaining cell pellet by various extraction methods. There have been very few attempts to study microbial polysaccharides with other approaches (NEU 1994).

Generally, due to the pure culture philosophy in microbiology, the production, the isolation, and the preparation of EPS were done with pure cultures only.

1.2
Change of Paradigms

The traditional interest in and meaning of EPS, as outlined above, is still justifiable. But in the last few years there has been a change of paradigms in microbiology. This refers to the following paradigms: germ theory, enrichment theory and selection theory (CALDWELL ET AL. 1997A,B). As a result, nowadays more and more microbiologists start to work with defined mixed or even fully complex cultures derived from certain natural or technical habitats. This trend is in opposition to the above paradigms. Although the traditional approach is important and necessary, if we are going to study natural systems we are confronted not with pure cultures but with very complex microbial communities (CALDWELL ET AL. 1996). As these complex communities produce complex mixtures of EPS we are also confronted with the challenge of analyzing these complex EPS matrices.

1.3
In Situ Methods

Apart from the high water content of up to 99%, biofilms consist mainly of EPS and cells. Several in situ techniques have been developed to investigate interfacial microbial communities. The major techniques include microelectrodes, gene probes, and new microscopic techniques. Microelectrodes were employed to investigate gradients within biofilm systems. In addition, they were used to study the physiology of biofilm communities (KÜHL ET AL. 1997; LASSEN ET AL. 1992; LÜBBERS 1992; REVSBECH AND JÖRGENSEN 1986). For the characterization of the cellular constituents of biofilm systems tremendous progress has been made due to the development of gene probes. In situ hybridization of bacteria with rRNA targeted oligonucleotide probes in natural samples has been extensively covered in various publications (AMANN ET AL. 1995; SAYLER AND LAYTON 1990; WARD 1989). For the EPS constituents of biofilm systems a similar in situ approach is urgently needed. This in situ approach should allow the investigation of the EPS non-destructively and in the fully hydrated form.

In this chapter we present some information on the possibilities to investigate the EPS in complex microbial communities. We will compare traditional approaches with new in situ methods which are now available as a result of new imaging techniques, e.g., confocal laser scanning microscopy (CLSM).

2
Destructive Analysis of EPS

2.1
Chemical Techniques

2.1.1
Extraction

The traditional approach to investigate polysaccharides is based on the isolation of the polysaccharides from a complex cell/EPS matrix. However, using these approaches the original structure of the cells, aggregates, or biofilm matrix is disrupted. Several methods have been suggested to isolate polysaccharides from biofilm systems (Nielsen and Jahn, this volume). Most publications are concerned with the extraction of polysaccharides from activated sludge. Early publications used a variety of approaches to extract the polymeric fraction of activated sludge. More recently, extraction by employing cation exchange resins seems to be the method of choice. Usually these extracts contained a high amount of protein, but lipids and nucleic acids were also found (Nielsen and Jahn, this volume). In addition to these investigations, there is a report on the isolation of EPS from sewer biofilms. This study also used cation exchange resin and reported that the polysaccharide fraction constituted only up to 5% of the total sample while the remainder was mainly humic substances and proteins (JAHN AND NIELSEN 1995). This of course raised questions about the significance and content of polysaccharides in natural biofilms. The findings are however in agreement with a report on river biofilms where humic substances have been suggested to represent a major constituent of lotic biofilms (NEU AND LAWRENCE 1997).

2.1.2
Analysis

The chemical analysis of polysaccharides can be divided between sugar analysis, linkage analysis, sequencing, and determination of anomeric configuration. The destructive chemical analysis of polysaccharides includes the following steps:

- Isolation by precipitation from the culture supernatant or extraction from the cell surface
- Purification by precipitation and size fractionation by gel chromatography (ASPINALL 1982)
- Release of single constituents by various types of hydrolyses
- Determination of charged compounds by high voltage electrophoresis
- Analysis of polysaccharide constituents (sugar and non-sugar) by HPLC or GC

Additional investigation of the polysaccharide may include methylation analysis to determine the linkage pattern of the carbohydrates, analysis of the products by GC-MS, specific degradation (LINDBERG ET AL. 1975), and isolation of di- or oligo-saccharides to confirm the structure. An indispensable technique to analyze the original polysaccharide molecule and to follow derivatization procedures of the polysaccharide molecule is ^{1}H NMR and ^{13}C NMR spectroscopy (PERLIN AND CASU 1982). Finally, the anomeric configuration (D or L form) of the carbohydrates has to be determined. If the polysaccharide is available in a pure form the three-dimensional structure may be investigated by X-ray crystallography (REES ET AL. 1982).

2.2
Electron Microscopy

The presence and significance of EPS in microbial adhesion and biofilm development was demonstrated by different electron microscopic techniques. An overview of a variety of methods may be found in a review article on ultrastructure techniques (ERDOS 1986). Scanning electron microscopy (SEM) in the traditional way includes fixation and dehydration to prepare the originally hydrated specimen for microscopic examination. The results usually show strands of polymeric material in between the microorganisms and between the micro-organisms and the surface to which they are attached to. These strands which are possibly precipitated due to the preparation procedure may represent the remains of EPS (COSTERTON ET AL. 1986; RICHARDS AND TURNER 1984).

Newer SEM techniques using a cryo-transfer chamber and cold stage were more gentle but revealed similar strands of polymers as well as "footprint" ma-terial (CHENU AND JAUNET 1992; NEU AND MARSHALL 1991; RICHARDS AND TURNER 1984). Another technique claiming the possibility of examining fully hydrated samples is environmental scanning electron microscopy (ESEM) (DANILATOS 1991). However, in reality, observation starts with fully hydrated samples, moves to partly hydrated samples, and ends with dehydrated samples. Highest resolution is achieved at the latter stage, whereby the final result may be similar to the cryo-SEM technique. Nevertheless, ESEM was used to examine biofilms as well as microbial aggregates (LAVOIE ET AL. 1995; SURMAN ET AL. 1996).

Transmission electron microscopy (TEM) also requires fixation and de-hydration as well as embedding and sectioning of the specimen. Thereby high

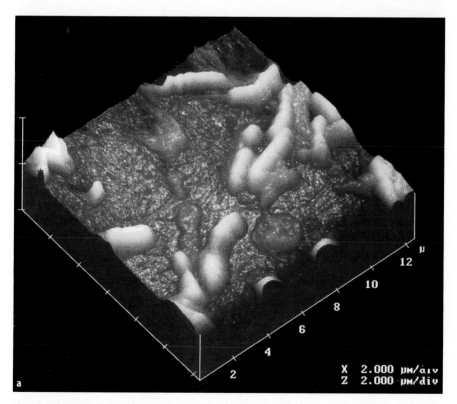

Fig. 1a, b. Atomic force microscopy (AFM) of biofilms (by courtesy of I. Beech, see also chap-ter in this volume): **a** AFM image of one-week old biofilm formed by *P. aeruginosa* showing bacterial cells and hydrated EPS

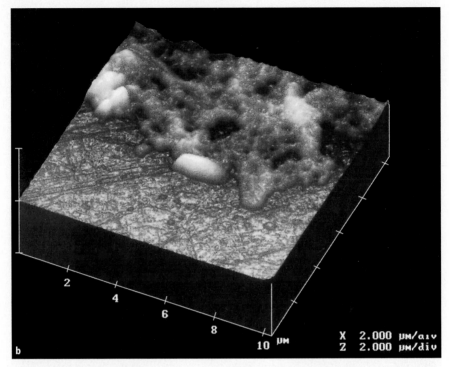

Fig. 1a, b (continued). **b** AFM image of a two-week old biofilm formed by mixed cultures of *P. aeruginosa* and *D. gigas*. Cells encased in EPS matrix can be clearly seen

resolution images of so-called fibrils became visible (COSTERTON ET AL. 1986). These TEM techniques were further developed by Gary Leppard who investigated EPS components of biofilms and marine snow. It has been suggested that EPS components have in situ a fibrillar structure which is dependent on the chemical composition of the polymer (LEPPARD 1986; LEPPARD ET AL. 1996).

The major disadvantage of all electron microscopic techniques lies in the creation and interpretation of artifacts caused by the various preparation steps. These artifacts may have been reduced by using new and fine tuned techniques. However, conclusions drawn from images of previously hydrated samples which have been fixed and dehydrated will still be subject to this criticism.

2.3
Scanning Probe Techniques

This suite of techniques includes scanning tunneling electron microscopy, atomic force microscopy, scanning ion-conductance microscopy, and scanning tunneling microscopy (BEECH 1996; DRAKE ET AL. 1989; HANSMA ET AL. 1989 A, B; MARTIN ET AL. 1988). In these cases a scanning probe provides a surface view of the material providing resolution at the atomic level. The few instances of its application to microbes have not revealed detailed structure, although there is potential for these methods to reveal structural surface information regarding the EPS "blanket" surrounding bacterial cells and biofilms (BREMER ET AL. 1992). An example of atomic force microscopy is shown in Fig. 1a, b (Beech and Tapper, this volume).

3
Non-Destructive Analysis of EPS

A short overview of the non-destructive techniques is given in Table 1.

3.1
Infrared Spectroscopy (FT-IR)

A variation of FT-IR spectroscopy, so-called attenuated total reflection/Fourier transform-infrared spectroscopy (ATR/FT-IR) is a non-destructive on-line technique (NICHOLS ET AL. 1985; CLARK 1990). This method allows the user to record the appearance of certain chemical groups on the surface of an internal reflection element (germanium crystal) exposed to the aqueous phase. ATR/FT-IR can be used to study adhesion and biofilm development of bacteria (NIVENS ET AL. 1993; SCHMITT ET AL. 1995) as well as the adsorption of isolated polysaccharides (BREMER AND GEESEY 1991A; NICHOLS ET AL. 1985; NIVENS ET AL. 1993, 1995; WHITE 1986). Furthermore, the technique has been successfully employed to investigate corrosion processes facilitated by bacteria and their exopolymers. By coating the germanium crystal with a thin copper film, biocorrosion in the μm range can be followed by ATR/FT-IR (BREMER AND GEESEY 1991B; GEESEY ET AL. 1987; JOLLEY ET AL. 1989). ATR/FT-IR was also combined

Table 1. Overview of non-destructive techniques for in-situ EPS analysis

Method	Information	Reference
ATR/FT-IR	Signal of chemical groups from: conditioning films, coatings, EPS, early biofilm events	BREMER AND GEESEY (1991A) GEESEY ET AL. (1987) JOLLEY ET AL. (1989) NICHOLS ET AL. (1985) NIVENS ET AL. (1993) SCHMITT ET AL. (1995) SCHMITT AND FLEMMING (1998) WHITE (1986)
NMR	Biomass, flow velocity, oxygen tension, cell distribution	LEWANDOWSKI ET AL. (1992) LEWANDOWSKI ET AL. (1993) LEWANDOWSKI ET AL. (1995) WILLIAMS ET AL. (1997)
CLSM	Wide range of probes available for: polysaccharides, proteins, nucleic acids	LAWRENCE ET AL. (1997) NEU AND LAWRENCE (1997) LAWRENCE ET AL. 1998A LAWRENCE ET AL. 1998B NEU (1999B)

with reflected differential interference contrast microscopy to obtain comple-
mentary data sets. This technique facilitates the simultaneous collection of information on chemical and structural features of adherent bacteria within a
1 µm layer at the surface of the germanium crystal (SUCI ET AL. 1997). Recently
the different FT-IR-spectroscopy techniques applicable to biofilm research have
been discussed. FT-IR may be employed as a measure for biofilms on an internal
reflection element; in the diffuse reflectance technique it is possible to study reflecting surfaces and the technique may be used to identify microorganisms
(SCHMITT AND FLEMMING 1998). An example of biofilm development on an ATR
crystal is presented in Fig. 2. Thus ATR/FT-IR has considerable potential for
fundamental biofilm research as well as for monitoring of biofilm formation.

3.2
Magnetic Resonance Spectroscopy (NMR)

As already mentioned, NMR spectroscopy is now an indispensable standard
technique for the investigation of polysaccharides (PERLIN AND CASU 1982).
The technique has several advantages. First of all, NMR spectroscopy is non-destructive, that means the sample is, after analysis, still available for subsequent
measurements. Another feature of NMR spectroscopy is the ease of determining the number of carbohydrate monomers in the repeating unit by simply
counting the signals in the anomeric region. Furthermore, application of certain chemical derivatizations and degradations can be easily controlled by NMR

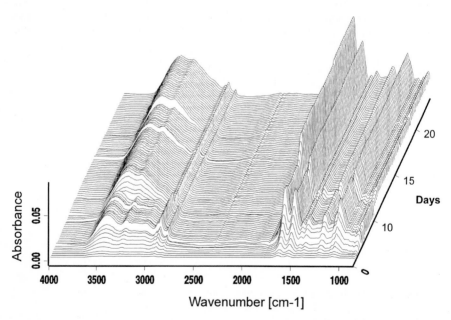

Fig. 2. Non destructive detection and analysis of biofilm growth as demonstrated with time
resolved FTIR-ATR cascade spectra in a flow through cell (by courtesy of J. SCHMITT AND H.-
C. FLEMMING)

spectroscopy. In addition, certain non-carbohydrate components can be identified from an NMR spectrum (see for example, NEU ET AL. 1992).

NMR spectroscopy can also be employed as a non-invasive technique to study intact biofilms (CLARK 1990). Lewandowski et al. employed NMR to investigate microbial biofilms demonstrating the simultaneous gathering of biomass distribution and flow velocity information by using ^{1}H NMR spectroscopy (LEWANDOWSKI ET AL. 1992, 1993). It was also shown that the flow velocity in a reactor was not affected by the presence of a biofilm. However, there was evidence for a flow field above the biofilm and another within the biofilm (LEWANDOWSKI ET AL. 1995). These findings are consistent with more recent biofilm models incorporating the presence of interstitial voids (LAWRENCE ET AL. 1991; COSTERTON ET AL. 1994; MASSOL-DEYA ET AL. 1995; NEU AND LAWRENCE 1997). The technique was suggested as a general tool in biotechnology to follow the physiological status of immobilized cells by using ^{13}C NMR, ^{32}P NMR as well as other NMR modes (LOHMEIER-VOGEL ET AL. 1990). Magnetic resonance imaging (^{1}H, ^{19}F) has been used to measure the spatial distribution of oxygen and mammalian cells in bioreactors (WILLIAMS ET AL. 1997). It is possible that ^{13}C NMR spectroscopy may have the potential to study EPS compounds in situ. However, to the authors knowledge this potential has not been assessed with complex biofilm communities.

3.3
Confocal Laser Scanning Microscopy (CLSM)

General technical and physical information about the CLSM method may be found in the second edition of "The handbook of biological confocal microscopy" (PAWLEY 1995). CLSM has become more and more available as a result of technical developments in light microscopy, video microscopy, confocal microscopy, available laser sources, and computer technology. The versatility of CLSM was further expanded by the development of a huge range of fluorescent labels for many different purposes (HAUGLAND 1996).

Comparison of various microscopy techniques showed that CLSM is for many biofilm examinations the optimal instrument (LAURENT ET AL. 1994; SURMAN ET AL. 1996). The applicability of CLSM to investigate biofilms has been discussed by several authors (LAWRENCE ET AL. 1997, 1998A,B; NEU 1999A). The major advantage is the possibility of non-destructive, three-dimensional, optical sectioning of fully hydrated, living biofilm systems.

4
CLSM Approach

4.1
Probes for Analysis of EPS

The in situ analysis of EPS relies on probes which are more or less specific for representative EPS molecules or characteristic chemical groups of EPS. The surface chemistry of bacterial cells and the EPS present in biofilms may be analyz-

Table 2. Fluorescent probes for in situ EPS analysis via epi-fluorescence microscopy (two-dimensional) and confocal laser scanning microscopy (three-dimensional)

Probe	Information	Reference
Lectin	General glycoconjugate distribution	HOLLOWAY AND COWAN (1997) LAWRENCE ET AL. (1998A) LAWRENCE ET AL. (1998B) MICHAEL AND SMITH (1995) NEU AND LAWRENCE (1997) NEU (1999B) WOLFAARDT ET AL. (1998)
Antibody	Specific EPS distribution	for complex systems not available, only against isolated EPS from pure cultures
Cell-impermeant protein specific	General protein distribution in "microbial footprints"	NEU AND MARSHALL (1991) PAUL AND JEFFREY (1985)
Cell-impermeant nucleic acid specific probes	Free/associated DNA equal to dead cell stain	HOLLOWAY AND COWAN (1997)
Charged probes	Charge distribution	WOLFAARDT ET AL. (1998)
Dextrans	Diffusion	KORBER ET AL. (1994) LAWRENCE ET AL. (1994)
Microspheres flow patterns	Hydrophobicity	LAWRENCE ET AL. (1998A) STOODLEY ET AL. (1994) ZITA AND HERMANSSON (1997)

ed using a variety of probes including general probes for polysaccharides, proteins, and nucleic acids. Furthermore, there are highly specific probes for EPS mono-, oligo-, and poly-mers such as lectins and antibodies. In addition, probes such as charged dextrans and ficolls may be used for very general EPS characterization, e.g., charge, hydrophobicity, or diffusion coefficients (see Table 2).

4.1.1
Polysaccharides

General fluorescent polysaccharide probes include stains like calcofluorwhite M2R and congo red. However, these stains are not really general polysaccharide stains as they have a specificity for all $(1 \rightarrow 4)$ and $(1 \rightarrow 3)$ beta-D-glucan polysaccharides (ALLISON AND SUTHERLAND 1984; WOOD 1980). In addition, the application of calcofluorwhite is dependent upon excitation from a UV laser source which may restrict its general use due to high cost. Congo red has been employed as a light microscopic stain but not yet examined with a CLSM setup (LAWRENCE ET AL. 1998A). Alcian blue, a copper-containing cationic phthalocyanin dye which binds to anionic molecules such as glycosaminoglycan, has been used as an indirect measure of EPS in biofilms (WETZEL ET AL. 1997) and may hold some promise as a general stain.

Lectins are carbohydrate binding proteins other than enzymes or antibodies. Fluor-conjugated lectins, characterized by their specific binding with glycoconjugates, can be used to map the chemical composition of the exopolymer materials associated with bacterial cells (CALDWELL ET AL. 1992a; NEU AND MARSHALL 1991). They have been mostly used in studies with pure cultures and visualized by electron or epi-fluorescence microscopy. Colloidal gold labeled lectins have been employed to examine microbial cells by transmission electron microscopy (MERKER AND SMIT 1988; MORIOKA ET AL. 1987; ONG ET AL. 1990; SANFORD ET AL. 1995; VASSE ET AL. 1994). Fluorescence labeled lectins in combination with epi-fluorescence microscopy were used in several studies characterizing the bacterial cell surface or cell appendages (HOOD AND SCHMIDT 1996; JONES ET AL. 1986; QUINTERO AND WEINER 1995; SIZEMORE ET AL. 1990; YAGODA-SHAGAM ET AL. 1988). Nevertheless, lectins in combination with CLSM have been applied in very few investigations of complex microbial biofilm systems. These reports are discussed later together with the "case studies".

Antibodies are a second specific group of probes. They have to be produced in animals or cell cultures against an isolated polysaccharide. This polysaccharide should be in a very pure form to ensure high specificity and little cross reactivity. However, the immunogenicity of polysaccharides is rather poor compared to that of proteins or lipopolysaccharides which results in a low antibody yield. In addition, the necessity of having a pure polysaccharide to produce antibodies, raises the question about their applicability to complex EPS matrices of biofilms. Nevertheless, antibodies against whole cell surfaces have been successfully applied in several epifluorescence and CLSM studies. They have been used in marine systems (ZAMBON ET AL. 1984), for in situ detection of bacteria in aquatic ecosystems (FAUDE AND HÖFLE 1997), to track down the presence of *Legionella* in biofilms (ROGERS AND KEEVIL 1992), to follow the survival of *Campylobacter* in microcosm experiments (BUSWELL ET AL. 1998) and in studies of the bacterial communities in the rhizosphere of plants (ASSMUS ET AL. 1997; GILBERT ET AL. 1998; HANSEN ET AL. 1997; SCHLOTER ET AL. 1993, 1997). A general assessment of the antibody approach is available (DAZZO AND WRIGHT 1996; SCHLOTER ET AL. 1995).

4.1.2
Proteins

There are general protein specific probes available such as Hoechst 2495. This has been used to label adhesive bacterial "footprints" which were left on surfaces after detachment of the cells (NEU AND MARSHALL 1991; PAUL AND JEFFREY 1985). More recently, new protein stains for measuring protein concentrations in solution or in gels have been developed (HAUGLAND 1996). However, these probes have not yet been tested in biofilms using CLSM.

If single purified protein fractions of constituents of biofilm matrices can be obtained, then the antibody approach would be applicable. However, the same problems noted for polysaccharide antibodies would remain. One always has to be aware that isolating the antigen of interest (cells or polymer) from a complex biofilm matrix may be very difficult or even impossible.

4.1.3
Nucleic Acids

In CLSM studies nucleic acids are usually stained inside microbial cells. This may be achieved with general nucleic acid specific probes applied to living, hydrated samples. Examples of such probes are the standard stains 4'6-diamidino-2-phenylindole (DAPI) and acridine orange (AO). AO is not really suitable for multi-channel CLSM as it will give a signal in several channels and thus cannot be combined with other probes. In the meantime new stains have been developed such as the SYTO series from one of the major suppliers of fluorescent chemicals (HAUGLAND 1996). This series is now available in several colors to be used as fluorescent stains with emissions in either the blue, green, or red part of the light spectrum. Other stains, e.g., SYTOX, YOYO-1, YO-PRO-1, and Pico-Green can only be used after fixation with aldehyde resulting in the loss of three-dimensional information (MARIE ET AL. 1996). Nevertheless, these stains should also allow the detection of free nucleic acid in complex biofilm matrices (HOLLOWAY AND COWEN 1997).

4.2
Physicochemical Characterization of EPS Matrix

4.2.1
Charge Distribution

The charge of the EPS matrix can be determined with commercially-available fluorescent dextrans bearing a defined charge (e.g., anionic, cationic, neutral). These probes have been used to study charge distribution in biofilms (KORBER ET AL. 1994; LAWRENCE ET AL. 1994). Furthermore, the binding of polyanionic and cationic dextrans indicated the presence of both positively and negatively charged micro zones within biofilms (WOLFAARDT ET AL. 1998).

4.2.2
Hydrophobicity/Hydrophilicity

Hydrophobic pockets within biofilms, e.g., on cell surfaces or within the EPS matrix, can be probed with either hydrophobic stains or hydrophobic microspheres. Similarly, surface modified hydrophilic beads may be used to probe for hydrophilic pockets. The microsphere approach was recently described for single cells but may be applicable to biofilm systems as well (ZITA AND HERMANSSON 1997). There are also a whole range of fluorescent stains available which may be suitable for labeling such areas (HAUGLAND 1996). For example, the lipophilic probe Nile Red may also be useful for assessing the distribution of specific residues in biological materials (LAMONT ET AL. 1987; WOLFAARDT ET AL. 1994B).

4.2.3
Permeability

Biofilm permeability and effective diffusion coefficients may be determined using the combination of fluorescent probes with known hydrated radii and CSLM. Probes that are most commonly used include size-fractionated dextrans or ficolls to examine diffusion and binding within biofilms. An in situ monitoring approach using size fractionated dextrans was developed by LAWRENCE ET AL. (1994) to determine effective diffusion coefficients for biofilm systems. DE BEER ET AL. (1997) used a micropipette injection method to examine the diffusion of a variety of probes including fluorescein, TRITC-IgG, and phycoerythrin in biofilms. The more standard FRAP (fluorescence recovery after photobleaching) approach may also be applied to biofilms. BIRMINGHAM ET AL. (1995) used FRAP to monitor diffusion and binding of dextrans in oral biofilms. Further details of the procedures are also outlined in other manuscripts (AXELROD ET AL. 1976; BLONK ET AL. 1993). Flow patterns in biofilms may be easily monitored using the approach described by STOODLEY ET AL. (1994) and LAWRENCE ET AL. (1998A). This involves the introduction of fluorescent beads and collection of averaged or serial images in which the changing position of the beads may be detected and measured.

4.3
EPS Bound and Associated Molecules

The EPS of biofilms is also an important site for the sorption or localization of carbon, metals (GEESEY ET AL. 1988; GEESEY AND JANG 1989), contaminants (SPÄTH ET AL. 1998; WOLFAARDT ET AL. 1994A, B) and enzymes (see CHROST 1991). The presence of these associated molecules is important to overall functioning of biofilms systems. However, very little research has been carried out using fluorescence and CSLM to visualize and quantify these moieties. Antibodies may be used to map the locations of specific enzymes, specific compounds such as pesticides, and a variety of other biologically relevant molecules. WOLFAARDT ET AL. (1995) demonstrated that autofluorescence could be used to localize the herbicide diclofop methyl in biofilm polymers and to follow its sorption and metabolism. There are a variety of ion sensitive probes such as fluorescent indicators for Zn and other metals such as the BAPTA series, Newport Green, and TSQ probes that may be used to indicate the presence of heavy metals in biofilms and within the EPS matrix.

5
Image Analyses and Three-Dimensional Data Presentation

CLSM is a multi-step procedure starting with preparation and staining of the sample, followed by scanning the optical sections and subsequently applying certain image analyzing procedures to the three-dimensional data sets. The latter steps are in many ways the most time consuming and difficult.

5.1
Analyses Techniques

A confocal laser scanning microscope can form the basis for quantitative measurements of various parameters through the application of digital image processing (for reviews see CALDWELL ET AL. 1992A; LAWRENCE ET AL. 1996, 1998 A). Quantitative image analyses involves the application of a sequence of computer operations including image acquisition, image processing, image segmentation or recognition, object measurement, and data output (GONZALEZ AND WINTZ 1977; RUSS 1990, 1995). Image processing may include histogram analysis, grey level transformation, normalization, contrast enhancement, application of median, lowpass, Gaussian, Laplacian filters, image subtraction, addition, multiplication, and erosion, and/or dilation of the object boundary. Lowpass and median filters smooth your image and reduce background noise. Application of shading correction to eliminate uneven illumination and other systemic errors may be necessary. AGARD (1984) and GORBY (1994) explained how CSLM images may be sharpened through mathematical removal of out-of-focus information. The UNIX-based program XCOSM available as freeware provides a good starting point for applications allowing mathematical removal of out-of-focus information.

Before measurements are made, the image must be defined through thresholding. Object recognition is however the most complicated aspect of image analysis. Differentiation between specific objects and autofluorescent material or individual cells, cell clumps, and varying cell morphologies is challenging for even the most advanced image analysis system. In most cases simple solutions are applied, for example cell clumps were excluded from the analysis (MÖLLER ET AL. 1995) or divided by two in the case of a diplococci monoculture (AN ET AL. 1995). However, pattern recognition has been used to determine fungal biomass, or separate cells based on shape factors (MORGAN ET AL. 1991; DUBUISSON ET AL. 1994). LAWRENCE ET AL. (1989) used cell size and motion to determine the numbers of bacteria and protists present in marine biofilms. BLOEM ET AL. (1995) is one of the most comprehensive studies and applies a range of image analysis approaches. Quantification of signal intensity is also a common application of image processing and analysis – for example, after hybridization of bacteria with rRNA-targeted probes to determine the ribosome content and estimate the actual metabolic activity in pure cultures or biofilms (DELONG ET AL. 1989; MÖLLER ET AL. 1995; POULSEN ET AL. 1993).

Image analyses packages are available through a number of sources, including the manufacturers Bio-Rad (Bio-Rad Microsciences, Hemel Hempstead, UK), Leica (Leica Microsystems, Heidelberg, Germany), KS Kontron (Carl Zeiss Vision, Eching, Germany), Quantimet (Leica Cambridge, Cambridge, UK) or GTFS (Ultimage, GTFS Inc., Santa Rosa, CA, USA); from the authors such as DUBUISSON ET AL. 1994; MÖLLER ET AL. 1995; MORGAN ET AL. 1991; VILES AND SIERACKI 1992; or over the internet via, for example, Cellstat (<url>http://www.lm.dtu.dk/cellstat/index.html</url>) or NIH image (<url>http://rsb.info.nih.gov/nih-image/</url>) The package ImageSpace (Molecular Dynamics, Sunnyvale, USA), operating on a Silicon Graphics platform provides an excellent analysis system including functional 3-D image analysis. However, none of the available packages meets the specific requirements of any end user, and thus care must be taken to assess the limitations of any package prior to purchase and application.

5.2
Three-Dimensional Imaging

CLSM allows the user to produce a three-dimensional image of an object with appreciable depth because the procedures create a series of optical thin sections in perfect register. The reader is referred to LAWRENCE ET AL. (1998A) for a discussion of the nature and limitations of the ideal z series. Critical considerations during z series collection are:

- Selection of the sectioning interval to prevent over or under sampling errors
- Minimize damage to the specimen by the laser beam
- Compensation for bleaching effects
- Reduce bleaching through application of fade retardant solutions
- Changing the fluor in use or staining approach (CALDWELL ET AL. 1992 B)
- Working within the total or optimal slice capacity (STEVENS 1994) of the objective lens and within its true working distance

Failure to consider these factors will degrade both the appearance and the analytical usefulness of a z series. The user should evaluate potential problems by varying laser beam intensity, frequency of section, number of scans, the angles of the scans through the biofilm, and comparing the sections and the resulting reconstructions of similar materials. Those approaches that result in similar views will be most effective in the system being studied.

5.3
Three-Dimensional Display

There are a variety of approaches to 3-D reconstructions. The user must select the number of sections to be used. A few sections, for example 12, may be used to create a smooth stereoscopic image without evidence of the individual sections. However, too many sections will create a projection that is confusing. The simplest approach for 3-D presentation is the creation of a stereo pair in which two extended focus images are made through the same location in the specimen and one is shifted 15 pixels or 5° relative to the other and then the two images aligned to make a stereo pair. Stereo pairs have been presented to show development of *Vibrio parahaemolyticus* biofilms, (LAWRENCE ET AL. 1991) and colonization and weathering of sulfide minerals (LAWRENCE ET AL. 1997). Red-green anaglyph projections may be used to create a 3-D effect which is achieved when they are examined using red green glasses. Examples of this type of projection may be found in AMANN ET AL. (1995), CALDWELL ET AL. (1992A) and NEU AND LAWRENCE (1997). Polarized double projection which also requires specialized glasses for observation or "simulated fluorescence" whereby the material is viewed as though it were illuminated from an oblique angle and the surface layer was fluorescent are other common options for 3-D presentation. Solid body or surface projections may also be animated allowing presentation of the entire data set.

Stereo projections may be color coded by depth, or by stain so that materials originating from the same source or depth appear in the same colour. NEU AND LAWRENCE (1997) showed the result of using a combination of nucleic acid stain, glycoconjugate staining (see Fig. 6a) and autofluorescence to examine the structure of mixed species biofilms in river microcosms.

6
Case Studies: In Situ Characterization of EPS

Previous studies have applied lectins to examine the EPS associated with pure cultures of various organisms and this lectin analysis has yielded useful information for the characterization of sugars and cells walls. These interactions have been evaluated using agglutination, flow cytometry, electron microscopy, and epifluorescence microscopy. As has been the case in a variety of studies, the lectins applied were commercially available and selected on the basis of other studies showing bacterial EPS to contain a variety of residues including mannose, glucose, galactose, glucosamine, *N*-acetyl-α-galactosaminyl, fucose, etc. However few authors have studied complex natural systems.

Preliminary CLSM data is available for cell surface and EPS characterization by employing fluorescent lectins. The principle proof of lectin binding has been demonstrated for a variety of complex samples. The examples include staining of fungal cell surfaces with FITC or TRITC conjugated *Triticum vulgaris* lectin. The images show a fully hydrated lotic aggregate colonized by fungi (Fig. 3a, b). Another example originates from a marine habitat. The marine *Thioploca* fila-

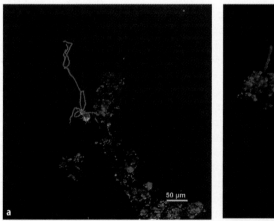
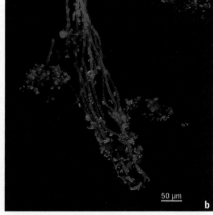

Fig. 3a, b. CLSM image of river aggregate with associated fungi stained with a specific lectin: **a** extended focus image of *Triticum vulgaris* FITC-lectin stained lotic aggregate with fungal hyphae; **b** extended focus image of *Triticum vulgaris* TRITC-lectin stained lotic aggregate with fungal hyphae

ments have been examined by using fluorescent labeled lectins. The filamentous cells are embedded in a sheath which allowed penetration of the lectins and staining of the cell surface (Fig. 4a, b). These images clearly demonstrate the specific differentiation between the target cell labeling and the associated components. Furthermore, by applying the same lectin with different fluorescence marker, e.g., FITC (green) or TRITC (red), the inverse staining can be demonstrated. Other examples from lotic habitats included biofilms and microbial aggregates. Complex biofilms grown in microcosms showed a specific labeling of algal "footprints" (COOKSEY AND COOKSEY 1986; NEU AND MARSHALL 1991; NEU 1992). This adhesive material which attached the algae to the surface did bind *Canavalia ensiformis* lectin (Fig. 5a). A final example of lectin cell surface labeling shows the resolution at the bacterial cell level. The image was taken from a biofilm reactor with a complex sulfate reducing microbial community (Fig. 5b). These examples from a range of habitats offer an outline of possible applications and show the potential as well as the specificity of the lectin binding analysis technique.

MICHAEL AND SMITH (1995) used lectins and epifluorescence microscopy to characterize the chemical nature and physical distribution of glycoconjugates (e.g., glycolipids, glycoproteins, polysaccharides) in marine biofilms. They described complex spatial patterns of receptors for fluorescently labeled lectins from *Canavalis ensiformis, Limulus polyphemus* and *Helix pomatia* in thin one- and three-day old marine biofilms. They noted that each lectin had localized discrete domains of binding on the biofilm surfaces.

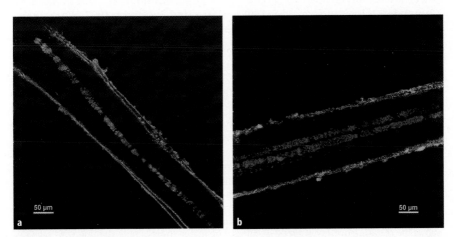

Fig. 4a, b. CLSM of marine Thioploca filaments stained with lectins: **a** extended focus image of Thioploca filaments in sheath labelled with *Ulex europaeus* FITC-lectin and *Lens culinaris* TRITC-lectin. Only the *Ulex europaeus* FITC-lectin did bind to the Thioploca cell surface (= green). However, both lectins did bind to material on the surface of the sheath in which the filaments were embedded (overlay of green and red = yellow); **b** inverse labelling of Thioploca filaments in sheath with the same two lectins having the opposite fluorescent labelling. *Ulex europaeus* TRITC-lectin and *Lens culinaris* FITC-lectin. Only the *Ulex europaeus* TRITC-lectin did bind to the Thioploca cell surface (= red)

ITC=tetramethylrhodamine
isothiocyanate
conjugate
(red)

FITC=
fluorescein
isothiocyanate conjugate
(green)

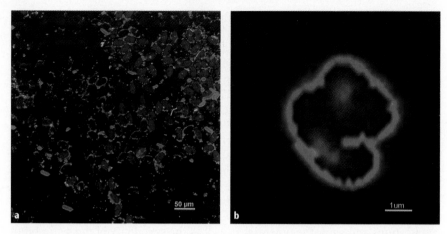

Fig. 5a, b. Lectin staining of cell surface and adhesive polymers: **a** polymeric "footprints" of algae. Biofilm with algae grown in a rotating annular reactor. The autofluorescence of algae is presented in blue. The algal cells were partly detached leaving algal "footprints" on the surface. The polymeric "footprints" material was stained with *Canavalia ensiformis* TRITC-lectin; **b** bacterial cell capsule. Zoomed image of bacteria from a complex anaerobic biofilm grown in a membrane biofilm reactor. The biofilm was stained with SYTO 9 for nucleic acids and *Canavalia ensiformis* TRITC-lectin to visualize the extracellular polymeric substances

Application of CSLM coupled with the putative specificity of lectins has recently been used to probe quantitatively the spatial relationships of EPS components within relatively thick complex biofilm communities (HOLLOWAY AND COWEN 1997; LAWRENCE ET AL. 1998 B; NEU AND LAWRENCE 1997; WOLFAARDT ET AL. 1998). The observations of these authors have indicated that sufficient quantities of specific residues were present to be easily visualized and localized within biofilm systems.

In the study of NEU AND LAWRENCE (1997) analysis of biofilm chemistry using a panel of FITC and TRITC conjugated lectins showed the complex chemical composition of these biofilms, indicating the presence of L-fucose, D-man, D-glc, D-galNAc, β-D-gal(1–4)-D-glcNAc and sialic acid residues. The spatial distribution of lectin binding residues was also highly heterogeneous as shown

Fig. 6a, b. Examples of complex microbial biofilm communities examined with CLSM: **a** complex biofilm grown in a river microcosm. Red-Green-Blue (RGB) overlay of a triple stained biofilm. The sample was probed with SYTO 9 = green as well as with the fluor conjugated lectins *Canavalia ensiformis*-FITC = red and *Arachis hypogaea*-TRITC = blue. The color wheel shows the resulting overlay of the RGB colors. The image shows single staining with each one of the probes as well as other combinations. It clearly demonstrates the variation of glycoconjugate distribution within the biofilm. Reprinted from FEMS Microbiology Ecology (by permission of Elsevier); **b** aggregate from the river Elbe (Germany). Simultaneous three-channel extended focus image of microbial community associated with suspended particulate matter. The aggregate was stained with *Triticum vulgaris* FITC-lectin (= green) and *Tetragonolobus purpureas* TRITC-lectin (= red). The double staining is shown as the overlay of green and red (= yellow). The algal autofluorescence was recorded in the far red channel (= blue)

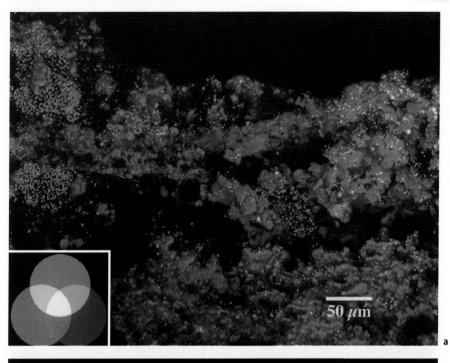

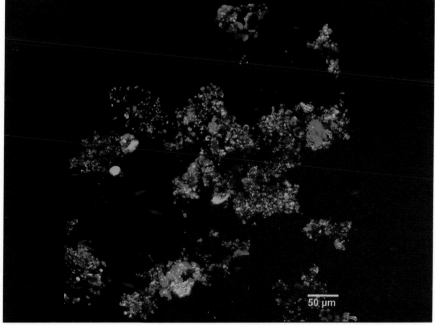

in Fig. 6a. This figure is a combination of images of three different staining techniques (*Canavalia ensiformis* lectin, *Arachis hypogaea* lectin, and SYTO 9) applied sequentially to the same biofilm location. The results indicated areas that bound none of the probes (black), all of the probes (white), and other possible combinations. For example, numerous microcolonies can be seen that stained with the nucleic acid stain but with neither of the lectin probes.

Similar biofilms were exposed to grazing by snails and ostracodes. The mature biofilms were examined by CLSM in three channel mode before and after grazing. The signals detected were autofluorescence, bacterial cells, and EPS distribution (LAWRENCE ET AL. 1998 B).

The nature of microbial EPS involved in the accumulation of chlorinated organics was studied in a nine-member biofilm community (WOLFAARDT ET AL. 1998). It was found that distribution of the herbicide diclofop methyl, *Ulex europaeus* Type I lectin glycoconjugate, a polyanionic dextran, and the hydrophobic dye Nile Red were identical. This indicated a degree of spatial organization and differentiation of EPS within this defined biofilm community.

Glycoconjugate distribution has also been demonstrated in formaldehyde fixed marine snow samples. The binding of *Canavalia ensiformis* lectin was shown for pure cultures and marine snow aggregates (HOLLOWAY AND COWEN 1997).

The fully hydrated architecture of microbial communities associated with suspended particulate matter from rivers was also examined by lectin analysis (Fig. 6b). By single channel scanning, simultaneous dual or triple channel scanning, the following signals were recorded: reflection, autofluorescence, cells, lectin glycoconjugates. The technique did not require any embedding or fixation which would disrupt the fragile structure of the aggregates thus providing clear impression of the architecture of these bioaggregates (NEU 1999 B).

In general, few if any studies have considered the difficulties inherent in the application of these protein fluor conjugates in complex microbial communities. In addition, questions regarding the specificity of the probe, non-specific binding, the application of positive and negative controls, the effect of the fluor conjugated to the lectin, concentration effects, and other environmental influences on the observed binding patterns. Thus although there is obvious potential in the method, definitive investigations in complex communities remain to be done.

7
Limitations of Non-Destructive Analyses

7.1
FT-IR/ATR

FT-IR/ATR can be used only if the biofilms develop on a clean or coated germanium or zinc selenide crystal. In addition, the depth of penetration of the IR signal is dependent on the refractive index of the optically thinner and denser medium, the selected wavelength, and the approaching angle of the beam. This results in some restrictions for the application of this technique to biofilms. But for that reason FT-IR/ATR is ideal for studying "conditioning films", elemental coatings or initial events of biofilm development. It is not the appropriate method for the analysis of biofilms having thicknesses of several hundred micrometers.

7.2
NMR

Studies in cell cultures indicate tremendous potential for the application of NMR to biofilms. However, for full exploitation of NMR as an in situ technique, considerable method development is still necessary.

7.3
CLSM

CLSM may be the most versatile analytical microscopic instrument for biofilm investigation. As a result of the development of cheaper hardware and software, CLSM may become one of the basic instruments in biofilm research. Nevertheless, there are some limitations to its use:

1. There is a physical limit due to the resolution of light microscopy.
2. The samples must be translucent for light to be able to excite the probes and to collect the emitted signal. This may be critical in sediment or samples associated with mineral precipitates.
3. The autofluorescence of the substratum may be very high (e.g., PVC) thereby making imaging of probes impossible.
4. Motility of organisms can create problems during collection of data. However, this may be the proof of imaging living samples.
5. There are uncertainties of binding specificity to EPS if antibodies or lectins are used. We have experienced very clear cut multi-labeling but also non-specific binding patterns. As a consequence, methodological development is still required to apply successfully and interpret the binding of fluorescent probes in CLSM.
6. The publication of four-channel and three-dimensional data on two-dimensional paper does not allow the full presentation and imagination of the datasets recorded. It is likely that in the near future video sequences collected by CLSM can be easily shown at conferences or in publications via the Internet.

Acknowledgements. The authors would like to acknowledge the financial support of Environment Canada and the Helmholtz Gemeinschaft Deutscher Forschungszentren. The work was also supported by the Canadian-German scientific and technological research cooperation (grants ENV46, ENV46/2). The supply of marine *Thioploca* samples by Bo Barker Jörgensen is highly appreciated. Special thanks are due to Ute Kuhlicke, George Swerhone, and Brij Verma for technical assistance on both sides of the Atlantic.

References

Agard DA (1984) Optical sectioning microscopy: cellular architecture in three dimensions. Annu Rev Biophys Bioeng 13:191–219

Allison DG, Sutherland IW (1984) A staining technique for attached bacteria and its correlation to extracellular carbohydrate production. J Microbiol Meth 2:93–99

Amann RI, Ludwig W, Schleifer K-H (1995) Phylogenetic identification and in situ detection of individual microbial cells without cultivation. Microbiol Rev 59:143–169

An YH, Friedman RJ, Draughn RA, Smith EA, Nicholson JH, John JF (1995) Rapid quantification of staphylococci adhered to titanium surfaces using image analyzed epifluorescence microscopy. J Microbiol Meth 24:29–40

Aspinall GO (1982) Isolation and fractionation of polysaccharides. In: Aspinall GO (ed) The polysaccharides, vol I. Academic Press, New York, pp 19–34

Assmus B, Schloter M, Kirchhof G, Hutzler P, Hartmann A (1997) Improved in situ tracking of rhizosphere bacteria using dual staining with fluorescence-labeled antibodies and rRNA-targeted oligonucleotides. Microb Ecol 33:32–40

Axelrod A, Koppel DE, Schlessinger J, Elsen E, Webb WW (1976) Mobility measurement by analysis of fluorescence photobleaching recovery kinetics. Biophys J 16:1055–1069

Beech IB (1996) The potential use of atomic force microscopy for studying corrosion of metals in the presence of bacterial films – an overview. Int Biodeter Biodegr 37:141–149

Birmingham JJ, Hughes NP, Treloar R (1995) Diffusion and binding measurements within oral biofilms using fluorescence photobleaching recovery methods. Phi Trans Soc Lond B Biol Sci 350:325–343

Bloem J, Veninga M, Sheperd J (1995) Fully automatic determination of soil bacterium numbers, cell volumes, and frequencies of dividing cells by confocal laser scanning microscopy and image analysis. Appl Environ Microbiol 61:926–936

Blonk JCG, Don A, van Aalst H, Birmingham JJ (1993) Fluorescence photobleaching recovery in the confocal scanning laser microscope. J Microsc 169:363–374

Bremer PJ, Geesey GG (1991a) An evaluation of biofilm development utilizing non-destructive attenuated total reflectance Fourier transform infrared spectroscopy. Biofouling 3:89–100

Bremer PJ, Geesey GG (1991b) Laboratory based model of microbiologically induced corrosion of copper. Appl Environ Microbiol 57:1956–1962

Bremer PJ, Geesey GG, Drake B (1992) Atomic force microscopy examination of the topography of a hydrated bacterial biofilm on a copper surface. Curr Microbiol 24:223–230

Buswell CM, Herlihy YM, Lawrence LM, McGuiggan JTM, Marsh PD, Keevil CW, Leach SA (1998) Extended survival and persistence of *Campylobacter* spp. in water and aquatic biofilms and their detection by immunofluorescence-antibody and -rRNA staining. Appl Environ Microbiol 64:733–741

Caldwell DE, Korber DR, Lawrence JR (1992a) Confocal laser microscopy and digital image analysis in microbial ecology. Adv Microbial Ecol 12:1–67

Caldwell DE, Korber DR, Lawrence JR (1992b) Imaging of bacterial cells by fluorescence exclusion using scanning confocal laser microscopy. J Microb Meth 15:249–261

Caldwell DE, Wolfaardt GM, Korber DR, Lawrence JR (1996) Cultivation of microbial consortia and communities. In: Hurst CJ, Knudson GR, McInerney MJ, Stetzenbach LD, Walker MV (eds) Manual for environmental microbiology. American Society for Microbiology, pp 79–90

Caldwell DE, Wolfaardt GM, Korber DR, Lawrence JR (1997a) Do bacterial communities transcend Darwinism? Adv Microb Ecol 15:105–191

Caldwell DE, Atuku E, Wilkie DC, Wivcharuk KP, Karthikeyan S, Korber DR, Schmid DF, Wolfaardt GM (1997b) Germ theory versus community theory in understanding and controlling the proliferation of biofilms. Adv Dent Res 11:4–13

Chenu C, Jaunet AM (1992) Cryoscanning electron microscopy of microbial extracellular polysaccharides and their association with minerals. Scanning 14:360–364

Chróst RJ (1991) Microbial enzymes in aquatic environments. Springer, Berlin Heidelberg New York

Clark DS (1990) Noninvasive techniques in studies of immobilized cells. In: Bont JAM de, Visser J, Mattiasson B, Tramper J (eds) Physiology of immobilized cells. Elsevier, Amsterdam, pp 603–613

Cooksey KE, Cooksey B (1986) Adhesion of fouling diatoms to surfaces: some biochemistry. In: Evans LV, Hoagland KD (eds) Algal fouling. Elsevier, Amsterdam, pp 41–53

Costerton JW, Nickel JC, Ladd TI (1986) Suitable methods for the comparative study of free-living and surface-associated bacterial populations. In: Pointdexter JS, Leadbetter ER (eds) Bacteria in nature. Plenum Press, New York, p 49–84

Costerton JW, Lewandowski Z, De Beer D, Caldwell D, Korber D, James G (1994) Biofilms, the customized microniche. J Bacteriol 176:2137–2142

Danilatos GD (1991) Review and outline of environmental SEM at present. J Microsc 162:391–X402

Dazzo FB, Wright SF (1996) Production of anti-microbial antibodies and their use in immunofluorescence microscopy. In. Akkermans ADL, van Elsas JD, de Bruijn FJ (eds) Molecular microbial ecology manual. Kluwer, Dordrecht, 4.1.2: pp 1–27

De Beer D, Stoodley P, Lewandowski Z (1997) Measurement of local diffusion coefficients in biofilms by microinjection and confocal microscopy. Biotechnol Bioeng 53:151–158

DeLong EF, Wickham GS, Pace NR (1989) Phylogenetic stains: ribosomal RNA-based probes for the identification of single cells. Science 243:1360–1363

Drake B, Prater CB, Weisenhorn AL, Gould SAC, Albrecht TR, Quate CF, Cannell DS, Hansma HG, Hansma PK (1989) Imaging crystals, polymers and processes in water with the atomic force microscope. Science 241:1586–1589

Dubuisson MP, Jain AK, Jain MK (1994) Segmentation and classification of bacterial culture images. J Microbiol Meth 19:279–295

Erdos GW (1986) Localization of carbohydrate-containing molecules. In: Aldrich HC, Todd WJ (eds) Ultrastructure techniques for microorganisms. Plenum Press, New York, pp 399–420

Faude UC, Höfle MG (1997) Development and application of monoclonal antibodies for in situ detection of indigenous bacterial strains in aquatic systems. Appl Environ Microbiol 63:4534–4542

Geesey GG, Jang L (1989) Interactions between metal ions and capsular polymers. In: Beveridge TJ, Doyle RJ (eds) Metal ions and bacteria. Wiley, New York, pp 325–357

Geesey GG, Iwaoka T, Griffith PR (1987) Characterization of interfacial phenomena occurring during exposure of a thin copper film to an aqueous suspension of an acidic polysaccharide. J Coll Interface Sci 120:370–376

Geesey GG, Jang L, Jolley JG, Hankins MR, Iwaoka T, Griffith PR (1988) Binding of metal ions by extracellular polymers of biofilm bacteria. Wat Sci Technol 20:161–165

Gilbert B, Assmuss B, Hartmann A, Frenzel P (1998) In situ localization of two methanotrophic strains in the rhizosphere of rice plants. FEMS Microbiol Ecol 25:117–128

Gonzalez RC, Wintz P (1977) Digital image processing. Addison-Wesley, Reading, Massachusetts.

Gorby GL (1994) Digital confocal microscopy allows measurements and three-dimensional multiple spectral reconstructions of *Neisseria gonorrhoe* / epithelial cell interactions in the human fallopian tube organ culture model. J Histochem Cytochem 42:297–306

Hansen M, Kragelund L, Nybroe O, Sörensen J (1997) Early colonization of barley roots by *Pseudomonas fluorescens* studied by immunofluorescence technique and confocal laser scanning microscopy. FEMS Microbiol Ecol 23:353–360

Hansma PK, Drake B, Mari O, Gould SAC, Prater CB (1989a) The scanning ion conductance microscope. Science 243:641–643

Hansma PK, Elinga VB, Mari O, Bracker CE (1989b) Scanning tunneling microscopy and atomic force microscopy: application to biology and technology. Science 241:209–216

Haugland RP (1996) Handbook of fluorescent probes and research chemicals. Molecular Probes, Eugene, Oregon

Holloway CF, Cowen JP (1997) Development of a scanning confocal laser microscopic technique to examine the structure and composition of marine snow. Limnol Oceanogr 42:1340–1352

Hood MA, Schmidt JM (1996) The examination of *Seliberia stellata* exopolymer using lectin assays. Microb Ecol 31:281–290

Jahn A, Nielsen PH (1995) Extraction of extracellular polymeric substances (EPS) from biofilms using a cationic exchange resin. Wat Sci Technol 32:157–164

Jolley JG, Geesey GG, Hankins MR, Wright RB, Wichlacz PL (1989) In situ, real-time FT-IR/CIR/ATR study of the biocorrosion of copper by gum arabic, alginic acid, bacterial culture supernatant and *Pseudomonas atlantica* exopolymer. Appl Spect 43:1062–1067

Jones AH, Lee C-C, Moncla BJ, Robinovitch MR, Birdsell DC (1986) Surface localization of sialic acid on *Actinomyces viscosus*. J Gen Microbiol 132:3381–3391

Korber DR, Caldwell DE, Costerton JW (1994) Structural analysis of native and pure-culture biofilms using scanning confocal laser microscopy. Proceedings of the National Association of Corrosion Engineers (NACE) Canadian Region Western Conference, Calgary, Ab, pp 347–353

Kühl M, Lassen C, Revsbech N-P (1997) A simple light meter for measurements of PAR (400 to 700 nm) with fiber-optic microprobes: application for P vs E_0 (PAR) measurements in a microbial mat. Aquat Microb Ecol 13:197–207

Lamont HC, Silvester WB, Torrey JG (1987) Nile red fluorescence demonstrates lipid in the envelope of vesicles from N_2-fixing cultures of *Frankia*. Can J Microbiol 34:656–660

Lassen C, Ploug H, Jörgensen BB (1992) A fibre optic scalar irradiance microsensor: application for spectral light measurements in sediments. FEMS Microbiol Ecol 86:247–254

Laurent M, Johannin G, Gilbert N, Lucas L, Cassio D, Petit PX, Fleury A (1994) Power and limits of laser scanning confocal microscopy. Biological Cell 80:229–240

Lavoie DM, Little BJ, Ray RI, Bennett RH, Lambert MW, Asper V, Baerwald RJ (1995) Environmental scanning electron microscopy of marine aggregates. J Microsc 178:101–106

Lawrence JR, Korber DR, Caldwell DE (1989) Computer-enhanced darkfield microscopy for the quantitative analysis of bacterial growth and behavior on surfaces. J Microbiol Meth 10:123–138

Lawrence JR, Korber DR, Hoyle BD, Costerton JW, Caldwell DE (1991) Optical sectioning of microbial biofilms. J Bact 173:6558–6567

Lawrence JR, Wolfaardt GM, Korber DR (1994) Monitoring diffusion in biofilm matrices using confocal laser microscopy. Appl Environ Microbiol 60:1166–1173

Lawrence JR, Korber DR, Wolfaardt GM, Caldwell DE (1996) Analytical imaging and microscopy techniques In: Hurst CJ, Knudson GR, McInerney MJ, Stetzenbach LD, Walker MV (eds) Manual of environmental microbiology. American Society for Microbiology Press, Washington, DC, pp 29–51

Lawrence JR, Kwong YTJ, Swerhone GDW (1997) Colonization and weathering of natural sulfide mineral assemblages by *Thiobacillus ferrooxidans*. Can J Microbiol 43:69–78

Lawrence JR, Wolfaardt GM, Neu TR (1998a) The study of biofilms using confocal laser scanning microscopy. In: Wilkinson MHF, Schut F (eds) Digital analysis of microbes. Imaging, morphometry, fluorometry and motility techniques and applications. Modern microbiological methods series. Wiley, Sussex, pp 431–465

Lawrence JR, Neu TR, Swerhone GDW (1998b) Application of multiple parameter imaging for the quantification of algal, bacterial and exopolymer components of microbial biofilms. J Microbiol Meth 32:253–261

Leppard GG (1986) The fibrillar matrix component of lacustrine biofilms. Wat Res 20:697–702

Leppard GG, Heissenberger A, Herndl GJ (1996) Ultrastructure of marine snow I. Transmission electron microscopy methodology. Mar Ecol Prog Ser 135:289–298

Lewandowski Z, Altobelli SA, Majors PD, Fukushima E (1992) NMR imaging of hydrodynamics near microbially colonized surfaces. Water Sci Technol 26:577–584

Lewandowski Z, Altobelli SA, Fukushima E (1993) NMR and microelectrode studies of hydrodynamics and kinetics in biofilms. Biotechnol Prog 9:40–45

Lewandowski Z, Stoodley P, Altobelli S (1995) Experimental and conceptual studies on mass transport in biofilms. Wat Sci Technol 31:153–162

Lindberg B, Lönngren J, Svensson S (1975) Specific degradation of polysaccharides. Adv Carbohydr Chem Biochem 31:185–240

Lohmeier-Vogel EM, McIntyre DD, Vogel HJ (1990) Nuclear magnetic resonance spectroscopy as an analytical tool in biotechnology. In: Bont JAM de, Visser J, Mattiasson B, Tramper J (eds) Physiology of immobilized cells. Elsevier, Amsterdam, pp 661–676

Lübbers DW (1992) Fluorescence based chemical sensors. Adv Biosensors 2:215–260

Marie D, Vaulot D, Partensky F (1996) Application of the novel nucleic acid dyes YOYO-1, YO-PRO-1, and PicoGreen for flow cytometric analysis of marine procaryotes. Appl Environ Microbiol 62:1649–1655

Martin Y, Williams CC, Wickramasinghe HK (1988) Tip techniques for microcharacterization of materials. Scanning Microsc 2:3–8

Massol-Deya AA, Whallon J, Hickey RF, Tiedje JM (1995) Channel structures in aerobic biofilms of fixed-film reactors treating contaminated groundwater. Appl Environ Microbiol 61:769–777

Merker RI, Smit J (1988) Characterization of the adhesive holdfast of marine and freshwater caulobacters. Appl Environ Microbiol 54:2078–2085

Michael T, Smith CM (1995) Lectins probe molecular films in biofouling: characterization of early films on non-living and living surfaces. Mar Ecol Prog Ser 119:229–236

Möller S, Kristensen CS, Poulsen LK, Carstensen JM, Molin S (1995) Bacterial growth on surfaces: automated image analysis for quantification of growth rate-related parameters. Appl Environ Microbiol 61:741–748

Morgan P, Cooper CJ, Battersby NS, Lee SA, Lewis ST, Machin TM, Graham SC, Watkinson RJ (1991) Automated image analysis method to determine fungal biomass in soils and on solid matrices. Soil Biol Biochem 23:609–616

Morioka H, Tachibana M, Suganuma A (1987) Ultrastructural localization of carbohydrates on thin sections of *Stahpylococcus aureus* with silver methenamine and wheat germ agglutinin-gold complex. J Bacteriol 169:1358–1362

Neu TR (1992) Microbial "footprints" and the general ability of microorganisms to label interfaces. Can J Microbiol 38:1005–1008

Neu TR (1994) The challenge to analyse extracellular polymers in biofilms. In: Stal LJ, Caumette P (eds) Microbial mats, structure, development and environmental significance. NATO ASI Series – vol G 35. Springer, Berlin Heidelberg New York, pp 221–227

Neu TR (1999a) Confocal laser scanning microscopy (CLSM) of biofilms. In: Flemming H-C (ed) Investigation of biofilms. Technomic, Lancaster (in press)

Neu TR (1999b) In situ confocal laser scanning microscopy (CLSM) of river snow (submitted)

Neu TR, Lawrence JR (1997) Development and structure of microbial biofilms in river water studied by confocal laser scanning microscopy. FEMS Microbiol Ecol 24:11–25

Neu TR, Marshall KC (1991) Microbial "footprints" – a new approach to adhesive polymers. Biofouling 3:101–112

Neu TR, Dengler T, Jann B, Poralla K (1992) Structural studies of an emulsion-stabilizing exopolysaccharide produced by an adhesive, hydrophobic *Rhodococcus* strain. J Gen Microbiol 138:2531–2537

Nichols PD, Henson JM, Guckert JB, Nivens DE, White DC (1985) Fourier transform-infrared spectroscopic methods for microbial ecology: analysis of bacteria, bacteria-polymer mixtures and biofilms. J Microbiol Meth 4:79–94

Nivens DE, Chambers JQ, Anderson TR, Tunlid A, Smit J, White DC (1993) Monitoring microbial adhesion and biofilm formation by attenuated total reflection/Fourier transform infrared spectrsoscopy. J Microbiol Meth 17:199–213

Nivens DE, Palmer RJ Jr, White DC (1995) Continuous non-destructive monitoring of microbial biofilms: a review of analytical techniques. J Indust Microbiol 15:263–276

Ong CJ, Wong MLY, Smit J (1990) Attachment of the adhesive holdfast organelle to the cellular stalk of *Caulobacter crescentus*. J Bacteriol 172:1448–1456

Paul JH, Jeffrey WH (1985) Evidence for separate adhesion mechanisms for hydrophilic and hydrophobic surfaces in *Vibrio proteolytica*. Appl Environ Microbiol 50:431–437

Pawley JB (1995) Handbook of biological confocal microscopy. Plenum Press, New York

Perlin AS, Casu B (1982) Spectroscopic methods. In: Aspinall GO (ed) The polysaccharides, vol I. Academic Press, New York, pp 133–196

Poulsen LK, Ballard G, Stahl DA (1993) Use of rRNA fluorescence in situ hybridization for measuring the activity of single cells in young and established biofilms. Appl Environ Microbiol 59:1354–1360

Quintero EJ, Weiner RM (1995) Evidence for the adhesive function of the exopolysaccharide of *Hyphomonas* strain MHS-3 in its attachment to surfaces. Appl Environ Microbiol 61:1897–1903

Rees DA, Morris ER, Thom D, Madden JK (1982) Shapes and interaction of carbohydrate chains. In: Aspinall GO (ed) The polysaccharides, vol I. Academic Press, New York, pp 185–290

Revsbech NP, Jörgensen BB (1986) Microelectrodes: their use in microbial ecology. Adv Microbial Ecol 9:293–352

Richards SR, Turner RJ (1984) A comparative study of techniques for the examination of biofilms by scanning electron microscopy. Wat Res 18:767–773

Rogers J, Keevil CW (1992) Immunogold and fluorescein immunolabelling of *Legionella pneumophila* within an aquatic biofilm visualised by using episcopic differential interference contrast microscopy. Appl Environ Microbiol 58:2326–2330

Russ JC (1990) Computer-assisted microscopy: the measurement and analysis of images. Plenum Press, New York

Russ JC (1995) The image processing handbook. CRC Press, Boca Raton

Sanford BA, Thomas VL, Mattingly SJ, Ramsay MA, Miller MM (1995) Lectin-biotin assay for slime present in in situ biofilm produced by *Staphylococcus epidermidis* using transmission electron microscopy (TEM). J Indust Microbiol 15:156–161

Sayler GS, Layton AC (1990) Environmental application of nucleic acid hybridization. Annu Rev Microbiol 44:625–648

Schloter M, Borlinghaus R, Bode W, Hartmann A (1993) Direct identification, and localization of *Azospirillum* in the rhizosphere of wheat using fluorescence-labelled monoclonal antibodies and confocal scanning laser microscopy. J Microsc 171:173–7

Schloter M, Assmuss B, Hartmann A (1995) The use of immunological methods to detect and identify bacteria in the environment. Biotech Adv 13:75–90

Schloter M, Wiehe W, Assmus B, Steindl H, Becke H, Höflich G, Hartmann A (1997) Root colonization of different plants by plant-growth-promoting *Rhizobium leguminosarum* bv. trifolii R39 studied with monospecific polyclonal antisera. Appl Environ Microbiol 63:2038–2046

Schmidt JE, Ahring BK (1994) Extracellular polymers in granular sludge from different anaerobic sludge blanket (UASB) reactors. Appl Microbiol Biotechnol 42:457–462

Schmitt J, Flemming H-C (1998) FTIR-spectroscopy in microbial and material analysis. Int Biodeter & Biodegr 41:1–11

Schmitt J, Nivens D, White DC, Flemming H-C (1995) Changes of biofilm properties in response to sorbed substances – an FTIR-ATR study. Water Sci Technol 32:149–155

Sizemore RK, Caldwell JJ, Kendrick AS (1990) Alternate Gram staining technique using a fluorescent lectin. Appl Environ Microbiol 56:2245–2247

Späth R, Flemming H-C, Würtz S (1998) Sorption properties of biofilms. Water Science Technol 37:207–210

Stevens JK (1994) Introduction to confocal three-dimensional volume investigation. In: Stevens JK, Mills LR, Trogadis JE (eds) Three dimensional confocal microscopy: volume investigation of biological systems. Academic Press, New York, pp 3–24

Stoodley P, De Beer D, Lewandowski Z (1994) Liquid flow in biofilm systems. Appl Environ Microbiol 60:2711–2716

Suci PA, Siedlecki KJ, Palmer RJ Jr, White DC, Geesey GG (1997) Combined light microscopy and attenuated total reflection Fourier transform infrared spectroscopy for integration of biofilm structure, distribution, and chemistry at solid-liquid interfaces. Appl Environ Microbiol 63:4600–4603

Surman SB, Walker JT, Goddard DT, Morton LHG, Keevil CW, Weaver W, Skinner A, Hanson K, Caldwell D, Kurtz J (1996) Comparison of microscopic techniques for the examination of biofilms. J Microbiol Meth 25:57–70

Vasse JM, Dazzo FB, Truchet GL (1994) Reexamination of capsule development and lectin-binding sites on *Rhizobium japonicum* 3I1B110 by glutaraldehyde/ruthenium red/uranyl acetate staining method. J Gen Microbiol 130:3037–3047

Viles CL, Sieracki ME (1992) Measurement of marine picoplankton cell size by using a cooled charge-coupled device camera with image-analyzed fluorescence microscopy. Appl Environ Microbiol 58:584–592

Ward DM (1989) Molecular probes for analysis of microbial communities. In: Characklis WG, Wilderer PA (eds) Structure and function of biofilms. Wiley, New York, pp 145–163

Wetzel, RG, Ward AK, Stock M (1997) Effects of natural dissolved organic matter on mucilaginous matrices of biofilm communities. Arch Hydrobiol 139:289–299

White DC (1986) Non-destructive biofilm analysis by Fourier transform spectroscopy (FT/IR). In: Megusar F, Gantar M (eds) Perspectives in microbial ecology. Slovene Society for Microbiology, pp 442–446

Williams SNO, Callies RM, Brindle KM (1997) Mapping of oxygen tension and cell distribution in a hollow-fiber bioreactor using magnetic resonance imaging. Biotechnol Bioeng 56:56–61

Wolfaardt GM, Lawrence JR, Headley JV, Robarts RD, Caldwell DE (1994a) Microbial exopolymers provide a mechanism for bioaccumulation of contaminants. Microb Ecol 27:279–291

Wolfaardt GM, Lawrence JR, Robarts RD, Caldwell DE (1994b) Multicellular organization in a degradative biofilm community. Appl Environ Microbiol 60:434–446

Wolfaardt GM, Lawrence JR, Robarts RD, Caldwell DE (1995) Bioaccumulation of the herbicide diclofop in extracellular polymers and its utilization by a biofilm community during starvation. Appl Environ Microbiol 61:152–158

Wolfaardt GM, Lawrence JR, Robarts RD, Caldwell DE (1998) In situ characterization of biofilm exopolymers involved in the accumulation of chlorinated compounds. Microb Ecol 35:213–223

Wood PJ (1980) Specificity in the interaction of direct dyes with polysaccharides. Carbohydr Res 85:271–287

Yagoda-Shagam J, Barton LL, Reed WP, Chiovetti (1988) Fluorescein isothiocyanate-labeled lectin analysis of the surface of the nitrogen-fixing bacterium *Azospirillum brasilense* by flow cytometry. Appl Environ Microbiol 54:1831–1837

Zambon JJ, Huber PS, Meyer AE, Slots J, Fornalik MS, Baier RE (1984) In situ identification of bacterial species in marine microfouling films by using immunofluorescence technique. Appl Environ Microbiol 4:1214–1220

Zita A, Hermansson M (1997) Determination of bacterial cell surface hydrophobicity of single cells in culture and in wastewater in situ. FEMS Microbiol Ecol 152:299–306

CHAPTER 3

Extraction of EPS

Per H. Nielsen · Andreas Jahn

Environmental Engineering Laboratory, Aalborg University, Sohngaardsholmsvej 57, DK-9000 Aalborg, Denmark, *E-mail: i5phn@civil.auc.dk*

Keywords. EPS, Composition, Sampling Extraction, Purification, Analysis

1
Introduction

There are a number of reasons for studying the composition and properties of extracellular polymeric substances (EPS) in microbial aggregates, such as mats or biofilms, activated sludge flocs, marine snow, etc. In all cases, these compounds are considered to be important for the activity of the microorganisms inhabiting the aggregates and for the physico-chemical properties of bioaggregates.

The approach to studying EPS from aggregates can be manifold. It is often necessary to extract or separate the EPS from the cell to perform a more detailed study. The amount or composition of EPS may be of interest, as well as their gel-forming properties. In some studies, a certain fraction of the EPS components are separated, purified, and examined for specific properties or chemical structure. These different aims are reflected in the numerous extraction protocols reported in the literature. Many different extraction mechanisms are used, either based on physical or chemical principles or a combination of these, and the selection of an appropriate method for a certain purpose must take place after careful evaluation of needs and available methods.

An important question always arises concerning the need to separate EPS and cell biomass quantitatively or if only certain compounds should be extracted from the EPS for further investigation. In the latter case, a homogenization of the entire sample can often be the best solution followed by a separation of cellular and extracellular macromolecules. This can for example be used for the extraction of lipopolysaccharide (JOHNSON AND PERRY 1976). If a quantitative separation is the object, it is important not to induce cellular lysis and keep the cell membrane structure as intact as possible. This is formulated in more general terms for the extraction of EPS from undefined cultures growing in aggregates (GEHR AND HENRY 1983). The procedure should (i) cause minimal cell lysis, (ii) not disrupt or alter biopolymers, and (iii) release all the EPS biopolymers. However, as will be discussed below, it is hardly possible to meet these requirements with the extraction methods known today.

The present chapter describes several parameters that are important when EPS are extracted from microorganisms, either from pure cultures or undefined aggregates in natural and technical systems. Focus will be on the extraction methods and the methods of validation the extraction performance.

1.1
Definition of EPS

Several definitions of EPS are reported in the literature. For cells in pure cultures, terms such as capsules, sheaths, and slimes are often used and basically defined by the nature of the polymer's association with the cells. Capsules are closely associated with the cells, while slimes do not have any direct contact with the cells. The way to separate the two fractions is usually by centrifugation, where the polymers in the supernatant are defined as slimes and the polymers bound to the pellet as capsular polymers. This illustrates that in many cases EPS are rather defined by the method used to separate/extract the EPS than by a theoretical consideration of the composition of the cell wall and the macro-molecules outside the cell wall.

The definitions of capsular and slime polymers in natural systems are often slightly different from the definition used in pure culture systems (DECHO 1994): slimes are used to describe a looser binding to the bioaggregate than the "condensed" gels (capsular polymers), but they are not dissolved. The dissolved polymers are called "colloidal".

However, as the physical state of the EPS gel is, to a large extent, governed by environmental parameters such as pH, available ions, etc., making the differen-tiation difficult, we use the terms "bound EPS" and "soluble EPS" in this chap-ter as has been done for some biofilm systems (HSIEH ET AL. 1994; NIELSEN ET AL. 1997):

- Bound EPS (sheaths, capsular polymers, condensed gel, loosely bound poly-mers, attached organic material)
- Soluble EPS (soluble macromolecules, colloids, slimes)

This means that all polymers outside the cell wall not directly anchored in the outer membrane/murein-protein-layer will be considered extracellular EPS material. We do not consider internal components of the cell wall such as LPS as extracellular, although LPS is associated with the EPS material. Moreover, a protein layer (S-layer) sometimes present around many bacterial cells is not considered EPS.

EPS (bound or soluble) does not only contain high-molecular-weight mu-cous secretions from the microorganisms, but also products of cellular lysis and hydrolysis of macromolecules. In addition, it can include compounds attached from the water phase. This is particularly important to note when bioaggregates from natural systems are examined. Lysis and hydrolysis products may, in com-bination with large amounts of adsorbed macromolecules, e.g., humic sub-stances, be integrated into the EPS matrix, contributing significantly to the phy-sico-chemical properties of the matrix. This is shown in Fig. 1, where it is also important to stress that EPS is composed by a number of different organic macromolecules of different origin. It further means that, in bioaggregates from natural systems, it is basically impossible to perform any extraction that ex-tracts only the exopolymers produced by the microbes present in the aggregate.

1.2
Composition of EPS

EPS from bioaggregates is a very heterogeneous material. In compositional ana-lyses of activated sludge flocs, anaerobic sludge granules, biofilms, or exopoly-mers from freshwater bacteria, a very heterogeneous composition of the extra-cellular polymers has been reported (Table 1). Main organic fractions were pro-teins and carbohydrates, many of these probably as glycoproteins (HORAN AND ECCLES 1986; MORGAN ET AL. 1991), and humic substances. Significant amounts of DNA and RNA were also found. In many publications it is assumed that the main part of substances other than carbohydrates are adsorbed from the water phase to the bioaggregates or are lysis products, and not products of the intact

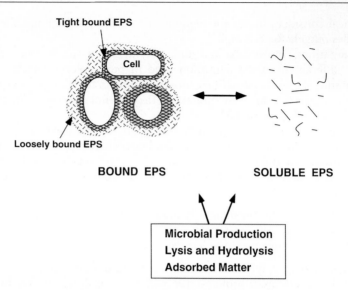

Fig. 1. Definition of cell biomass and extracellular polymeric substances (EPS). The EPS may arise from microbial production, lysis and hydrolysis, or from attachment

bacterial metabolism. However, data from pure cultures support the observations that many bacteria produce a range of compounds in the EPS matrix (BROWN AND LESTER 1980; JAHN AND NIELSEN 1995). It should be noted that in many protocols for EPS extraction, a step to remove protein and DNA from the sample is included, because these compounds are considered, often erroneously, as contamination with cellular material produced during the extraction procedure.

1.3
Extraction Procedure

A typical scheme of sampling, handling and analysis of bound EPS in bioaggregates is as follows:

1. Sampling and pretreatment: sampling of biological aggregates (e.g., activated sludge or biofilms) in environmental samples or sampling from bioreactors – pretreatment often includes a washing step and homogenization of the sample; the samples might be stored before further handling.
2. Extraction: the EPS components are extracted by an appropriate extraction procedure.
3. Purification: in some cases the extracted EPS are purified before further analysis.
4. Analysis: the EPS are usually analyzed for macromolecular composition (e.g., polysaccharides and proteins), and in some cases a more detailed investigation of the chemical composition or other characteristics is performed.

In the following sections, each of theses steps will be described in more detail.

Table 1. EPS composition and extraction efficiency with different physical and chemical extraction methods used in different systems. In brackets are given the extraction efficiencies reported for various substances. The carbon-extraction efficiency was calculated by summing up the organic carbon content (TOC) of the substances in the extract (protein: 0.375; carbohydrate/uronic acids: 0.378; DNA: 0.25; Humic substances: 0.28 (all in g C g⁻¹ VSS); these values were determined in a Beckmann model 700 carbon analyser with BSA, glucose, salmon testes DNA and humus as standards, respectively). Conversion factors from the total were calculated from VS(S) (volatile (suspended) solids) using a carbon-content of 0.55 (HENZE ET AL. 1995). From suspended solids (SS) VSS was calculated as $0.7 \times$ SS. n.d. = not determined

Method	System	Protein	DNA	Carbo-hydrate	Uronic acids	Humic sub-stances	Unit in original publication	Extraction effi-ciency (% org. C extracted from total org. C)	Reference
High-speed centrifugation	*Klebsiella aerogenes*	54.0	9.1	89.4	26.8	n.d.	mg g⁻¹ SS	17	BROWN AND LESTER 1980
Sonication/centrifugation	Activated sludge	279.9	67.34	60.54	n.d.	n.d.	mg l⁻¹	10	URBAIN ET AL. 1993
Heating 70 °C	UASB sludge	80	n.d.	13	n.d.	n.d.	mg g⁻¹ VSS	6.3	SCHMIDT AND AHRING 1994
Heating 80 °C	Activated sludge	121	n.d.	8	2.2	n.d.	mg g⁻¹ VS	9	FRØLUND ET AL. 1996
Steaming, 10 min	Activated sludge	77.1	3.7	15.8	n.d.	n.d.	mg g⁻¹ SS	10	BROWN AND LESTER 1980
NaOH (pH 11)	Activated sludge	96	n.d.	22	3.1	n.d.	mg g⁻¹ VS	8	FRØLUND ET AL. 1996
EDTA	Freshwater bio-film bacterium	94 (3)	19 (8)	110 (20)	14 (9)	n.d.	mg l⁻¹	3	PLATT ET AL. 1985
Dowex/shear	Activated sludge	243	n.d.	48	6.1	126	mg g⁻¹ VS	27	FRØLUND ET AL. 1996
Dowex/shear	Sewer biofilm	154 (14.1)	12	12 (3.3)	6 (19.3)	293 (67.6)	mg g⁻¹ TOC	15	JAHN AND NIELSEN, 1995
Dowex/shear	*Pseudomonas putida*	214 (15.7)	15	56 (29.8)	2 (40.8)	n.d.	mg g⁻¹ TOC	11	JAHN AND NIELSEN 1995
Cold aqueous technique	Anaerobic sludges	30	n.d.	10	n.d.	n.d.	mg g⁻¹ VSS	3	JIA ET AL. 1996

2
Sampling and Pretreatment

2.1
Sampling and Storage

Transport of environmental samples to the laboratory and further storage should take place at 0–4 °C to prevent any changes occurring, e.g., by exoenzymatic activity. Biofilms and activated sludge can be stored for 1–2 days without significant changes in the extracted EPS composition (Nielsen, unpublished results).

Freezing and thawing of the samples are generally not recommended, as cells may lyse and release periplasmic proteins (LALL ET AL. 1989). However, some authors recommend freeze-drying of the samples (biofilms in marine sediments) as fast as possible for later extraction of polysaccharides and subsequent analysis (UNDERWOOD ET AL. 1995). No detailed studies on the effect of freezing and thawing on extraction of EPS from aggregates have to the best of our knowledge been reported in the literature.

2.2
Homogenization

In order to obtain an optimal extraction, samples should be well homogenized without disruption of the cells. Different homogenizers are available, and a microscopic check of the particle size after homogenization is recommended.

2.3
Washing

Washing of the aggregate before extraction is often conducted to remove soluble EPS. It is important to select a washing buffer carrying an ionic strength and composition not too different from the sample, otherwise some bound EPS components might desorb and thus be washed away from the EPS matrix. If, for instance, the sample is mixed with deionized water, it can be regarded as a simple form of extraction (GAUDY AND WOLFE 1962).

3
Extraction of EPS

No universal extraction method exists for a quantitative extraction of bound EPS compounds from microorganisms growing in suspension or in aggregates. A number of different methods have been applied in studies on pure cultures or on undefined cultures, mainly related to activated sludge and biofilms. Only a few of the methods have been thoroughly evaluated to obtain an optimal extraction procedure with a high extraction efficiency without unwanted cell lysis and disruption of macromolecules. The extraction methods include various physical or chemical methods or combinations thereof. It is important to note that almost all methods rely on a certain water solubility of the components extracted, where the more hydrophobic compounds cannot be expected to be extracted.

3.1
Selection of an Appropriate Extraction Method

The extraction procedure must be selected for each case, considering the specific needs and constraints. In some studies, a certain fraction, e.g., the polysaccharides, are extracted for a more detailed chemical or structural analysis. In that case, lysis of the bacteria may not be a problem, if only the impurities can be removed before further analysis. In other cases, a quantitative extraction of all EPS is desired, if possible, and here it is of critical importance that no lysis takes place.

The best extraction method will depend on the type of interactions that keep the EPS components together in the matrix. The main forces involved in the binding of the polymers in the EPS matrix is van der Waal forces, electrostatic interactions, hydrogen bonds, hydrophobic interactions (CHRISTENSEN AND CHARACKLIS 1990), and in some cases, covalent bonds as disulfide bonds in glycoproteins (EMERSON AND GHIORSE 1993). The dominating forces may be different from one EPS matrix to another, so various methods must often be tested. Many chemical extraction methods rely on a breakage of the electrostatic interactions, thereby promoting an extraction of water soluble compounds. Less focus has been on the hydrophobic components, probably because it is difficult not to destroy the cells with the procedures.

It is particularly difficult to test the extraction performance in undefined cultures such as biofilms and flocs. As no universal extraction method exists, we recommend that extraction is only performed after running some comparative methods and initially optimizing and standardizing a selected extraction technique. This can be done by varying the variable (extraction time, shear, temperature, or chemical) and by recovery of added standards. This is very important (FRØLUND ET AL. 1996), because the EPS yield depended on the extraction time and shear rate, and moreover, the dependence was different for various EPS components. Furthermore, an evaluation of lysis and polymer disruption must be included.

3.2
Physical Methods

With the physical methods a shear is applied to extract EPS by centrifugation, mixing or shaking, sonication, or heat treatment (Tables 2 and 3). In general, the extraction yield is lower when using a physical treatment than when using combined chemical and physical treatments (see below). The extent of lysis or disruption of macromolecules has been evaluated only in a few studies, but, particularly, heat treatment may cause significant lysis and disruption (see below).

Centrifugation is often used to separate the soluble (slime) EPS fraction from cell biomass (TROCH ET AL. 1992; HEBBAR ET AL. 1992). The shear applied by centrifugation may only to a very limited extent extract bound EPS. High speed

Table 2. Physical extraction methods for defined cultures

Method	System	Intracellular leakage	Reference
Low-speed centrifugation (5000–10,000 × g)	*Azospirillum brasilense*	n.i.	TROCH ET AL. 1992
	Pseudomonas putida/fluorescens	n.i.	CONTI ET AL. 1994
	Pseudomonas alcaligenes	n.i.	TITUS ET AL. 1995
High-speed centrifugation (30,000–48,000 × g)	*Klebsiella aerogenes*	Protein/DNA	BROWN AND LESTER 1980
	Freshwater/marine isolates	n.i.	KENNEDY AND SUTHERLAND 1987
	Different pure cultures	n.i.	PAVONI ET AL. 1972
	Pseudomonas aeruginosa	n.i.	PAVONI ET AL. 1972
			BUCKMIRE 1984
High-speed centrifugation (113,000 × g)	Marine *Pseudomonas* sp.	n.i.	WRANGSTADH ET AL. 1986
Ultra Turrax (20,000 rpm, 60 s)	*Rhodopseudomonas capsulata*	n.i.	OMAR ET AL. 1983
Ultrasonication (10 min, 120 V, 18 W)	*Klebsiella aerogenes*	Protein/DNA	BROWN AND LESTER 1980
Ultrasonication (10 min, 120 V, 18 W)/	*Klebsiella aerogenes*	Protein/DNA	BROWN AND LESTER 1980
High-speed centrifugation		Protein	RUDD ET AL. 1982
Boiling (30 min)	*Proteus vulgaris*	n.i.	RAHMAN ET AL. 1997
Steam (autoclave, 10 min)	*Klebsiella aerogenes*	Protein/DNA	BROWN AND LESTER 1980

n.i. = not investigated.

Table 3. Physical extraction methods for undefined cultures. n.i. = not investigated

Method	System	Intracellular leakage	Reference
Passing through a syringe needle/PBS (phosphate-buffered saline) buffer	Anaerobic sludge	n.i.	DeBeer et al. 1996
Filtration (0.4 μm)	Activated sludge	n.i.	Hejzlar and Chudoba 1986
Repeated homogenization with a hand homogenizer	Methanogenic granular sludge	DNA, protein, polysaccharide	Grotenhuis et al. 1991
Heating (70 °C, 1–8 h)	Anaerobic sludge	n.i.	Schmidt and Ahring 1994
Heating (80 °C, 1 h)	Activated sludge	n.i.	Horan and Eccles 1986
	Anaerobic/activated sludge	n.i.	Morgan et al. 1990
	Anaerobic/granular sludge	n.i.	Forster and Quarmby 1995
Boiling (1 h)	Activated sludge	n.i.	Beccari et al. 1980
Steaming treatment (autoclave, 10 min)	Activated sludge	Protein/DNA	Brown and Lester 1980
		Protein/polysaccharide ratio	Karapanagiotis et al. 1989
Sonication (up to 60 W min^{-1})	Activated sludge	n.i.	King and Forster 1990
Sonication, 60 W, varying time	UASB granules	n.i.	Quarmby and Forster 1995
Sonication, 18 W, 120 V, 10 min	Activated sludge	Protein/DNA	Brown and Lester 1980
Sonication. 37 W, 60 s, centrifugation (20,000 × g)	Activated sludge	DNA, cell count	Jorand et al. 1995
Sonication, two 15 s periods, 50 W, 20 KHz/centrifugation (33,000 × g)	Activated sludge	n.i.	Urbain et al. 1993
Standard blender, 1 min/centrifugation (13,000 × g)	Activated sludge	DNA	Gehr and Henry 1983

n.i. = not investigated.

centrifugation was proposed (BROWN AND LESTER 1980) as the most effective method for EPS extraction from *K. aerogenes*. However, centrifugation does not provide any significant extraction of bound EPS from microbial aggregates from natural systems, as the components are usually strongly bound (e.g., NOVAK AND HAUGAN 1981; FRØLUND ET AL. 1996). Centrifugation is, however, almost always used after any extraction procedure to separate the extracted EPS from cells and other particles.

Shaking, stirring, or pumping the bacterial matrix through needles by using syringes to obtain a certain shear are other common physical methods used. A high shear level may be provided by sonication under defined conditions, which has been used to extract EPS from activated sludge and granular sludge. It should be noticed that it is essential to use defined and well-described shear conditions in order to get reproducible results. Also heat, either at 70–80 °C or boiling, typically for 1 h, has been used in several studies, sometimes combined with shear provided by shaking (Table 2).

3.3
Chemical Methods

The chemical treatments include addition of various chemicals to the bacterial sample that can break different linkages in the EPS matrix, facilitating a release of EPS to the water (Tables 4 and 5).

Alkaline treatment by addition of NaOH causes many charged groups, such as carboxylic groups in proteins and polysaccharides, to be ionized because their isoelectric points are generally below pH 4–6. It results in a strong repulsion between EPS within the EPS gel and provides a higher water solubility of the compounds. Alkaline hydrolysis of many polymers may take place (HANCOCK AND POXTON 1988). Also the covalent disulfide bindings in glycoproteins may be broken at pH values above 9 (ZAYAS 1997), promoting an extraction of these compounds. Disulfide bindings can also be reduced by adding β-mercaptoethanol or other compounds (EMERSON AND GHIORSE 1993). The pH used in various studies using alkaline treatment varies from 9 to 13, or it is indicated as addition of 1–9 N NaOH.

The repulsion among the components in the EPS matrix and the water solubility can also be increased by an exchange of divalent cations with monovalent ions. The divalent cations, mainly Ca^{2+} and Mg^{2+}, are very important for the crosslinking of charged compounds in the EPS matrix and by removing these, the EPS matrix tends to fall apart. The divalent cations can be removed by using a resin, e.g., Dowex, or by using a complexing agent such as EDTA or EGTA. Both methods have been used in biofilm systems, activated sludge and pure bacterial cultures (Table 2). There are some drawbacks in using EDTA. It can remove divalent cations from the cell wall, leading to destabilization of cell wall and release of components such as LPS (JOHNSON AND PERRY 1976) and possibly also contamination with cellular macromolecules. It may also interfere in the protein determination (see below).

Another way to carry out a cation exchange is by using a high concentration of NaCl. It has been used for the extraction of adhesive exopolymers from *P.*

Table 4. Chemical extraction methods for defined cultures. n.i. = not investigated

Method	System	Intracellular leakage	Reference
NaOH (2 N), 2–3 h	*Escherichia coli*	n.i.	SATO AND OSE 1980
EDTA (3.72 g l^{-1}), 15 min stirring	periphytic marine bacteria	n.i.	LABARE ET AL. 1989
Zwittergent (0.1 %) in 50 mmol l^{-1} citrate, pH 4.5, 42 °C, 30 min	*Klebsiella pneumoniae*	DNA	DOMENICO ET AL. 1989
Pyridine acetate (0.1 mol l^{-1}, pH 5.0), 1 h	*Escherichia coli*	n.i.	PELKONEN ET AL. 1988
NaCl (0.9 %)	*Pseudomonas aeruginosa*	n.i.	MAY AND CHAKRABARTY 1994
NaCl (0.5 mol l^{-1}), 60 min	*Pseudomonas* sp. NCMB 2021	n.i.	CHRISTENSEN ET AL. 1985
NaCl (1.0 mol l^{-1})	*Pseudomonas putida/ Pseudomonas fluorescens*	n.i.	READ AND COSTERTON 1987
Deflocculating enzyme/ 0.5 N NaOH, stirring for 24 h	*Pseudomonas* sp.	n.i.	TAGO AND AIDA 1977

n.i. = not investigated.

Table 5. Chemical extraction methods for undefined cultures

Method	System	Intracellular leakage	Reference
K_2HPO_4 (0.4 mol l^{-1}), 15 min	Activated sludge	DNA	GEHR AND HENRY 1983
NaOH (1.5 mol l^{-1}, 2 h)	Digested activated sludge	Protein/polysaccharide ratio	KARAPANAGIOTIS ET AL. 1989
NaOH (2 N, 2–3h)	Activated sludge	n.i.	SATO AND OSE 1980
NaOH (2 N, 5 h)	Activated sludge	Protein, DNA	BROWN AND LESTER 1980
Dowex (1:3 vol, 1 h)	Digested activated sludge	Protein/polysaccharide ratio	KARAPANAGIOTIS ET AL. 1989
Dowex (1:1 vol)/formaldehyde (1%)	Activated sludge	n.i.	RUDD ET AL. 1983
Ethanolic extraction, 1–16 days	Activated sludge	n.i.	FORSTER AND CLARKE 1983
Phenol (45% (w/v)), 70 °C, 15 min	Digested sewage sludge	Protein/polysaccharide ratio	KARAPANAGIOTIS ET AL. 1989
2-step extraction (1) 30 s homogenization with NH_4OH (1 N)/ (2) EDTA (2% (w/v)), pH 10, 3 h, 4 °C	Activated sludge	n.i.	SATO AND OSE 1984

n.i. = not investigated.

putida and *P. fluorescens* (READ AND COSTERTON 1987; CHRISTENSEN ET AL. 1985). In some investigations, the combination of high pH and ion exchange has been applied by using NH_4OH and EDTA (SATO AND OSE 1984).

Enzymatic digestion may be used to destabilize bioaggregates and enhance extraction. A deflocculation enzyme, prepared from the fluid culture of a *Pseudomonas sp.* isolated from activated sludge, has been used to deflocculate cells before further extraction (TAGO AND AIDA 1977).

The more hydrophobic part of the EPS is usually not extracted by specific procedures. In some cases, however, specific hydrophobic components are extracted from pure cultures by use of detergents, e.g., to isolate capsular antigen from *E. coli* (JANN ET AL. 1980; SCHMIDT AND JANN 1982) or to purify exopolysaccharides from *Klebsiella pneumoniae* (DOMENICO ET AL. 1989). Very few studies are being published on the hydrophobic extracellular fraction in environmental samples, probably because it is very difficult to assess whether cellular lysis takes place. Ethanol was used to extract lipids from activated sludge (FORSTER AND CLARKE 1983), but it is unknown to what extent cell lipids were extracted as well.

3.4
Combination of Physical and Chemical Methods

Many of the chemical methods reported have been used without applying a defined shear during the extraction, e.g., by shaking a few times. However, it appears that the chemical extraction becomes more reproducible and effective when it is combined with defined shear. The shear is typically provided by heat, sonication or stirring under defined conditions (Tables 6 and 7).

Alkaline treatment has been combined with heat treatment (70°C) to extract capsular EPS from *R. trifolii* (BREEDVELD ET AL. 1990). Ion exchange by a Dowex extraction has been used in combination with shear (stirring) to extract EPS from activated sludge and biofilms (FRØLUND ET AL. 1996; JAHN AND NIELSEN 1995). As ion exchange is controlled by diffusion, it is very important to standardize the experimental conditions such as shear, time and temperature. NaCl, formaldehyde, and ultrasonication were used to extract EPS from anaerobic sludge (JIA ET AL. 1996). The formaldehyde was added in order to tighten the cells to minimize cell lysis during the procedure. However, formaldehyde changes the properties of many EPS components and interferes in a possible later analysis of carbohydrates by the phenol-sulphuric acid method (UNDERWOOD ET AL. 1995).

4
Contamination by Intracellular Macromolecules

The questions concerning cell lysis have been described in detail only for some of the extraction methods mentioned above. Moreover, it is difficult to compare the results because different methods for evaluation of lysis have been used and because it is doubtful if all the methods actually measure cell lysis.

Table 6. Combined physical and chemical extraction methods for defined cultures. n.i. = not investigated; Glucose6PDH = glucose-6-phosphate dehydrogenase

Method	System	Intracellular leakage	Reference
Deionized water/blending (2 min)	*Sphaerotilus natans*	n.i.	GAUDY AND WOLFE 1962
NaOH (pH 11.5)/heating 40 °C, time unknown)	*Rhizobacteria*	n.i.	HEBBAR ET AL. 1992
NaOH (1 N)/heating (70 °C, 15 min)	*Rhizobium trifolii*	n.i.	BREEDVELD ET AL. 1990
Saline/sonication (60 W, 2 min)	*Staphylococcus epidermidis*	n.i.	EVANS ET AL. 1994
EDTA (10 mmol l⁻¹)/Waring blender (60 s)	Freshwater sediment bacterium	Glucose6PDH	PLATT ET AL. 1985
NaCl (0.1 mol l⁻¹)/EDTA 0.01 mol l⁻¹, 30 min stirring	*Clostridium acetobutylicum*	n.i.	JUNELLES ET AL. 1989
Dowex (240 g g⁻¹ organic carbon/shear (600 rpm, 1–2 h)	*Pseudomonas putida* (biofilm and suspended culture)	Glucose6PDH	JAHN AND NIELSEN 1995
Ethanol/high-speed centrifugation (48,000 × g)	Bacteroids *Bradyrhizobium japonicum*	n.i.	STREETER ET AL. 1994
Cetyltrimethylammonium bromide (1 % (w/v)/NaCl (0.04 mol l⁻¹))/heating (50 °C, 2 h)	*Rhodopseudomonas capsulata*	n.i.	OMAR ET AL. 1983
Hexadecyltrimethylammonium bromide (0.1% (w/v)/CaCl₂ (1 mol l⁻¹) precipitation	*Escherichia coli*	n.i.	SCHMIDT AND JANN 1982 JANN ET AL. 1980
NaCl (0.9%)/heating (50 °C, 2 h)	*Rhodopseudomonas capsulata*	n.i.	OMAR ET AL. 1983
K₂HPO₄ (0.02 mol l⁻¹)/blending (2 min)	*Zoogloea*	n.i.	FARRAH AND UNZ 1976

Table 7. Combined physical and chemical extraction methods for undefined cultures.

Method	System	Intracellular leakage	Reference
Dowex/shear (65–80 g g⁻¹ VS, 900 rpm, 1–2 h)	Activated sludge	Glucose6PDH	FRØLUND ET AL. 1996
Dowex/shear (240 g g⁻¹ organic matter, 600 rpm, 1–2 h)	Sewer biofilm	Glucose6PDH	JAHN AND NIELSEN 1995
NaCl (2.5 % (w/v))/EDTA (100 mmol l⁻¹), for 15 min, Vortex mixing	Intertidal sediments	n. i.	UNDERWOOD ET AL. 1995
NaCl (8.5 % (w/v))/formaldehyde (0.22 % (v/v))/ sonication (40 W for 3 min on ice) = "cold aqueous technique"	Anaerobic sludge	n. i.	JIA ET AL. 1996; original published in SUTHERLAND AND WILKINSON 1971
Phenol extraction (20 % (v/v)), 50 °C, 45 min, intermittent sonication	Methanogenic granules	n. i.	VEIGA ET AL. 1997

n.i. = not investigated; Glucose6PDH = glucose-6-phosphate dehydrogenase; VS = volatile solids.

The extent of cell lysis during extraction is difficult to evaluate in undefined cultures. In many studies, the accumulation of protein and nucleic acids in the crude extract has been taken as an indication of lysis. However, now it has been recognized that the EPS matrix usually contains large amounts of protein, nucleic acids, and probably also glycoprotein (GEHR AND HENRY 1983; URBAIN ET AL. 1993; FRØLUND ET AL. 1996), so the presence of these compounds in the extract is very difficult to use as an indicator for cell lysis. Instead, other methods must be applied, e.g., substances that are truly intracellular and do not accumulate in the EPS matrix. ATP has been used in addition to DNA as intracellular marker, but it is difficult to obtain good accuracy of the ATP measurements (GROTENHUIS ET AL. 1991). Another promising method is the appearance of intracellular enzymes, i.e., glucose-6-phosphate dehydrogenase (G6PDH) in the extract, indicating the extent of lysis (PLATT ET AL. 1985). A certain background activity may be present in the EPS matrix of flocs and biofilms (FRØLUND ET AL. 1995, 1996). Cell counts can only be used to see if cells are destroyed, but not to assess whether they leak intracellular material (FRØLUND ET AL. 1996).

Despite the problems with the analysis of cell lysis, boiling and addition of NaOH seem to cause severe cell lysis in activated sludge (KARAPANAGIOTIS ET AL. 1989). The cation exchange method with Dowex resin did not show strong lysis if short extraction times (less than 2 h) were conducted (FRØLUND ET AL. 1996).

An interesting question arises when extracting EPS from bioaggregates. How large is the cell fraction, and is some cell lysis important for the composition of the extracted EPS? In activated sludge as well as many natural biofilms, the cell biomass represents only 10–20% of the total organic matter (WANNER 1994; FRØLUND ET AL. 1996; JAHN AND NIELSEN 1998; MÜNCH AND POLLARD 1997). In such systems it can be argued that some lysis does not make much of a difference on EPS amount and composition (FRØLUND ET AL. 1996). Of course this will depend on the system investigated because higher values for cell biomass have been recorded in some defined biofilm cultures (see references in NIELSEN ET AL. 1997).

5
Disruption of Macromolecules

Disruption of macromolecules during extraction may take place. It is critical if the chemical structure or some macromolecular properties are to be investigated, but not if only the total amount of EPS is determined. Boiling and alkaline treatment have been reported to cause disruption of macromolecules (KARAPANAGIOTIS ET AL. 1989). In particular NaOH causes changes in the polymer composition by hydrolysis at pH values above 9 (HANCOCK AND POXTON 1988). Also deacylation of acylated alginates may take place, resulting in increased solubility (and decreased hydrophobicity). High pH also breaks disulfide bindings in glycoproteins (ZAYAS 1997), and uronic acids are degraded (HAUG ET AL. 1967). Possible disruption has not been investigated in detail for extraction by heating or sonication, while the cation exchange method with Dowex does not cause disruption (KARAPANAGIOTIS ET AL. 1989; FRØLUND ET AL. 1996).

It is also important to realize that the intrinsic enzymatic activities such as proteolytic/lipolytic/sugar cleaving activities can take place during extraction. These enzymes may be stable in the matrix and change the properties and composition of the macromolecules during extraction. So whenever possible, all procedures should be performed on ice. A protease inhibitor mix can be added when starting an extraction procedure to prevent protein degradation. Changes in the macromolecular composition of extract from undefined cultures can be evaluated using gel permeation chromatography (KARAPANAGIOTIS ET AL. 1989) or HPSEC on the crude extract (FRØLUND AND KEIDING 1994; FRØLUND ET AL. 1996).

6
Extraction Efficiency

Extraction efficiency can be defined in two ways: as the total amount of EPS extracted from all the organic matter in a certain sample, or as the total amount of EPS extracted from the total EPS pool in a certain sample. The second definition is the most correct definition, but since the total amount of EPS is usually unknown, this definition is rarely used. By using the first definition, it does not reveal anything about the total amount of EPS, but only how much EPS a certain method can extract of the total organic matter.

The extraction efficiency, in this section indicated as the amount of organic matter extracted, differs from method to method and from macromolecule to macromolecule, as described above. Some comparative work has been published on extraction of EPS from activated sludge (BROWN AND LESTER 1980; FRØLUND ET AL. 1996). For routine extraction of biofilms or flocs with a high yield of particularly protein and humic substances, the best choice seems at present to be to use a cation exchange resin (Dowex), combined with stirring or sonication under defined conditions as suggested by Rudd and co-workers (RUDD ET AL. 1983; KARAPANAGIOTIS ET AL. 1989) and by Frølund and co-workers (FRØLUND ET AL. 1996). An example of extraction yields from different methods is given in Table 8. Large variations can be observed from one study to another, reflecting use of many different extraction methods and different analysis methods. In some studies simple things such as units are incompatible. In order to compare extraction yields obtained in different studies it is important to refer to a common measurement, e.g., the amount of organic matter (volatile matter).

It is still uncertain which part of the exopolymers is extracted by the various methods. Many of the hydrophobic compounds, together with some polysaccharides, are not extracted by the commonly used methods. This might explain why some tightly bound exopolymers associated with cell clusters are not extracted from activated sludge (FRØLUND ET AL. 1996). Also in pure culture studies some capsular polysaccharides can be very difficult to extract and rough methods must be used, e.g., pH of 11.5 (HEBBAR ET AL. 1992). Such capsular material is probably not extracted by use of the methods normally applied for biofilms and activated sludge, and can hardly be extracted without damaging cells and polymers. It is important to realize that use of even well-standardized extraction procedures is still qualitative in nature, and perhaps only a minor part of the EPS is extracted.

Table 8. Extraction check list

Sampling/storage	Transport samples on ice to laboratory Check change in the sample during different storage times/conditions Do not freeze the sample
Washing	Use buffer with similar ionic strength/conductivity as sample Check loss of EPS/cellular material
Homogenization	Homogenize and check particle size after homogenization procedure
Extraction	Optimize and standardize extraction parameters (i.e. extraction time, stirring intensity, centrifugation, G-forces, chemical concentration, temperature) Evaluate cellular lysis by measuring intracellular enzyme activity Evaluate disruption of macromolecules by gel filtration or HPSEC Evaluate extraction efficiency after number of bacteria/size measurement, based on either TOC or protein; try to make a mass balance
Purification	Check crude extract for macromolecular composition When using further purification steps be aware of selective purification of macromolecules, depending on hydrophobicity/hydrophilicity
Analysis	Investigate possible interference in commonly used methods: e.g., protein/humic substances

In many studies it is important to know how much of the EPS is extracted (definition 2 above). Because no direct quantitative methods for separation of cell biomass and EPS are available, indirect methods must be used. One possibility is to measure the total biovolume by measuring the number and sizes of bacteria present. By using conversion factors from biovolume to TOC or protein, it is possible to calculate the total contribution from the cellular biomass fraction. By measuring the total content of protein or TOC in the samples, the EPS fraction can be calculated by simple subtraction as shown in Fig. 2 (JAHN AND NIELSEN 1998; MÜNCH AND POLLARD 1997). Protein accounts for about 50% of the cell dry weight (RUSSELL AND COOK 1995). Other cellular molecules like DNA or polysaccharide cannot be used because the cellular content varies. The DNA content can vary up to four times, depending on the growth conditions (MASON AND EGLI 1993), while polysaccharides can vary because they are used as storage material. TOC can also be problematic if much organic debris, e.g., fibers, are present in the sample. A cell count and cell size analysis is recommended to be included into the development of any extraction protocol because the total amount of EPS and also the risk of cellular lysis can at least be roughly estimated. In addition, the extraction efficiencies for different types of macromolecules can be found. This may vary according to the applied extraction procedure.

Different attempts have been made to estimate the EPS fraction in various bioaggregates by using biovolume determination described above, and reported values vary from 10% to 90% of the total organic matter (see references in

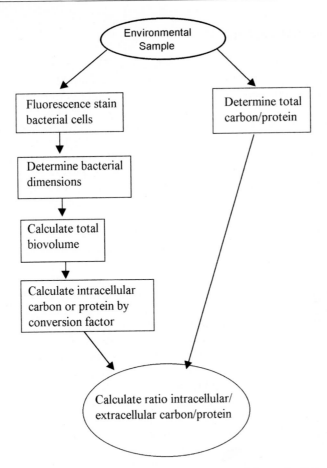

Fig. 2. Estimation of cell biomass and EPS by microscopy and determination of total carbon/protein

NIELSEN ET AL. 1997). In natural systems it is often more that 70% (JAHN AND NIELSEN 1998). Considering the usually very low extraction yields from the total organic matter recorded in many studies, 3–27% (Table 8), the extraction yields in respect of the total EPS amount are in general very small. Extraction efficiencies of EPS in biofilms in technical systems have been found in the range of 3–68%, depending on the compound (JAHN AND NIELSEN 1995, 1997). Or, in other words, 32–97% of the EPS was not extracted and characterized by the method applied.

If EPS are extracted from pure cultures, the efficiency also depends on bacterial species, growth conditions, and growth phase. Therefore, if the extraction efficiency is important, it must be determined for each species. In addition, the different bacteria can be very different in their ability to lyse during extraction (Jahn, unpublished results).

7
Purification and Analysis

In many studies of EPS no further purification of the crude extract is conducted before analysis of the EPS composition and amount. This is common in investigations of activated sludge, granular sludge, and biofilms in technical systems (RUDD ET AL. 1983; KARAPANAGIOTIS ET AL. 1989; FRØLUND ET AL. 1996; JAHN AND NIELSEN 1995). In some studies, however, purification before any further analysis always takes place (MORGAN ET AL. 1990, 1991). They use heat extraction, precipitate macromolecules in alcohol/acetone overnight, and rinse and dehydrate in acetone/petroleum ether before analysis of components. This purification step will remove various macromolecules from the EPS matrix, which should be considered when results from different investigations are compared.

If the aim is to perform a more detailed analysis of the extracted EPS components, more purification of the biopolymers is needed. It can, for example, include precipitation of polysaccharides in cold alcohols, removal of protein by protease treatment or phenol extraction, or use of gel filtration chromatography (DOMENICO ET AL. 1989). These steps are well described in the literature about purification of biomolecules and will not be further described here.

The analysis of the amount and composition of the extracted EPS is important in many studies. The composition of the EPS matrix in biofilms and activated sludge flocs has been reported to be very complex, containing protein, polysaccharides, nucleic acids, lipids, various heteropolymers, and humic substances (Tables 1 and 8). If cell biomass is to be estimated, e.g., to estimate the extraction efficiency, the selected analytical methods must be suitable for extracted EPS as well as the total bioaggregate. Most methods are colorimetric and developed for use in "clean" systems and thus very susceptible to presence of unknown interfering substances. Lack of a proper reference material stresses that a good evaluation of the methods must be conducted, e.g., by recovery of internal standards. Furthermore, the methods always use simple standards, e.g., glucose for polysaccharides or BSA for protein, which are not always representative of the actual composition (HORAN AND ECCLES 1986; FRØLUND ET AL. 1996).

It is important to choose extraction methods that do not interfere with the subsequent analysis. Examples, where interferences have been documented, are formaldehyde, which interferes with analysis of carbohydrates (phenol-sulfuric acid method (UNDERWOOD ET AL. 1995) and EDTA, which strongly interferes with the common colorimetric analysis for protein, the Lowry method (PETERSON 1979).

8
Conclusion

Extraction of EPS from bioaggregates includes sampling of bioaggregates, pretreatment, extraction, and analysis. Often a purification step is also included. No universal extraction method exists due to a variation in the properties of EPS components in various bioaggregates, but also due to different aims for performing an extraction. Before an extraction procedure is selected, it is important to consider that the methods are not quantitative and that often only a minor part of the total amount of EPS is extracted. Furthermore, cell lysis and polymer disruption may take place during the extraction. It is important to test several methods for a particular case and to optimize the effect of important extraction parameters such as time, temperature, and shear. If necessary, methods for evaluation of cell lysis and polymer disruption should be included. An overview of the entire extraction protocol is given in Table 8.

References

Beccari M, Mappelli P, Tandoi V (1980) Relationship between bulking and physicochemical-biological properties of activated sludge. Biotechn Bioeng 12:969–979

Breedveld MW, Zevenhuizen TM, Zehnder AJB (1990) Osmotically induced oligo- and polysaccharide synthesis by *Rhizobium meliloti* Su47. J Gen Microbiol 136:2511–2519

Brown MJ, Lester JN (1980) Comparison of bacterial extracellular polymer extraction methods. Appl Environ Microbiol 40:179–185

Buckmire FLA (1984) Influence of nutrient media on the characteristics of the exopolysaccharide produced by three mucoid *Pseudomonas aeruginosa* strains. Microbios 41:49–63

Christensen BE, Characklis WG (1990) Physical and chemical properties of biofilms. In: Characklis WG, Marshall KC (eds) Biofilms. Wiley, New York, pp 93–130

Christensen BE, Kjosbakken J, Smidsrød O (1985) Partial chemical and physical characterization of two extracellular polysaccharides produced by marine, periphytic *Pseudomonas* sp. strain NCMB 2021. Appl Environ Microbiol 50:837–845

Conti E, Flaibani A, O'Regan M, Sutherland IW (1994) Alginate from *Pseudomonas fluorescens* and *P. putida*: production and properties. Microbiol 140:1125–1132

deBeer D, O'Flaharty V, Thaveesri J, Lens P, Verstraete W (1996) Distribution of extracellular polysaccharides and flotation of anaerobic sludge. Appl Microbiol Biotechnol 46:197–201

Decho AW (1994) Molecular-scale events influencing the macroscale cohesiveness of exopolymers. In: Krumbein WE, Paterson DM, Stal LJ (eds) Biostabilization of sediments. BIS-Verlag, Oldenburg, Germany, pp 135–148

Domenico P, Diesrich DL, Cunha BA (1989) Quantitative extraction and purification of exopolysaccharides from *Klebsiella pneumoniae*. J Microbiol Methods 9:211–219

Emerson D, Ghiorse WC (1993) Role of disulphide bonds in maintaining the structural integrity of the sheath of *Leptothrix discophora* SP-6. J Bacteriol 175:7819–7827

Evans E, Brown MRW, Gilbert P (1994) Iron chelator, exopolysaccharide and protease production in *Staphylococcus epidermidis*: a comparative study of the effects of specific growth rate in biofilm and planktonic culture. Microbiology 140:153–157

Farrah SR, Unz RF (1976) Isolation of exocellular polymer from *Zoogloea* strains MP6 and 106 and from activated sludge. Appl Environ Microbiol 32:33–37

Forster CF, Clarke AR (1983) The production of polymer from activated sludge by ethanolic extraction and its relation to treatment plant operation. Wat Pollut Contr 82:430–433

Forster CF, Quarmby J (1995) The physical characteristics of anaerobic granular sludges in relation to their internal architecture. Ant v Leeuwenhoek 67:103–110

Frølund B, Keiding K (1994) Implementation of an HPLC polystyrene divinylbenzene column for separation of activated sludge exopolymers. Appl Microbiol Biotechnol 41:708–716

Frølund B, Griebe T, Nielsen PH (1995) Enzymatic activity in the activated sludge floc matrix. Appl Microbiol Biotechnol 43:755–761

Frølund B, Palmgren R, Keiding K, Nielsen PH (1996) Extraction of extracellular polymers from activated sludge using a cation exchange resin. Wat Res 30:1749–1758

Gaudy E, Wolfe RS (1962) Composition of an extracellular polysaccharide produced by *Sphaerotilus natans*. Appl Microbiol 10:200–205

Gehr R, Henry JG (1983) Removal of extracellular material – techniques and pitfalls. Wat Res 12:1743–1748

Grotenhuis JTC, Smit M, van Lammeren AAM, Stams AJM, Zehnder AJB (1991) Localization and quantification of extracellular polymers in methanogenic granular sludge. Appl Microbiol Biotechnol 36:115–119

Hancock I, Poxton I (1988) Bacterial cell surface techniques. Wiley, New York

Haug A, Larsen B, Smidsrød O (1967) Alkaline degradation of alginate. Acta Chem Scand 21:2859–2870

Hebbar KP, Gueniot B, Heyraud A, Colin-Morel P, Heulin T, Blandreau J, Rinaudo M (1992) Characterization of exopolysaccharides produced by rhizobacteria. Appl Microbiol Biotechnol 38:248–253

Hejzlar J, Chudoba J (1986) Microbial polymers in the aquatic environment I. Production by activated sludge microorganisms under different conditions. Wat Res 10:1209–1216

Henze M, Harremoes P, Jansen J, Arvin E (1995) Wastewater treatment. Biological and chemical processes. Springer, Berlin Heidelberg New York

Horan NJ, Eccles CR (1986) Purification and characterization of extracellular polysaccharide from activated sludges. Wat Res 20:1427–1432

Hsieh KM, Murgel GA, Lion LW, Shuler ML (1994) Interactions of microbial biofilms with toxic trace metals 1. Observation and modeling of cell growth, attachment, and production of extracellular polymer. Biotechnol Bioeng 44:219–231

Jahn A, Nielsen PH (1995) Extraction of extracellular polymeric substances (EPS) from biofilms using a cation exchange resin. Wat Sci Tech 32:157–164

Jahn A, Nielsen PH (1998) Cell biomass and exopolymer composition in sewer biofilms. Wat Sci Tech 37:17–24

Jann K, Jann B, Schmidt MA, Vann WF (1980) Structure of the *Escherichia coli* K2 capsular antigen, a teichoic acid-like polymer. J Bacteriol 143:1108–1115

Jia XS, Furumai H, Fang HHP (1996) Yields of biomass and extracellular polymers in four anaerobic sludges. Env Tech 17:283–291

Johnson KG, Perry MB (1976) Improved techniques for the preparation of bacterial lipopolysaccharides. Can J Microbiol 22:29–34

Jorand F, Zartarian F, Thomas F, Block JC, Bottero JY, Villemin G, Urbain V, Manem J (1995) Chemical and structural (2D) linkage between bacteria within activated sludge flocs. Wat Res 29:1639–1647

Junelles AM, Kanouni AE, Petitdemange H, Gay R (1989) Influence of acetic and butyric acid addition on polysaccharide formation by *Clostridium acetobutylicum*. J Ind Microbiol 4:121–125

Karapanagiotis NK, Rudd T, Sterritt RM, Lester JN (1989) Extraction and characterisation of extracellular polymers in digested sewage sludge. J Chem Tech Biotechnol 44:107–120

Kennedy AFD, Sutherland IW (1987) Analysis of bacterial exopolysaccharides. Biotechnol Appl Biochem 9:12–19

King RO, Forster CF (1990) Effects of sonication on activated sludge. Enz Microbiol Technol 12:109–115

Labare MP, Guthrie K, Weiner RM (1989) Polysaccharide exopolymer adhesives from periphytic marine bacteria. J Adhesion Sci Technol 3:213–223

Lall SD, Eribo BE, Jay JM (1989) Comparison of four methods for extracting periplasmic proteins. J Microbiol Methods 9:195–199

Mason CA, Egli T (1993) Dynamics of microbial growth in the decelerating and stationary phase of batch culture. In: Kjelleberg S (ed) Starvation in bacteria. Plenum Press, New York, pp 81–102

May TB, Chakrabarty AM (1994) Isolation and assay of *Pseudomonas aeruginosa* alginate. Meth Enzymol 235:295–304

Morgan JW, Evison LM, Forster CF (1991) An examination into the composition of extracellular polymers extracted from anaerobic sludges. TransIChemE Part B:69:231–236

Morgan JW, Forster CF, Evison L (1990) A comparative study of the nature of biopolymers extracted from anaerobic and activated sludges. Wat Res 24:743–750

Münch E, Pollard PC (1997) Measuring bacterial biomass-COD in wastewater containing particulate matter. Wat Res 31:2550–2556

Nielsen PH, Jahn A, Palmgren R (1997) Conceptual model for production and composition of exopolymers in biofilms. Wat Sci Tech 36:11–19

Novak JT, Haugan B-E (1981) Polymer extraction from activated sludge. J Wat Pollut Control Fed 53:1420–1424

Omar AS, Weckesser J, Mayer H (1983) Different polysaccharides in the external layers (capsule and slime) of the cell envelope of *Rhodopseudomonas capsulata* Sp11. Arch Microbiol 136:291–296

Pavoni JL, Tenney MW, Echelberger WF (1972) Bacterial exocellular polymers and biological flocculation. J Wat Pollut Control Fed 44:414–431

Pelkonen S, Häyrinen J, Finne J (1988) Polyacrylamide gel electrophoresis of capsular polysaccharides of *Escherichia coli* and other bacteria. J Bacteriol 170:2646–2653

Peterson GL (1979): Review of the folin phenol quantitation method of Lowry, Roseborough, Farr and Randall. Anal Biochem 100:201–220

Platt RM, Geesey GG, Davis JD, White DC (1985) Isolation and partial chemical analysis of firmly bound exopolysaccharide from adherent cells of a freshwater sediment bacterium. Can J Microbiol 31:675–680

Quarmby J, Forster CF (1995) An examination of the structure of UASB granules. Wat Res 11:2449–2454

Rahman MM, Guard-Petter J, Asokan K, Carlson RW (1997) The structure of the capsular polysaccharide from a swarming strain of pathogenic *Proteus vulgaris*. Carbohydrate Res 301:213–220

Read RR, Costerton JW (1987) Purification and characterization of adhesive exopolysaccharides from *Pseudomonas putida* and *Pseudomonas fluorescens*. Can J Microbiol 33:1080–1090

Rudd T, Sterritt RM, Lester JN (1982) The use of extraction methods for the quantification of extracellular polymer production by *Klebsiella pneumoniae* under varying cultural conditions. Eur J Appl Microbiol Biotechnol 16:23–27

Rudd T, Sterritt RM, Lester JN (1983) Extraction of extracellular polymers from activated sludge. Biotechnol Lett 5:327–332

Russell JB, Cook GM (1995) Energetics and bacterial growth: balance on anabolic and catabolic reactions. Microbiol Rev 59:48–62

Sato T, Ose Y (1980) Floc forming substances extracted from activated sludge by sodium hydroxide solution. Wat Res 14:333–338

Sato T, Ose Y (1984) Floc-forming substances extracted from activated sludge with ammonium hydroxide and EDTA solutions. Wat Sci Tech 17:517–528

Schmidt JE, Ahring BK (1994) Extracellular polymers in granular sludge from different upflow anaerobic sludge blanket (UASB) reactors. Appl Microbiol Biotchnol 42:457–462

Schmidt MA, Jann K (1982) Phospholipid substitution of capsular (K) polysaccharide antigens from *Escherichia coli* causing extraintestinal infections. FEMS Microbiol Lett 14:69–74

Streeter JG, Salminen SO, Beuerlein JE, Schmidt WH (1994) Factors influencing the synthesis of polysaccharides by *Bradyrhizobium japonicum* bacteroids in field-grown soybean nodules. Appl Environ Microbiol 60:2939–2943

Sutherland IW, Wilkinson JF (1971) Chemical extraction methods of microbial cells. In: Methods in microbiology, vol 5B. Academic Press, London, chap 5

Tago Y, Aida K (1977) Exocellular mucopolysaccharide closely related to bacterial floc formation. Appl Environ Microbiol 34:306–314

Titus S, Gaonkar SN, Srivastava RB, Karande AA (1995) Exopolymer production by a fouling bacterium *Pseudomonas alcaligenes*. Ind J Med Sci 24:45–48

Troch P, Philip-Hollingworth S, Orgambide G, Dazzo FB, Vanderleyden J (1992) Analysis of extracellular polysaccharides isolated from *Azospirillum brasilense* wild type and mutant strains. Symbiosis 13:229–241

Underwood GJC, Paterson DM, Parkes RJ (1995) The measurement of microbial carbohydrate exopolymers from intertidal sediments. Limnol Oceanogr 40:1243–1253

Urbain V, Block JC, Manem J (1993) Bioflocculation in activated sludge: an analytic approach. Wat Res 27:829–838

Veiga MC, Mahendra KJ, Wu W-M, Hollingsworth RI, Zeikus JG (1997) Composition and role of extracellular polymers in methanogenic granules. Appl Environ Microbiol 63:403–407

Wanner J (1994) Activated sludge bulking and foaming control. Technomic Publishing

Wrangstadh M, Conway PL, Kjelleberg S (1986) The production and release of an extracellular polysaccharide during starvation of a marine *Pseudomonas* sp. and the effect thereof on adhesion. Arch Microbiol 145:220–227

Zayas JF (1997) Functionality of proteins in food. Springer, Berlin Heidelberg New York

CHAPTER 4

Biofilm Exopolysaccharides

Ian W. Sutherland

Institute of Cell and Molecular Biology, Edinburgh University, Mayfield Road, Edinburgh EH9 3JH, UK, *E-mail: I.W.Sutherland@ed.ac.uk*

Keywords. Polysaccharide, Transition, Order, Synergism, Gelation

1
Introduction

The presence of exopolysaccharides as essential components of biofilms can be demonstrated by chemical analysis or by direct examination of the biofilm using microscopy or electron microscopy. The development of specific staining methods has revealed both the attachment of cells and the presence of micro-colonies surrounded by extensive amounts of exopolysaccharide (e.g. ALLISON AND SUTHERLAND 1984). In such laboratory experiments of biofilm elaboration on glass surfaces, there was good correlation between the physical appearance of exopolysaccharide and the amount of carbohydrate-containing material attached to the glass. Scanning electron microscopy (SEM) studies have also revealed material thought to be polysaccharide in nature (Fig. 1). An SEM examination of bacteria attached to sand grains has revealed strands of polysaccharide or glycoprotein-like material thought to attach the cells to the solid surface

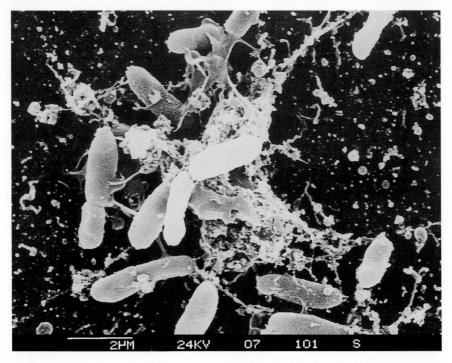

Fig. 1. SEM – Pseudomonas strain S61 attached to glass by adhesive exopolysaccharide (EPS) following 15 h growth in yeast extract medium (bar = 2 μm)

(WEISE AND REINHEIMER 1978). The strands were of varying thickness and were associated with various types of cells. However, this material cannot be chemically analysed and may also be subject to artefacts from specimen preparation. On the other hand, use has been made of ruthenium red, a dye which interacts strongly with polysaccharides and can be visualised under the transmission electron microscope. Thus, natural biofilms revealed dense populations of bacterial cells enmeshed in fibrous structures (GEESEY ET AL. 1977). Although other reports using this technique have also indicated the presence of fibrous material presumed to be polysaccharide, the actual appearance depended on the method used for sample preparation.

It is probable that the exopolysaccharides found in biofilms vary greatly in their composition and in their physical properties. They are, however, unlikely to differ much from the extensive range of exopolysaccharides from non-biofilm species which have now been characterised (SUTHERLAND 1997). It remains to be seen whether any are truly biofilm-specific, being synthesised only under biofilm as opposed to planktonic conditions. Laboratory studies using artificial biofilms have only tested a limited number of microbial components, one of which has commonly been an alginate-producing *Pseudomonas aeruginosa* strain. This has perhaps resulted in the misconception that all biofilm polysaccharides are similar and that they all resemble algal alginate. The major differences between bacterial and algal alginates lies in their monosaccharide sequences and in the presence of O-acetyl groups in the former. Even though they are all polyanionic macromolecules composed of D-mannuronic acid and L-guluronic acid, their physical properties are very different. It is probably necessary to look at each microbial species and each exopolysaccharide as distinct entities with unique properties whether in the planktonic or biofilm mode of growth.

We have attempted to determine whether any exopolysaccharides are truly biofilm specific (SUTHERLAND 1995). Studies in our laboratory have indicated that both planktonically prepared and biofilm exopolysaccharides could be degraded by the same specific endoglycanases and the resultant patterns of oligosaccharides appeared similar (HUGHES ET AL. 1998A). Similarly, COSTERTON ET AL. (1981) used antibodies against planktonically-synthesised polysaccharide to demonstrate interaction with material in a biofilm matrix. As yet, the cryptic capability of biofilm micro-organisms to synthesise novel polymers only in the biofilm mode has not been successfully demonstrated. It is, however, known to be present in a number of other bacterial species such as *Rhizobium meliloti*, *Escherichia coli*, etc. (e.g. GLAZEBROOK AND WALKER 1989; GRANT ET AL. 1969) and to be expressed under suitable physiological conditions. It has also been reported that alginate synthesis in *P. aeruginosa* can be induced following attachment, presumably through activation of one of the complex "switches" employed by these bacteria to regulate polysaccharide synthesis. Many laboratory studies have employed artificial biofilms in which the key microbial components are alginate-producing *P. aeruginosa* strains. However, bacterial alginates synthesised by diverse bacterial species and strains differ in their chemical structure. Consequently, they vary greatly from their algal equivalents in their physical properties. This is true even though their monosaccharide constituents are identical. It has also to be understood that biofilm exopolysaccharides interact with other exopolysaccharides, with proteins and other macromolecules, lipids and with ions. An example of this can be seen in the adhesion of *Staphylococcus epidermidis* to inert surfaces; adhesion and cell accumulation have been distinguished as separate functions. For the latter to take place, a 140 kDa protein produced only under sessile conditions is required (HUSSAIN ET AL. 1997). A synergistic effect, possibly involving polysaccharide-polysaccharide interaction, has also been observed when two exopolysaccharide-producing Enterobacterial strains are allowed to form a mixed biofilm. Either following pre-treatment with exopolysaccharide or exposure to the mixed culture, adhesion of the bacterial cells in the form of a biofilm is enhanced (SKILLMAN ET AL. 1997). It remains to be seen whether other bacteria are as complex in their biofilm-associated activities but, clearly, a variety of cell constituents may be involved in the adhesion and accumulation process and exopolysaccharide represents just one of these factors. In those interactions in which the mutual exclusion of incompatible molecules occurs, a resultant increase in the effective concentration of each polymer is observed. This may result in localised gelation or other effects. Conversely, some polysaccharides bind strongly to one another when energetically favourable segments of regular structures interact. Such interchain associations may be favoured by the presence of low molecular mass hydrophilic molecules (SUTHERLAND 1983).

Physical studies on a considerable number of exopolysaccharides have shown that the great majority of these macromolecules can exist either in ordered or disordered forms, the latter being favoured at elevated temperatures and (frequently) at extremely low ionic concentrations. Thus, in the natural environment, these polymers will almost always be found in some sort of ordered configuration (SUTHERLAND 1990). It is also possible that some microbial exopolysaccharides resemble crystalline cellulose in their behaviour. Each crystal of cellulose contains numerous glycan chains in parallel orientation (TEERI 1997). The reducing ends are at one terminus while the non-reducing ends are at the other. Despite the high degree of crystallinity, the structure is not uniform and amorphous regions may be present as well as highly crystalline regions. Many bacterial exopolysaccharides adopt a double helical configuration in the ordered form and the association between double helices is facilitated by ions and by water molecules. In this, they can be regarded as resembling the double helix of DNA. At the molecular level the physical properties of the polysaccharides, and hence of the biofilms, are dependent on these interactions and may be very greatly influenced by the presence or absence of free anionic groups derived from uronic acids, phosphate groups, pyruvate ketals or succinyl half-esters. Hydrogen bonding involving exposed hydroxyl groups can also be significant. Localised hydrophobic regions, either deriving from O-acetyl groups or 6-deoxyhexoses such as L-rhamnose or L-fucose, may also exert considerable influence. One very unusual polymer is the linear glucosaminoglycan from biofilm-forming *S. epidermidis*. Its structure contains 80–85% N-acetylated residues, while the remainder are non-acetylated and hence provide one of the few basic biofilm exopolysaccharides so far reported (MACK ET AL. 1996). An associated and structurally related polymer showed greater acetylation and was anionic due to the presence of ester-linked succinyl residues and also phosphate groups.

If the exopolysaccharides are to be effective in maintaining the biofilm structure, they will do so through formation of a network of cross-linked linear macromolecules. Binary networks may be independent of each other but may either be interpenetrating or phase separated. ROSS-MURPHY AND SHATWELL (1993) have distinguished three categories of networks: those in which covalent cross-linking leads to the formation of junctions; physical junctions which can be readily disrupted through alteration of temperature; and entanglement networks which result from local chain entanglements. It is also possible to envisage a single polymer network in which a second component is soluble and diffuses through the polysaccharide mesh without any direct interaction. When there are more types of polysaccharide present, the network structure will inevitably become even more complicated.

2
Polysaccharide Composition and Structure

Most microbial exopolysaccharides are highly soluble in water or dilute salt solutions and, if present in the form of capsules, are attached to the microbial cells through covalent bonds to other surface polymers. On the other hand, some exopolysaccharides are virtually insoluble in water or form very rigid gels when in the ordered form and among these are several which are commonly found in biofilms. Structurally, these polysaccharides usually contain a backbone in which there is a predominance of either 1,3 or 1,4 linkages in either the α- or more commonly the β-configuration. Once synthesised, polysaccharides of this type are likely to be extremely effective molecules in maintaining the integrity of biofilms and their components. The actual composition of microbial polysaccharides shows an almost infinite range of possibilities. Some are simple homopolysaccharides containing a single linkage type as exemplified by the 1,3 linked glucans mutan from *Streptococcus mutans* (Fig. 2) and curdlan from *Agrobacterium radiobacter* and other species (Fig. 2), or the amino sugar-containing 1,6-β-glucan from *S. epidermidis* (Fig. 3) (MACK ET AL. 1996). More complex polymers may be composed of three to five different monosaccharides and repeat units of hexa-, hepta -or octasaccharides. The complexity is increased further by the presence of various acyl adornments. An example of a complex polysaccharide structure can be seen in colanic acid. This exopolysaccharide is secreted by many strains of *Escherichia coli* and also other enteric species such as *Enterobacter cloacae*. The repeat unit of this polymer is a hexasaccharide composed of four sugars in which both acetyl and pyruvate groups are present (Fig. 4) (LAWSON ET AL. 1969).

There are few reports on chemical composition of natural biofilm exopolysaccharide material. However GLOAGUEN ET AL. (1995) used NMR spectroscopy to identify monosaccharides and oligosaccharide fragments from polymer synthesised predominantly by the cyanobacterium *Mastigocladus laminosus*. Similarly, SUTHERLAND (1996) examined the composition of a viscous terrestrial biofilm produced by cyanobacteria and green algae. In these studies, plenty

Curdlan is synthesised by *Agrobacterium radiobacter* and other related bacteria

→ [-3-β-D-Glc*p*-(1→3)-β-D-Glc*p*-(1→3)-β-D-Glc*p*-(1-]-→

Mutan is produced by *Streptococcus mutans*

→ [-3-α-D-Glc*p*-(1→3)-α-D-Glc*p*-(1→3)-α-D-Glc*p*-(1-]-→

Fig. 2. The linear homopolysaccharides (glucans) curdlan and mutan. Curdlan is synthesised by *Agrobacterium radiobacter* and other related bacteria

Polysaccharide I

→ [6-β-D-Glc*p*NAc-(1→ 6)-β-D-Glc*p*NAc-(1→ 6)-β-D-Glc*p*-N-(1→6)-β-D-Glc*p*NAc-(1→ 6)-β-D-Glc*p*NAc-(1→

Ratio: Glc*p*NAc : Glc*p*N   c. 5:1 (Cationic)

Polysaccharide II

→ [6-β-D-Glc*p*NAc-(1→ 6)-β-D-Glc*p*NAc-(1→6)-β- D-Glc*p*-N-(1→6)-β-D-Glc*p*NAc-(1→ 6)-β-D-Glc*p*NAc-(1→

Ratio: Glc*p*NAc : Glc*p*N   c. 16:1   Also contains: phosphate and succinyl half esters (Anionic)

Fig. 3. *Staphylococcus epidermidis* biofilm exopolysaccharides (results of MACK ET AL. 1996)

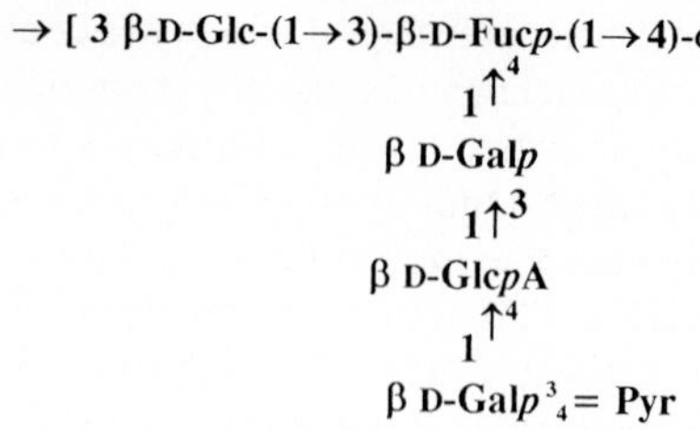

$$\rightarrow [3 \ \beta\text{-D-Glc-}(1\rightarrow 3)\text{-}\beta\text{-D-Fuc}p\text{-}(1\rightarrow 4)\text{-}\alpha\text{-L-Fuc}p\text{-}(1] \rightarrow$$

$$1 \uparrow^4$$
$$\beta \text{ D-Gal}p$$
$$1 \uparrow^3$$
$$\beta \text{ D-Glc}p\text{A}$$
$$1 \uparrow^4$$
$$\beta \text{ D-Gal}p^3_4 = \text{Pyr}$$

Typically the polymer carries an O-acetyl group on the 2 or 3 position of the unsubstituted fucose residue of each hexasaccharide repeating unit.

Fig. 4. The structure of the exopolysaccharide colanic acid from strains of *Escherichia coli, Salmonella typhimurium (enterica)* and *Enterobacter cloacae*

of material was available but usually this is not the case. The other problem is that such mixed cultures cannot be reproduced in the laboratory to yield identical products.

One group of bacterial species synthesises a family of closely related polymers, which have been designated galactoglucans. The most complex are those found associated with various *Rhizobium* spp., and possess an acylated octasaccharide repeat unit in which pyruvate, acetate and succinate may be found. Simpler structures composed of disaccharide repeat units also belong to this group (Table 1). The galactoglucan molecule forms a 2-fold helix in which acetyl and pyruvyl substituents are located on the periphery (CHANDRA-SEKARAN ET AL. 1994). The location of acyl groups on the outside of the polymer chain is not unusual and enables such substituents to interact with other macromolecules and with cations. Even monovalent cations such as potassium play a role in stabilising the helical structure of the polysaccharide.

"Mutan" from *S. mutans* is well characterised as being predominantly water-insoluble and also is resistant to the activity of enzymes secreted by associated microbial components of oral biofilms (BIRKHED ET AL. 1979). It owes its insolubility and effective hydrophobicity to its 1,3-α-linked glucan structure. The corresponding β-linked D-glucan, curdlan, is synthesised by *A. radiobacter* and other soil micro-organisms where it may well play a similar role in the adhesion of the bacterial cells to particulate matter or to plant material. A number of physical studies have shown that curdlan normally forms a tightly bound *triple helix*. Within the highly ordered, microcrystalline interior of the polymer strands the environment can be effectively hydrophobic and is thus protected from access by smaller water-soluble molecules and from enzymic or even acid attack (KANZAWA ET AL. 1989).

Strains of *Enterobacter agglomerans* form a complex group of bacteria, which are difficult to differentiate, and which are widely found in the environment. They are opportunistic bacteria which may be found as contaminants of medicaments although they are probably not normally of clinical significance (GRIMONT AND GRIMONT 1992). Several strains of this bacterial species, along with other Enterobacterial species, have also been isolated as biofilm compo-

Table 1. The composition of bacterial galactoglucans, demonstrating the variety of substituents which can be found in a polysaccharide containing only two monosaccharides

Galactogluco-polysaccharides Source	Glucose	Galactose	pyruvate	Acetate	Succinate	Glc → Gal	Gal → Glc
Achromobacter sp.	1	1[a]	081–0.99	–	–	–	–
Agrobacterium radiobacter	0.9	1	0.83	–	–	β 1.3	α 1.3
Burkholderia cepacia	1	1	1	0	0	β 1.3	α 1.3
Escherichia coli K37	1	2	1	+	–	β 1.3	α 1.3
Pseudomonas fluorescens	1	1	0.5	0	1	–	–
Pseudomonas marginalis	1	1[a]	1	+	0	β 1.3	α 1.3
Pseudoinonas putida	1	1	1	+		–	–
Rhizobium melfloti YE2	1	1	1	0.9	0	β 1.3	α 1.3
Zoogloea ramigera 1	2	1	+	(Ill defined)	0		
Z. ramigera. 2	11	3	1.5		+		

[a] Pyruvylated sugar.

nents from food manufacturing plants. Recent work by HUGHES (1997) in our laboratory has shown that for the two strains that we have studied most intensively, the exopolysaccharides are very poorly soluble in water and also in alkali or dimethylsulphoxide. The polymers are also, unusually, composed entirely of neutral monosaccharides and are free from acyl substituents. Preliminary studies have revealed that both polysaccharides contain a high proportion of 1,3-linked residues. This indicates that these polymers are likely to be similar in some of their physical properties to the 1,3-linked D-glucans mutan or curdlan. They would therefore provide a very effective matrix for bacterial adhesion in biofilms and might be able to exclude many water-soluble molecules from the interior of the matrix. This may make eradication of the bacteria somewhat difficult when standard disinfection procedures are applied.

It is not known whether, as well as the potential action of glycanases, esterases capable of removing acetyl, succinyl or other substituents may be present in the microbial complexes found in biofilms. Acetyl esterases have recently been reported in *Erwinia chrysanthemi* (SHEVCHIK AND HUGOUVIEUX-COTTE-PATTAT 1997) but their only action was on pectin. Removal of ester groups can have marked effects on the chain-chain interactions and the physical properties of bacterial exopolysaccharides. Typically, amorphous structures may be altered to yield highly microcrystalline structures. This was observed following chemical removal of acetyl groups from *Enterobacter* XM6 exopolysaccharide (NISBET ET AL. 1984). Similarly, removal of ester-linked acyl groups from the exopolysaccharide "gellan" synthesised by *Sphingomonas* spp. caused a change from weak rubbery gels to hard brittle gels. This indicated that they resembled agar in some of their properties (e.g. CHANDRASEKARAN AND RADHA 1995). Another polymer recently shown to exhibit behaviour of this type was "beijeran" from *Azotobacter beijerinckia*. When deacetylated, this D-galacturonic acid-containing polysaccharide composed solely of $1 \rightarrow 3$ linked monosaccharides, became much more crystalline and formed insoluble gels in the presence of Ca^{2+} (OGAWA ET AL. 1997). If such behaviour occurred within biofilms it would assist in maintaining their attachment and integrity. It would also mean that while glycanases are likely to destroy the biofilm structure and to release the microbial cells, other enzymes would actually have the potential to enhance the biofilm state. As yet, no esterase activity against O-acyl substituents on neutral sugars in bacterial exopolysaccharides has been demonstrated. However, phage-induced enzymes deacetylating the enterobacterial Vi antigen and Salmonella lipopolysaccharide have been found (KWIATKOWSKI ET AL. 1975; IWASHITA AND KANEGASAKI 1976). As these esterases proved to be capable of removing acetyl groups from other substrates such as poly-D-galacturonic acid, it is possible that such non-specific enzymes do exist and would be capable of affecting the structure of regions within biofilms.

As many studies on biofilms have employed alginate-synthesising species the question arises as to how adhesive are bacterial alginates? They could be very effective adhesins capable of ensuring that the bacterial cells from which they are produced are very firmly attached to surfaces. Alternatively, they might be relatively easily removed from solid or cellular substrates. The answer may well depend on the bacterial species. *Azotobacter vinelandii* alginates most closely resemble the commercial products extracted from marine algae and share some of their physical properties. In these alginates, the presence of poly-L-guluronic acid sequences (polyG blocks) provides much greater effective gel strength than do the corresponding poly-D-mannuronic acid (polyM) blocks. In the proton form, algal alginate gels were almost six times stronger than *P. aeruginosa* alginate gels in which there were polyM blocks of equivalent length to the polyG blocks in the algal material (DRAGET ET AL. 1994). Thus a biofilm in which *A. vinelandii* alginate or similar material was present, as might be the case on the soil surface, would be much more likely to have a high rigidity than a *P. aeruginosa* biofilm whether the latter was in the proton or Ca^{2+} form. PRINCE (1996) suggested that in respiratory infections, alginate from *P. aeruginosa* played a more important function in interbacterial adhesion than it did in binding the bacteria to host tissues. It was also suggested that the polysaccharide facilitated microcolony and biofilm formation.

3
Promotion of Order and Gelation

Polysaccharides are probably present in most biofilms in a hydrated, semi-solid state. Many microbial polysaccharides are relatively water-soluble, easily removed and probably fairly ineffective in attaching the bacterial cells to the solid substratum. Depending on the composition and linkages present in the polymer chains, these can adopt an appropriate conformation or secondary structure. Frequently the conformation takes the form of a helix and the helices can in turn aggregate and form networks. In some systems, no ordered conformation has yet been demonstrated. Between these extremes are the many polysaccharides that can exhibit some degree of order either in the presence or absence of ions and thus allow the bacteria producing them to form biofilms which adhere with varying degrees of tenacity. Most exopolysaccharides in solution undergo a change from order to disorder on heating or on removal of ions. This change may be reflected by increasing solubility and eventual dissolution from a biofilm, but at ambient temperatures the exopolysaccharides will be ordered and may even show synergistic effects with other exopolysaccharides. Thus, two exopolysaccharides may form a gel at polysaccharide concentrations that do not do so individually. Probably many exopolysaccharides possess transition temperatures well above those likely to be encountered in biofilm situations, but others may exhibit transition from order to disorder at 30 °C or below. This is exemplified by the *Enterobacter* XM6 exopolysaccharide which forms highly crystalline gels below 30 °C and also melts at the same temperature (NISBET ET AL. 1984). The corresponding polysaccharides synthesised by *Enterobacter aerogenes* type 54 strains differ only in the presence of O-acetyl groups on every tetrasaccharide or octasaccharide, but fail to show either ordered structure or a distinct transition (ATKINS ET AL. 1987). The failure to gel of this and other polysaccharides, including several of those from *Sphingomonas paucimobilis* strains, is almost certainly due to shielding of the carboxylate groups by side-chains (LEE AND CHANDRASEKARAN 1991; CHANDRASEKARAN ET AL. 1988) or by acyl substituents. It has been suggested by MORRIS ET AL.

(1989) that cation-dependent gelation of acid exopolysaccharides from *Rhizobium leguminosarum* provided a non-specific mechanism for adhesion to plant roots. Similarly, gelation of biofilm polymers may require cations for effective adhesion to occur.

Even for those exopolysaccharides which are polyanionic in nature due to the presence of either uronic acid residues or acyl substituents such as pyruvate ketals or succinyl half-ester groups, much depends on the conformation adopted by the polysaccharide. In some of these microbial exopolysaccharides the charged groups will be on the exterior of the adopted structure, where they will certainly bind cations but are less likely to play a major role in forming any tight binding of adjacent strands. In marked contrast to this, many alginates possess sequences of L-guluronosyl residues (poly-L-guluronic acid) which can bind Ca^{2+} very effectively within the "egg-box" structure that is adopted. Sequences of poly-guluronic acid are also present in *A. vinelandii* alginates but not in those from *Pseudomonas* spp. Nevertheless, *P. aeruginosa* alginate in the nonacetylated form does bind cations with some selectivity for Ca^{2+} (GEDDIE AND SUTHERLAND 1994). This binding is insufficient to form the strong gels observed with algal material although it allows some weak gel formation, but requires a polysaccharide concentration of ca. 2% (DRAGET ET AL. 1994). The low gel strength is probably due to the small number of stable intermolecular bonds. Typically a *P. aeruginosa* alginate with 7–40% guluronic acid content did not form gels with Ca^{2+} but did so in the proton form. Exchange of protons with cations caused a sol-gel transition. It is probable that cations are not evenly distributed in their association with polysaccharide chains. THU ET AL. (1997) recently used synchrotron-induced X-ray emission to demonstrate the spatial distribution of calcium ions in alginate gels. They confirmed that distribution of the Ca^{2+} was inhomogeneous. It was highest at the surface and lower in the gel interior. This agreed with earlier findings that demonstrated inhomogeneity of polymer concentration when calcium-induced alginate gels were studied (SKJÅK-BRÆK ET AL. 1989). It also suggested that a similar situation would exist in biofilm polysaccharides. Some of the carboxylic acid groups would be in the proton form while others would be associated with a range of counterions. The actual density of cations would depend on both the available anionic groups on the polymer and the availability of cations from the biofilm environment. Competition between different biofilm polysaccharides (and other components) for the available cations must undoubtedly occur.

Some polysaccharides, such as the galactoglucans found in some *R. meliloti* strains and in various other bacteria, always remain in a disordered conformation which may be characterised by a high degree of flexibility. This probably causes loosely attached microbial cells which can be easily removed from the biofilm. Polysaccharides of this type lack intra-chain hydrogen bonds and the pyruvate ketals, providing the only anionic portion of this Rhizobium polymer, are found along with acetyl groups on the periphery of the helix (CESÀRO ET AL. 1992; CHANDRASEKARAN ET AL. 1994). A polymer of very similar composition and with some structural similarities is formed by *Zoogloea ramigera* and was recently shown to play a role in floc formation (NAKAMURA ET AL. 1987; TROYANO ET AL. 1996).

4
Ionic Interactions

As has already been pointed out, for some polysaccharides the presence of multivalent cations is essential for the formation of ionic bridges and maintenance of ordered structures. The polymers could also serve a useful role in trapping and maintaining a reservoir of ions to supplement those which might be deficient under oligotrophic growth conditions. The ionic complement of any biofilm polysaccharide, which is usually polyanionic in nature, could be expected to change with the nature of the fluid perfusing the biofilm. Studies on the ion binding capacity of xanthan revealed that a typical preparation in which both glucuronosyl residues and ketal bound pyruvate were available to bind ions, contained ca. 3% by weight of Na^+, K^+ and Ca^{2+} (SHATWELL 1989). Loss of the pyruvate residues reduced the quantity of ion bound by almost 50%. Ion binding depends very much on the polysaccharide structure and on the configuration that it adopts. Thus some exopolysaccharides may resemble algal alginates in highly specific binding of cations such as Ca^{2+} and Sr^{2+}, but more commonly binding would be less specific. The presence of the O-acetyl groups in bacterial alginates had a strongly inhibitory effect on cation binding and greatly reduced the selectivity of cation binding (GEDDIE AND SUTHERLAND 1994). Nevertheless, P. aeruginosa alginate in the non-acetylated form bound cations with some selectivity for Ca^{2+}. Recent studies from LOAËC ET AL. (1997) demonstrated that the polymer from a strain of Alteromonas macleodii possessed affinity for lead, cadmium and zinc. Lead was preferentially absorbed, while there was competition between zinc and cadmium for the same binding site. Removal of acetate from an exopolysaccharide apparently does not always enhance ion binding. In a study using an undefined Pseudomonas polysaccharide, MARQUÉS ET AL. (1990) demonstrated that uranium uptake reached 96 mg g^{-1} polymer. This value was reduced to almost half in deacetylated polysaccharide.

While some exopolysaccharides bind cations and alter their ordered structure, others are precipitated by a range of metallic ions. Multivalent cations including Sn^{2+}, Al^{3+}, Fe^{3+} and Th^{4+} may all, under certain conditions, precipitate polymers including those of Bradyrhizobium species (CORZO ET AL. 1994). Similar results were observed with various polysaccharides produced by marine bacteria and originating from biofilms on rock surfaces (SUTHERLAND 1983). Ion concentration and pH play an important role in these reactions which may have secondary effects on biofilm structure and function.

5
Synergistic Effects

When two or more microbial strains form a biofilm, the presence of exopolysaccharides may not only assist in establishing the biofilm but also promote greater biofilm growth than the comparable single species biofilms. The reason for this is unknown but it does suggest that the exopolysaccharide helps to promote not only adhesion and growth of the cells which synthesise it, but also those of other microbial species. While this phenomenon has been observed for some pairs of bacterial strains, others show no enhancement. In some mixed biofilms, it is clear that the presence of more than one exopolysaccharide leads to increased adhesive properties (L.C. Skillman, unpublished results). What is not yet known is the effect of partial or complete removal of one of the polymers from such a complex matrix.

An examination of mixtures of polysaccharides and their effect on tobramycin sensitivity of *P. aeruginosa* was made by ALLISON AND MATTHEWS (1992). Although they found that mixtures of polymer solutions were much more viscous than pure solutions, the greatest effect was observed when Ca^{2+} was also added to the system. This enhanced gelation of the mixture of bacterial alginate, mucin and *Burkholderia cepacia* polysaccharide. It also greatly reduced the effect of tobramycin. To what extent such studies mimic the situation where these bacteria form a biofilm in the human lung is open to question, but they do serve to indicate the protective effect of certain bacterial exopolysaccharides against some antimicrobial agents.

6
Other Functions of Biofilm Exopolysaccharides

Clearly, one of the major roles of biofilm exopolysaccharides is to provide an adhesive matrix that both promotes microbial cell adhesion to inert surfaces and also enables the biofilm cells to remain attached to that surface and in close proximity to one another. In a study of freshwater isolates forming biofilms, ALLISON AND SUTHERLAND (1987) found that mutants devoid of exopolysaccharide still attached but failed to form microcolonies. This observation suggests that the role of the polymer came after the initial attachment and was essential for microcolony formation and cell:cell adhesion. Growth conditions that affected exopolysaccharide production in the wild type bacteria also affected microcolony formation but not attachment. In many respects, the involvement of exopolysaccharide in biofilm development resembles the process of fruiting body formation in certain Myxobacteria (DWORKIN AND KAISER 1993). In both processes there is extensive production of exopolysaccharide and structural development. In the Myxobacteria, there is accompanying morphogenesis which is absent in biofilm development. The polysaccharide synthesised in vegetative Myxobacterial cultures appeared to be very similar if not identical to that isolated from those fruiting bodies which have been examined (SUTHERLAND 1979; SUTHERLAND AND THOMSON 1975). It probably has a dual function of surface lubrication in the motile vegetative cells and as an adhesive holding the cells (and later microcysts) together and enhancing cell:cell communication.

The intergeneric coaggregation which occurs in oral systems involves exopolysaccharides as well as other surface macromolecules. Some of the aggregation can be inhibited by monosaccharides and disaccharides, or by proteolytic treatment of the cells. Thus lectin-like mechanisms are probably involved. The systems are complex as bacterial cells, which do not coaggregate per se, may form a complex with a third cell type. It requires synthesis of both adhesins and carbohydrate receptors. The structures of three streptococcal carbohydrate receptors have been reported to be polysaccharides with hexasaccharide or heptasaccharide repeat units (KOLENBRANDER 1991) (Fig. 5). These molecules appear to permit the streptococci to coaggregate with other bacteria drawn from several different genera. This model also demonstrates a system in which the exopolysaccharides appear to play a role in adhesion which also involves considerable cell:cell interaction and may be closer to the Myxobacterial model in this respect.

There are however other possible roles, some of which have already been postulated above. The matrix maintains a high degree of hydration in the immediate vicinity of the microbial cells and may thus assist in their survival under conditions of desiccation. Such conditions would be found in those surface biofilms which are exposed to frequent cycles of dehydration and rehydration. This protective role has been confirmed in the laboratory by HUGHES ET AL. (1998 B) using monospecies biofilms with a variety of enteric species. Protection against antimicrobial agents and toxic metal cations is also a possibility. The reduced uptake of biocides caused by the biofilm exopolysaccharide may also lead to development of resistance against such agents. Here, it should be noted that some antimicrobial compounds at sub-lethal concentrations actually promote exopolysaccharides synthesis. Bacterial growth is inhibited and considerably more exopolysaccharide is formed than in the absence of the agents. Particularly interesting is the observation that this phenomenon can be observed with a range of agents which differ considerably in their mode of antimicrobial action (HUGHES ET AL. 1998B). It does suggest that in industrial environments in which there is regular use of disinfectants etc., failure to eliminate the micro-organisms present in a biofilm may increase their resistance to further disinfection through the enhancement of exopolysaccharide synthesis. The exopolysaccharide may also interact with the antimicrobial agent if the polymer is polyanionic while the agent is basic. Thus, a study of gentamycin penetration into solutions of alginate prepared from *P. aeruginosa* and gellan from *Sphingomonas elodea* (KUMON ET AL. 1994) indicated that polymer concentration and the presence of Ca^{2+} greatly influenced the result. Positively

-[α-L-Rha*p*-(1→2)-α-L-Rha*p*-(1→3)-α-D-Gal*p*-(1→3)-β-D-Gal*p*(1→4)-β-D-Glc*p*-(1→3)-α/β-D-Gal*p*-]-

Streptococcus oralis H1

Each hexasaccharide repeat unit carries a phosphate substituent

-[α-D-Gal*p*NAc-(1→3)-β-L-Rha*p*-(1→4)-β-D-Glc*p*-(1→6)-β-D-Gal*f*-(1→6)-β-D-Gal*p*NAc-(1→3)-α-D-Gal*p*-]$_n$-

→PO$_4$'

Streptococcus oralis 34

-[→6)-α-D-Gal*p*NAc-(1→3)-β-L-Rha*p*-(1→4)-β-D-Glc*p*-(1→6)-β- D-Gal*f*-(1→6)-β-D-Gal*p*-(1→3)-α-D-Gal*p*NAc-]$_n$-
$$\underset{\text{α-L-Rha}p}{_1\uparrow^2}$$
→PO$_4$'

Streptococcus oralis J22

Fig. 5. The polysaccharides involved in multicellular adhesion biofilms of *Streptoccus* and other oral species

charged, hydrophilic antimicrobial compounds including aminoglycosides and polypeptides were inhibited, whereas quinolones, macrolides and β-lactams were unaffected. Alternatively selective or sacrificial destruction of exopolysaccharide might leave insufficient concentrations of the antimicrobial agent to kill the microbial cells. This would be more applicable to treatment of biofilms with disinfectants in either the home or in industrial environments. L. C. Skillman (unpublished results) has recently demonstrated that removal of the exopolysaccharide through the use of highly specific enzymes results in increased susceptibility to a range of antimicrobial agents. In a recent study, *P. aeruginosa* biofilms were grown in media which caused different amounts of exopolysaccharide synthesis (SAMRAKANDI ET AL. 1997). Chlorine disinfection efficacy was lower for biofilms prepared in the medium with higher exopolysaccharide production, whereas monochloramine was equally effective under both conditions tested. This was possibly due to its better penetration of the biofilm. Clarithromycin, a macrolide antibiotic which is not directly bactericidal for *P. aeruginosa*, nevertheless caused removal of biofilms of this bacterium from Teflon surfaces (KONDOH ET AL. 1996). The action was considered to be a result of inhibition of polysaccharide synthesis. It was suggested that this antibiotic might be useful for the treatment of biofilm infections resulting from the insertion of biomedical devices.

The ability of certain exopolysaccharides to adsorb normally toxic cations such as Cd, Zn, Pb, Cu and Sr would suggest that the presence of such polymers might protect other microbial cells in a mixed culture biofilm. NORBERG AND PERSSON (1984) and KUHN AND PFISTER (1989) have demonstrated the absorption of these toxic ions by *Zoogloea ramigera* exopolysaccharide. As well as such a protective role through the entrapment of ions, exopolysaccharides in biofilms might indirectly store nutrients. Macromolecules enmeshed in the biofilm matrix could later be degraded by extracellular enzymes and converted to usable oligomers or monomers.

There are various other roles that could possibly be ascribed to exopolysaccharides, although firm evidence often awaits the development of suitable methods for assay or detection. Many capsulate micro-organisms owe their pathogenicity to the presence of the exopolysaccharide capsule surrounding the cell. In biofilms containing pathogenic species, exopolysaccharides might act as a functional barrier against cells of the immune system and increase resistance to antibody-mediated mechanisms. The polymer might also interfere with killing by phagocytes by scavenging oxygen radicals produced by them or by acting as a barrier to cells of the immune system, to antibodies and to complement. Similarly, grazing protozoa might in some cases be prevented from utilising cells within the biofilm matrix. The presence of the exopolysaccharides in biofilms might also affect both excretion of microbial products and the transfer of plasmids. Even although exopolysaccharides are seldom themselves carbon or energy reserves, they could assist biofilm-bound microbial cells to bind or retain nutrients. Proteins bound in this way could then be hydrolysed later and serve as a source of amino acids and peptides for microbial growth. The conservation of enzymes within the biofilm, assisted by the exopolysaccharides also reduces the requirement for repeated synthesis of extracellular enzymes used in degradation of macromolecular nutrients. It also enables other, associated bacteria to benefit from release of these nutrients even though they are unable to degrade the polymers – an example of co-metabolism.

7
Dissolution of Physical Structures

Within the complex structure and heterogeneity of the biofilm there may be various factors that lead to solubilisation of either the very strong gel type or the more soluble types of adhesive. Changes may also occur either at the gross or molecular level. Some changes may lead to very rapid loss of biofilm integrity (e.g. HUGHES 1997). Many of the microbial constituents of the biofilm will, during growth and metabolism, secrete products which are strong chelating agents. The gluconic acid produced by *P. aeruginosa* or *Pseudomonas putida* when high glucose concentrations are present provides one such example. Removal of multivalent cations such as Ca^{2+}, even within a restricted part of the biofilm, might be expected to lead to either local or more general dissolution of the biofilm through solubilisation of the exopolysaccharide. Excess Na^+ might exchange with Ca^{2+} and phosphate would also leach Ca^{2+} from alginate (KEVEKORDES 1996). Alternatively, local proton gradients might convert the salt form of the exopolysaccharide to the proton form, again altering its properties. In the case of alginate, conversion to the proton form might enhance its rigidity. This would also be true for some other bacterial exopolysaccharides. The deacylated exopolysaccharide from *S. elodea* formed gels with a large range of monovalent and divalent cations but the gel strength depended on the ion and on its concentration. Generally higher concentrations of monovalent ions are needed to yield gels equivalent in strength to those induced by divalent counterions. Gels of this polysaccharide in the proton form resembled alginate in that they were even stronger than those formed with divalent cations (LARWOOD ET AL. 1996). Some other exopolysaccharides might behave similarly, but for most of these macromolecules there would probably be no such effect.

Changes in the ionic mix associated with a biofilm will inevitably lead to alterations in the conformation and strength of biofilm exopolysaccharides and their contribution to structure. Information from algal alginate gels may again be relevant in this context. Polysaccharide of low mass leached first, especially alginate in which there was less guluronic acid and more mannuronic acid (DRAGET ET AL. 1996). If the same happened in microbial biofilms, any change in ionic profile might result in changes to the composition of the exopolysaccharide mixture present, some of the polymers remaining in situ while others were relatively rapidly removed.

Enzymes may provide a major factor in dissolving the exopolysaccharide structure of biofilms. Many studies on exopolysaccharide synthesis have now revealed that either glycanases or polysaccharide lyases are gene products associated with biosynthesis of the exopolysaccharide itself. The genes for the enzymes were found to form part of operonic systems regulating synthesis, polymerisation and excretion of the exopolysaccharide (e.g. GLUCKSMAN ET AL. 1993; SUTHERLAND AND KENNEDY 1996). Enzymes present in the exopolysaccharide-synthesising bacteria might only be released slowly as cells lyse, but might nevertheless rapidly reduce the polymer mass. In the case of alginates from *P. fluorescens* or *P. putida*, the mass fell approximately 50% per 24 h at 30°C in planktonic cultures due to the action of endogenous poly-D-mannuronate-specific lyases and would probably be similarly reduced in biofilm (CONTI ET AL. 1994). The effect of such action would be to reduce greatly the aqueous solution viscosity and the binding properties of the exopolysaccharide. Considerable alteration to biofilm might thus be expected.

Other enzymes derive from bacteriophage infection and provide a means of enabling the phage to reach receptors on the cell surface. The enzymes are either integral components of the viral particle or are soluble proteins released during cell lysis on phage maturation. Most of the enzymes are endo-acting glycanases or polysaccharide lyases causing very rapid, random cleavage of the exopolysaccharide chains with resultant loss of viscosity and integrity (Sutherland, this volume). Exopolysaccharides will however, protect the microbial cells against some phage attack when the viral particles lack depolymerases and are unable to penetrate the polymer matrix.

Bacterial cells on the surface of a biofilm could be subject to high velocity attack by *Bdellovibrio*. This action would not destroy any exopolysaccharide present but would certainly cause considerable perturbation at the biofilm surface as capsules and slime provide the bacterial cell with no protection against penetration by the *Bdellovibrio* as was recently elegantly demonstrated by KOVAL AND BAYER (1997). If intracellular glycanases were present in the cell, they would possibly be released by such action and the *Bdellovibrio* might thus indirectly lead to some localised enzymic attack on the exopolysaccharide. Many micro-organisms are also capable of synthesising biosurfactants (emulsifiers) (ROSENBERG AND RON 1997). These clearly might play a significant role in the localised dissolution of biofilm material. It has been suggested that one role for biosurfactants is to enhance desorption of the microbial cells from hydrophobic surfaces which no longer contain usable carbon sources. The biosurfactants may also greatly affect attachment to liopophilic matrix material. Bioemulsans from bacterial species such as *Acinetobacter calcoaceticus* are macromolecules which enhance adhesion to surfaces and form strong monomolecular layers (ROSENBERG 1988). They are thus probably less affected by turbulence and changes in solute than are other exopolysaccharides.

Application of a range of agents has been shown to inhibit adhesion of *P. aeruginosa* (MAI ET AL. 1993). The agents tested included alginase, monoclonal antibodies and antibiotics. All showed some effect on bacterial adhesion. In some cases, synergism was seen between two types of agent. This was most noticeable for combinations of alginase and ciprofloxacin or the antibiotic with monoclonal antibodies.

8
Conclusions

Clearly our direct knowledge of the role of exopolysaccharides in biofilms is still rather limited. It is probable that each mixture of microbial species should be regarded as distinct, with its own unique microenvironments. The variance between exopolysaccharides of different structure and different physical properties in complex mixtures provides an enormous number of possible microenvironments and requires much more detailed study. One should certainly not extrapolate from highly charged polysaccharides such as bacterial alginates to neutral or less highly charged macromolecules and expect to get identical or even similar results. Nor should one ignore the interactions that undoubtedly occur between exopolysaccharides and other macromolecules within the biofilm. Hopefully, novel techniques, improved methods of microanalysis and greater use of microprobes may provide a clearer picture of the exact function(s) of these vital biofilm constituents.

References

Allison DG, Matthews MJ (1992) Effect of polysaccharide interactions on antibiotic susceptibility of *Pseudomonas aeruginosa*. J Appl Bacteriol 73:484–488

Allison DG, Sutherland IW (1984) A staining technique for attached bacteria and its correlation to extracellular carbohydrate production. J Microbiol Methods 2:93–99

Allison DG, Sutherland IW (1987) Role of exopolysaccharides in adhesion of freshwater bacteria. J Gen Microbiol 133:1319–1327

Atkins EDT, Attwool PT, Miles MJ, Morris VJ, O'Neill MA, Sutherland IW (1987) Effect of acetylation on the molecular interactions and gelling propoerties of a bacterial polysaccharide. Int J Biol Macromol 9:115–117

Birkhed D, Rosell K, Granath K (1979) Structure of extracellular water-soluble polysaccharides synthesised from sucrose by oral strains of *Streptococcus mutans, Streptococcus salivarius, Streptococcus sanguis* and *Actinomyces viscosus*. Arch Oral Biol 24:53–61

Cesàro A, Tomasi G, Gamini A, Vidotto S, Navarini L (1992) Solution conformation and properties of the galactoglucan from *Rhizobium meliloti* strain YE2 (Sl). Carbohydr Res 231:117–135

Chandrasekaran R, Radha A (1995) Molecular architectures and functional-properties of gellan gum and related polysaccharides. Trends Food Sci Technol 6:143–148

Chandrasekaran R, Puigjaner LC, Joyce KL, Arnott S (1988) Cation interactions in gellan: an X-ray study of the potassium salt. Carbohydr Res 181:23–40

Chandrasekaran R, Radha A, Lee EJ, Zhang M (1994) Molecular architecture of araban, galactoglucan and welan. Carbohydr Polymers 25:235–244

Conti E, Flaibani A, O'Regan M, Sutherland IW (1994) Alginate from *Pseudomonas fluorescens and Pseudomonas putida*: production and properties. Microbiol 140:1128–1132

Corzo J, Leon-Barrios M, Hernando-Rico V, Gutierrez-Navarro AM (1994) Precipitation of metallic cations by the acidic exopolysaccharides from *Bradyrhizobium japonicum*. Appl Environ Microbiol 60:531–4536

Costerton JW, Irvin RT, Cheng K (1981) Bacterial biofilms in nature and disease. Ann Rev Microbiol 35:399–424

Draget KI, Skjåk-Bræk G, Smidsrød O (1994) Alginic acid gels: the effect of alginate chemical composition and molecular weight. Carbohydr Polymers 25:31–38

Draget KI, Skjåk-Bræk G, Christensen BE, Gaserød O, Smidsrød O (1996) Swelling and partial solubilization of alginic acid gel beads in acidic buffer. Carbohydr Polymers 29:209–215

Dworkin M, Kaiser D (1993) Myxobacteria II. American Society for Microbiology, Washington

Geddie JL, Sutherland IW (1994) The effect of acetylation on cation binding by algal and bacterial alginates. Biotech Appl Biochem 20:117–129

Geesey GG, Richardson WT, Yeomans HG, Irvin RT, Costerton JW (1977) Microscopic exami-
    nation of natural sessile bacterial populations from an alpine stream. Can J Microbiol
    23:1733–1736
Glazebrook J, Walker GC (1989) A novel exopolysaccharide can function in place of the cal-
    cofluor-binding exopolysaccharide in nodulation of alfalfa by *Rhizobium meliloti*. Cell
    56:661–672
Gloaguen V, Wieruszeski J-M, Strecker G, Hoffmann L, Morvan H (1995) Identification by
    NMR spectroscopy of oligosaccharides obtained by acidolysis of the capsular polysac-
    charides of a thermal biomass. Int J Biol Macromol 17:387–393
Glucksman MA, Reuber TL, Walker GC (1993) Genes needed for the modification, polyme-
    rization, export and processing of succinoglycan by *Rhizobium meliloti*: a model for suc-
    cinoglycan biosynthesis. J Bacteriol 175:7045–7055
Grant WD, Sutherland IW, Wilkinson JF (1969) Exopolysaccharide colanic acid and its occur-
    rence in the *Enterobacteriaceae*. J Bacteriol 100:1187–1193
Grimont F, Grimont PAD (1992) The genus *Enterobacter*. In: Balows A, Trüper HG, Dworkin
    M, Harder W, Schleifer K-H (eds) The prokaryotes, 2nd edn, vol III, chap 148. Springer,
    Berlin Heidelberg New York, pp 2797–2815
Hughes KA (1997) Bacterial biofilms and their exopolysaccharides. PhD thesis, Edinburgh
    University
Hughes KA, Sutherland IW, Clark J, Jones MV (1998a) Bacteriophage and associated poly-
    saccharide depolymerases – novel tools for study of bacterial biofilms. J Appl Microbiol
    85:583–590
Hughes KA, Sutherland IW, Jones MV (1998b) The function of biofilm exopolysaccharides:
    increased exopolysaccharide production following exposure to antimicrobial agents and
    enhanced desiccation resistance. Microbiol 144:3039–3047
Hussain M, Herrmann M, von Eiff C, Perdreau-Remington F, Peters GA (1997) A 140-kilodal-
    ton extracellular protein is essential for the accumulation of *Staphylococcus epidermidis*
    strains on surfaces. Infect Immun 65:519–524
Iwashita S, Kanegasaki S (1976) Deacetylation reaction catalyzed by *Salmonella* phage. J Biol
    Chem 251:5361–5365
Kanzawa Y, Harada A, Koreeda A, Harada T, Okuyama K (1989) Difference of molecular asso-
    ciation in two types of curdlan gel. Carbohydr Polymers 10:299–313
Kevekordes K (1996) Using light scattering measurements to study the effects of monovalent
    and divalent cations on alginate aggregates. J Exp Bot 47:677–682
Kolenbrander PE (1991) In: Dworkin M (ed) Microbial cell:cell interactions, chap 10.
    American Society for Microbiology, Washington, USA, pp 303–329
Kondoh K, Hashiba M, Baba S (1996) Inhibitory activity of clarithromycin on biofilm syn-
    thesis with *Pseudomonas aeruginosa*. Acta Oto-laryngologica S525:56–60
Koval SF, Bayer ME (1997) Bacterial capsules – no barrier against *Bdellovibrio*. Microbiol
    143:749–753
Kuhn SP, Pfister RM (1989) Adsorption of mixed metals and cadmium by calcium alginate
    immobilised *Zoogloea ramigera*. Appl Microbiol 31:613X–618
Kumon H, Tomochika K-I, Matunga T, Ogawa M, Ohmori H (1994) A sandwich cup method
    for the penetration assay of antimicrobial agents through *Pseudomonas* exopolysaccha-
    ride. Microbiol Immunol 38:615–619
Kwiatkowski B, Beilharz H, Stirm S (1975) Disruption of Vi bacteriophage III and localization
    of its deacetylase activity. J Gen Virol 29:267–280
Larwood, VL, Howlin BJ, Webb GA (1996) Solvation effects on the conformational behav-
    iour of gellan and calcium ion binding to gellan double helices. J Mol Modeling
    2:175–182
Lawson CJ, McCleary CW, Nakada HI, Rees DA, Sutherland IW, Wilkinson JF (1969) Structural
    studies of colanic acid from *Escherichia coli* by using methylation and base-catalysed
    fragmentation. Biochem J 115:947–958
Lee EJ, Chandrasekaran R (1991) X-ray and computer modelling studies on gellan related po-
    lymers: molecular structures of welan, S657 and rhamsan. Carbohydr Res 214:11–24

Loaëc M, Olier R, Guezennec JG (1997) Uptake of lead, cadmium and zinc by a novel bacterial exopolysaccharide. Wat Res 31:1171–1179

Mack D, Fischer W, Krokotsch A, Leopold K, Hartmann R, Egge H, Laufs R (1996) The intercellular adhesin involved in biofilm accumulation of *Staphylococcus epidermidis* is a linear β-1,6-linked glucosaminoglycan: purification and structural analysis. J Bacteriol 178:175–183

Mai GT, McCormack JG, Seow WK, Pier GB, Jackson LA, Thong YH (1993) Inhibition of adherence of mucoid *Pseudomonas aeruginosa* by alginase. Infect Immun 61:4338–4343

Marqués AM, Bonet R, Simon-Pujol MD, Fuste MC, Congegado F (1990) Removal of uranium by an exopolysaccharide from *Pseudomonas* sp. Appl Microbiol Biotechnol 34: 429–431

Morris VJ, Brownsey GJ, Harris JE, Gunning AP, Stevens BJH, Johnston AWB (1989) Cation-dependent gelation of the acidic extracellular polysaccharides of *Rhizobium leguminosarum*: a non-specific mechanism for the attachment of bacteria to plant roots. Carbohydr Res 191:315–320

Nakamura T, Koo SJ, Pradipasena P, Rha C, Sinskey AJ (1987) Solution properties of polysaccharide flocculant produced by *Zoogloea ramigera* 115. Biotech Sep 399–413

Nisbet BA, Sutherland IW, Bradshaw IJ, Kerr M, Morris ER, Shepperson WA (1984) XM6 a new gel-forming bacterial polysaccharide. Carbohydr Polymers 4:377–394

Norberg AB, Persson H (1984) Accumulation of heavy metals by *Zoogloea ramigera*. Biotechnol Bioeng 26:239–246

Ogawa K, Yui T, Nakata K, Kakuta M, Misaki A (1997) X-ray study of beijeran sodium salts, a new galacturonic acid-containing exopolysaccharide. Carbohydr Res 300:41–45

Prince A (1996) *Pseudomonas aeruginosa*: versatile attachment mechanisms. In: Fletcher M (ed) Bacterial adhesion: molecular and ecological diversity. Wiley, New York, pp 183–199

Rosenberg E (1988) Microbial surfactants. Crit Rev Biotechnol 3:109–132

Rosenberg E, Ron EZ (1997) Bioemulsans – microbial polymeric emulsifiers. Current Opinion in Biotechnol 8:313–316

Ross-Murphy SB, Shatwell KP (1993) Polysaccharide strong and weak gels. Biorheol 30: 217–227

Samrakandi MM, Roques C, Michel G (1997) Influence of trophic conditions on exopolysaccharide production – bacterial biofilm susceptibility to chlorine and monochloramine. Can J Microbiol 43:751–758

Shatwell KP (1989) The influence of acetyl and pyruvate groups on the solution and interaction properties of xanthan. PhD thesis, Edinburgh University

Shevchik WE, Hugouvieux-Cotte-Pattat N (1997) Identification of a bacterial pectin acetyl esterase in *Erwinia chrysanthemi* 3937. Mol Microbiol 24:1285–1301

Skillman LC, Sutherland IW, Jones MV (1997) Cooperative biofilm formation between two species of *Enterobacteriaceae*. In: Wimpenny JWT, Handley P, Gilbert P, Lappin-Scott HM, Jones MV (eds) Biofilms: community interactions and control. Bioline Publications, Cardiff, pp 119–128

Skjåk-Bræk G, Grasdalen H, Draget K, Smidsrød O (1989) Inhomogeneous calcium alginate beads. In: Crescenzi V, Dea ICM, Paoletti S, Sutherland IW (eds) Biomedical and biotechnological advances in industrial polysaccharides. Gordon and Breach, New York, pp 345–362

Sutherland IW (1979) Polysaccharides produced by *Cystobacter, Archangium, Sorangium* and *Stigmatella* species. J Gen Microbiol 111:211–216

Sutherland IW (1983) Microbial exopolysaccharides – their role in microbial adhesion in aqueous systems. Crit Rev Microbiol 10:173–201

Sutherland, IW (1990) Biotechnology of microbial exopolysaccarides. Cambridge University Press, Cambridge

Sutherland, IW (1995) Biofilm specific polysaccharides – do they exist? In: Wimpenny JWT, Handley P, Gilbert P, Lappin-Scott HM (eds) The life and death of biofilm. Bioline Publications, Cardiff, pp 103–106

Sutherland IW (1996) A natural terrestrial biofilm. J Ind Microbiol 17:281–283

Sutherland IW (1997) Microbial biofilm exopolysaccharides – superglues or Velcro? In: Wimpenny JWT, Handley P, Gilbert P, Lappin-Scott HM, Jones MV (eds) Biofilms: community interactions and control. Bioline Publications, Cardiff, pp 33–39

Sutherland IW, Kennedy L (1996) Polysaccharide lyases from gellan-producing *Sphingomonas* spp. Microbiol 142:867–872

Sutherland IW, Thomson S (1975) Comparison of polysaccharides produced by *Myxococcus* strains. J Gen Microbiol 89:124–132

Teeri TT (1997) Crystalline cellulose degradation: new insight into the function of cellobiohydrolases. Trends Biotechnol 15:160–167

Thu B, Skjåk-Bræk G, Micali F, Vittur F, Rizzo R (1997) The spatial distribution of calcium in alginate gel beads analysed by synchrotron-radiation induced X-ray emission (SRIXE). Carbohydr Res 297:101–105

Troyano E, Lee SP, Rha CK, Sinskey AJ (1996) Presence of acetate and succinate in the exopolysaccharide produced by *Zoogloea ramigera* 115SLR. Carbohydr Polymers 31:35–40

Weise W, Reinheimer G (1978) Scanning electron microscopy and epifluorescence investigation of bacterial colonization of marine sand sediments. Microb Ecol 4:175–188

CHAPTER 5

Regulation of Matrix Polymer in Biofilm Formation and Dispersion

David G. Davies

Department of Biological Sciences, Science III, Binghamton University, Binghamton, N.Y. 13902, USA, *E-mail: dgdavies@binghamton.edu, Tel.: 607-777-2006*

Keywords. *Pseudomonas aeruginosa*, Alginate, Biosynthesis, Regulation, Autoinducer, Biofilm

1
Alginate Biosynthesis in Pseudomonas aeruginosa

Biofilms are biological films that develop and persist at interfaces in aqueous environments in natural and manmade ecosystems. These biological films are composed of microorganisms embedded in a gelatinous matrix composed of one or more organic polymers which are secreted by the resident microorganisms.

Biofilm bacteria have been shown to predominate both in numbers and in metabolic activity in natural (GEESEY ET AL. 1977; COSTERTON ET AL. 1994), industrial (BOIVIN AND COSTERTON 1991), and medical (KHOURY ET AL. 1992) environments. The involvement of extracellular polymers in bacterial biofilms has been documented for both aquatic (JONES ET AL. 1969; SUTHERLAND 1980) and marine bacteria (FLOODGATE 1972), and the association of exopolysaccharides with attached bacteria has been demonstrated using electron microscopy (GEESEY ET AL. 1977; DEMPSEY 1981) and light microscopy (ZOBELL 1943; ALLISON AND SUTHERLAND 1984). The presence of extracellular polymers is considered necessary for the development and persistence of the microbial biofilm (WARDELL ET AL. 1983; ALLISON AND SUTHERLAND 1987). Analyses of biofilm bacteria isolated from both freshwater and marine environments have shown that the polymers they produce are composed largely of acidic polysaccharides (FLETCHER 1980; SUTHERLAND 1980; CHRISTENSEN AND CHARAKLIS 1990).

The acidic polysaccharide alginate has been shown to be involved in biofilm formation by the Gram-negative bacterium *Pseudomonas aeruginosa* in pulmonary infections (DOGGETT 1969). This extracellular substance is composed of β-1,4-linked D-mannuronic acid and its C-5 epimer, L-guluronic acid (Fig. 1).

P. aeruginosa is commonly isolated from biofilms in the environment, industrial water systems and from infections. LINKER AND JONES (1966) reported that this organism has the ability to produce alginate. Alginate overproducing strains of *P. aeruginosa* generates copious quantities of the acidic polysaccharide and as a result produces distinctive mucoid colonies. The term mucoid when applied to *P. aeruginosa* is restricted to those strains producing large slimy colonies during 24 h incubation on common agar-based media and whose appearance results from the overproduction of alginate as a major component of the bacterial glycocalyx (GOVAN 1990). Mucoid variants of *P. aeruginosa* are most commonly isolated from the respiratory tract infections that accompany the genetic disease, cystic fibrosis (DOGGETT ET AL. 1977).

Cystic fibrosis (CF), is the most prevalent lethal genetic disease among people of European descent. It is inherited as an autosomal recessive trait at a rate of 1 in 2000 live births among Caucasians (MAY ET AL. 1991). Clinical manifestations of the disease include chronic pulmonary disease with a persistent cough, elevated sweat electrolytes and pancreatic abnormalities that result in problems with nutrient absorption and recurrent diarrhea (CHARTRAND AND MARKS 1983). CF is caused by a genetic defect resulting in an abnormal electrolyte transport and mucous secretion from exocrine glands and secretory epithelia (MCPHERSON AND GOODCHILD 1988).

During initial phases of CF lung infections, *P. aeruginosa* isolates produce nonmucoid colonies which are not associated with alginate production. However, over time and with the initiation of antibiotic therapy, the nonmucoid form begins to over-produce alginate (DOGGETT ET AL. 1966). The emergent mucoid form predominates in the CF-damaged lung and is the major pathogen isolated from sputum cultures of patients in advanced stages of the disease. The mucoid form has also been identified in approximately 2% of non-CF *P. aeruginosa* infections (DOGGETT 1969).

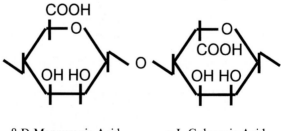

β-D-Mannuronic Acid   α-L-Guluronic Acid

Fig. 1. Structure of the constituent uronic acids of alginate

An interesting aspect of the mucoid phenotype is that it is unstable when cells are grown in laboratory culture; such bacteria have an unusually high reversion rate to the nonmucoid phenotype (GOVAN 1975; GOVAN ET AL. 1979). It is because of the high frequency of spontaneous mutations to nonmucoidy that it has been difficult to characterize and map structural gene mutations in the alginate pathway.

LINKER AND JONES (1964) were the first to characterize the alginate produced by *P. aeruginosa*. This alginate was later found to be similar to the commercially useful polymer obtained from marine algae (LINKER AND JONES 1966; LIN AND HASSID 1966A) and to the polysaccharide later identified with the capsular material of the bacterium *Azotobacter vinelandii* (GORIN AND SPENCER 1966).

Preliminary studies of the alginate biosynthesis pathway were made by LIN AND HASSID (1966 A, B), following investigations performed on the exopolysaccharide produced by the alga *Fucus gardneri* and by PINDAR AND BUCKE (1975), in studies on the capsular polysaccharide of *A. vinelandii*. This seminal work demonstrated that the first alginate precursor was fructose 6-phosphate which was recruited from the carbohydrate pool of the Entner-Doudoroff pathway and fructose 1,6-bisphosphate aldolase. Similar recruitment of this precursor has since been shown to take place in *P. aeruginosa* (MAY ET AL. 1991).

The first alginate biosynthesis enzymes from *P. aeruginosa* were reported by PIGGOTT ET AL. (1981). These investigators were able to demonstrate the presence of very low levels of phosphomannose isomerase (PMI), GDP-mannose pyrophosphorylase (GMP), and GDP-mannose dehydrogenase (GMD). PADGETT AND PHIBBS (1986) detected another alginate biosynthesis enzyme, phosphomannomutase (PMM).

In continuing studies into the regulation of alginate biosynthesis, the research group headed by Dr. A.M. Chakrabarty, at the University of Illinois at Chicago, was able to isolate a number of stable alginate mutants derived from the lung of a CF patient (Fig. 2).

The original isolate (8821) was mucoid on solid, rich, nutrient medium and possessed a high background reversion rate. One of the revertant mutants was discovered to have a stable nonmucoid phenotype and was designated strain 8822. This strain was mutagenized with ethyl methanesulfonate (EMS) to produce the stable mucoid derivative, strain 8830. Again, following EMS mutagenesis, a fourth strain was produced, 8852 which is nonmucoid but can be complemented back to mucoidy by the addition of AlgR1, a DNA binding protein. To determine that the mutational changes resulting from these procedures were not the result of lesions in the carbohydrate metabolism genes, each mutant was tested for its ability to grow on minimal medium supplemented with different carbon sources. No growth aberrations were noted in the presence of glucose, gluconate, glycerol, mannitol, fructose, glutamate, or succinate (DARZINS AND CHAKRABARTY 1984).

Other useful mucoid strains have since been made in a number of laboratories. Using these strains and their derivatives, it has been possible to isolate individual structural gene mutants defective in alginate synthesis. This work (particularly in the case of strain 8830) has been instrumental in developing our current understanding of the genes involved in alginate biosynthesis in *P. aeru-*

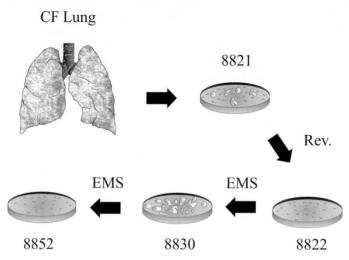

Fig. 2. Production of stable *Pseudomonas aeruginosa* mutants of a strain isolated from the lung of a cystic fibrosis patient

ginosa (BANERJEE ET AL. 1983; DARZINS ET AL. 1985A; ROYCHOUDHURY ET AL. 1989; SÁ-CORREIA ET AL. 1987). This has been due to the increased activity of the alginate genes that occurs only in mucoid strains.

A schematic representation of the alginate biosynthesis pathway of *P. aeruginosa* is diagrammed in Fig. 3. The first enzyme to be isolated and identified was the gene product from *algA* (DARZINS ET AL. 1985B). This is the first enzyme required for alginate synthesis and has been found to be bifunctional, having phosphomannose isomerase (PMI) activity, as well as GDP-mannose pyrophosphorylase (GMP) activity (SÁ-CORREIA ET AL. 1987). PMI is responsible for interconversion of fructose 6-P and mannose 6-P. Mannose 6-P is converted to mannose 1-P by phosphomannomutase (PMM). GDP-mannose pyrophosphorylase converts mannose 1-P to GDP-mannose, and GDP-mannose dehydrogenase converts GDP-mannose to GDP-mannuronic acid (GMD). The remaining steps of the *P. aeruginosa* alginate pathway have not been fully elucidated. Following the activity of GMD, acetylation, polymerization, epimerization, and export are known to occur.

Acetylation is carried out by *algF* (FRANKLIN AND OHMAN 1993; SHINA-BARGER ET AL. 1993). This enzyme is believed to be active only on mannuronic acid, and *O*-acetyl modification is proposed to regulate the degree of epimerization by shielding mannuronic acid groups from the epimerase enzyme. Acetylation of mannuronic acid residues has also been shown to be protective against enzymatic cleavage by alginate lyases (GACESA 1992). Acetylation of alginate is also known to be dependent upon the products of *algI* and *algJ*, although the specific functions of these gene products have not been fully elucidated (FRANKLIN AND OHMAN 1996). Epimerization interconverts D-mannuronic acid and L-guluronic acid, a function carried out by the *algG* gene product (FRANKLIN ET AL. 1994). The incorporation of L-guluronic acid into alginate

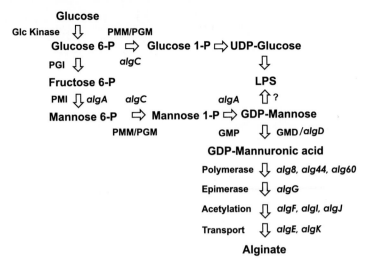

Fig. 3. Bacterial alginate biosynthesis pathway in *P. aeruginosa*. The gene encoding each enzyme is given in *italics*. Abbreviations for enzymes: PGI, phosphoglucose isomerase; PMI, phosphomannose isomerase; PMM, phosphomannomutase; PGM, phosphoglucomutase; GMP, GDP-mannose pyrophosphorylase; GMD, GDP-mannose dehydrogenase. The remaining steps in the production of alginate include polymerization, epimerization, acetylation, and export. A number of remaining alginate genes are of unknown function

may result in changes in the rheological properties of the finished polymer, as well as provide a protective role against cleavage by alginate lyase (BOYD AND CHAKRABARTY 1994). Alginate produced by *P. aeruginosa* 8830 has been shown to be composed of 95% mannuronic acid and 5% guluronic acid (ABRAHAMSON ET AL. 1996). None of the guluronate residues have been shown to be arranged into poly-guluronate blocks. *O*-Acetylation has been found on approximately 69% of the mannuronate residues; guluronate residues are not acetylated.

The polymerization and export of alginate are believed to be carried out by the gene products of *alg8, alg44, alg60, algE,* and *algK.* The polymerization of alginate is presumed to occur prior to acetylation and export, although the temporal and spatial aspects of the elongation reaction have yet to be fully described. It is currently believed that Alg8, Alg44, and possibly Alg60 (AlgX) are subunits involved in the polymerization process (MAHARAJ ET AL. 1993; REHM AND VALLA 1997). The *algE* gene product is a porin-like protein which is detectable only in mucoid strains of *P. aeruginosa* (REHM ET AL. 1994). The mature form of AlgE is located within the bacterial outer membrane and is presumed to be involved in the translocation of polymannuronate. The *algK* gene product is less well characterized but is also believed to be involved in the translocation of alginate (AARONS ET AL. 1997).

The production of alginate by *P. aeruginosa* is known to be regulated by transcriptional activators as well as by a σ^{54} sigma factor recognition sequence. A schematic representation of the known alginate genes is shown in Fig. 4. The

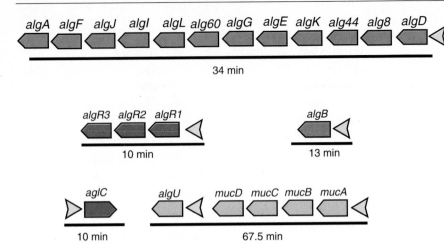

Fig. 4. Genes involved in alginate biosynthesis and regulation. The function of the gene products are as follows: AlgD, GDP-mannose dehydrogenase; Alg8 and Alg44, probable membrane proteins involved in alginate polymerization; AlgK, poorly characterized, presumed to be involved in transport; AlgE, probable outer membrane porin involved in transport; AlgG, epimerase; Alg60, required gene product, possibly involved in polymerization; AlgL, alginate lyase; AlgI, involved in acetylation; AlgJ, involved in alginate acetylation; AlgF, alginate acetylase; AlgA, phosphomannose isomerase/GDP-mannose pyrophosphorylase; AlgC, phosphomannomutase/phosphoglucomutase. AlgR1, AlgR2 and AlgR3 are positive response regulators for the alginate biosynthesis genes; MucA, MucB, MucC, MucD, and AlgU, regulate the switch to mucoidy. AlgB has been shown to enhance alginate production. *Triangle* indicates promoter region and direction of transcription. For a complete discussion of the alginate genes, refer to the text

DNA binding protein AlgR1 has been shown to be a positive transcriptional activator which interacts with the *algD* promoter region at three separate regions upstream from the mRNA start site (MOHR ET AL. 1992). Similar binding sites have been postulated for the *algC* promoter region (MAY ET AL. 1991).

The *algD* promoter regulates transcription of *algD, alg8, alg44, algK, algE, algG, alg60, algL, algI, algJ, algF,* and *algA* (BOYD ET AL. 1993; AARONS ET AL. 1997; CHITNIS AND OHMAN 1990; FRANKLIN AND OHMAN 1996; MAY ET AL. 1991). The *algC* promoter is known to regulate only the *algC* gene (ZIELINSKI ET AL. 1991). Other regulatory proteins have been found. AlgR2 has been shown to increase *algD* promoter activity when in the presence of AlgR1 (KATO ET AL. 1989). AlgR3, a highly basic regulatory protein, has not been assigned a known function; however, it has been shown to have significant sequence homology (44% identity) with a sea urchin histone protein believed to regulate DNA supercoiling in response to changes in osmolarity (KATO ET AL. 1990; HULTON ET AL. 1990).

AlgR1, AlgR2, and AlgR3 appear to regulate the level of transcription of the alginate biosynthesis genes. In addition to these response regulators, studies concerning the molecular mechanisms for conversion to mucoidy have defined a cluster of genes identified as *algU, mucA,* and *mucB.* The mucoid condition of *P. aeruginosa* has been shown to be dependent upon the presence of AlgU (also known as AlgT), a σ^{54}-like protein which interacts with RNA polymerase core to activate the promoters of the alginate biosynthesis genes (MARTIN ET AL. 1993a; DEVRIES AND OHMAN 1994). The activity of *algU* is controlled by two downstream accessory genes, *mucA* and *mucB* (also described as *algN*) (MARTIN ET AL. 1993b, c). MucA and MucB act to suppress AlgU activity, and mutations in these genes can cause conversion to mucoidy by relieving AlgU from negative regulation (MARTIN ET AL. 1994).

In laboratory studies it has been shown that the mucoid strain *P. aeruginosa* 8830 is capable of regulation of alginate production, even though there appeared to be constitutive expression of the alginate operons. When strain 8830 was grown in L-broth medium in the laboratory of Dr. A.M. Chakrabarty, it failed to produce detectable levels of alginate, despite the addition of a wide variety of supplemental carbon sources. However, when utilizing the same substrates on solid medium, the organism was able to produce large amounts of alginate (DARZINS AND CHAKRABARTY 1984). Strain 8830 has also been shown to be responsive to medium osmolarity, demonstrating maximum up-regulation when grown on 0.3–0.4 mol l^{-1} NaCl (A.M. Chakrabarty, personal communication).

Alginate biosynthesis in strains of *P. aeruginosa* isolated from environments other than the CF lung has not been well characterized. Transcription of their alginate biosynthesis genes is usually not detectable (MAY ET AL. 1991). It is for these reasons, and because of the unusual stability of its mucoid phenotype (DARZINS AND CHAKRABARTY 1984), that *P. aeruginosa* 8830 has been used to evaluate alginate biosynthesis in a biofilm mode of growth.

While *P. aeruginosa* 8830 is derived from a medical isolate, the alginate genes that it carries are presumed to be under similar control as those that affect alginate biosynthesis in environmental strains of *P. aeruginosa* and perhaps in other bacteria. Past investigations have suggested that the production of alginate by *P. aeruginosa* was unique to the environment of the CF lung (FIALHO ET AL. 1990). Yet it is unlikely that these bacteria, which are found predominantly in the soil and water, developed the ability to produce alginate in order to inhabit the lungs of humans having the genetic disorder leading to CF. It is also unlikely that infection of the CF lung by *P. aeruginosa* results from contact with other patients having CF. DERETIC ET AL. (1993) have suggested that the mucoid phenotype of *P. aeruginosa* is probably present in environments other than in chronically infected patients. Recent investigations have shown that alginate genes are widespread within the group *Pseudomonas*, even though they are not normally expressed (WALLACE ET AL. 1994). Examples of 23 *Pseudomonas* species were probed for the presence of *P. aeruginosa* alginate genes. Of these, all group I pseudomonads except *P. stutzeri* were shown to contain homologous sequences to the gene probes: *algA*, *algD* and *algR* (FIALHO ET AL. 1990). The presence of these principal alginate genes indicated that these organisms were capable of producing alginate, and that the pathway for alginate synthesis by these organisms incorporated similar elements to those found in *P. aeruginosa* 8830.

Additional studies by GROBE ET AL. (1995) demonstrated the presence of mucoid *Pseudomonas* species isolated from the surfaces of various technical water systems. In their study, 10 out of 81 *Pseudomonas* isolates demonstrated a mucoid phenotype when grown on *Pseudomonas* Isolation Agar. This work further indicated that attached *Pseudomonas* species in the environment possess the biosynthetic capability to produce alginate.

2
Regulation of Alginate Biosynthetic Genes During Biofilm Development

While *P. aeruginosa* is known to form biofilms on natural surfaces and CF lung epithelial cells (CHRISTENSEN AND CHARACKLIS 1990; COSTERTON ET AL. 1987), very little is known concerning the role of biofilm development on the triggering of exopolysaccharide formation. It is generally assumed that the initiation step in biofilm development is the attachment of the microorganism to the substratum. Surface activation of bacterial genes has been reported previously by DAGOSTINO ET AL. (1991) in *Pseudomonas* S9 growing on polystyrene microtiter plates. In that study transposon mutagenesis was used to insert promoterless *lacZ* genes into recipient organisms, giving some the ability to display β-galactosidase activity at a surface but not in liquid or on agar media. The authors, however, did not identify the specific target genes controlling *lacZ* activation. In a separate study by BELAS ET AL. (1986) the *laf* gene was activated in *Vibrio parahaemolyticus* on agar medium but not in liquid. Activation was, however, shown to occur in a highly viscous liquid. In that study the authors concluded that gene activation was a consequence of medium viscosity.

Two studies performed at the Center for Biofilm Engineering (DAVIES ET AL. 1993; DAVIES AND GEESEY 1995) were directed at investigating the activation of both the *algC* and the *algD* operons in *P. aeruginosa* 8830 during biofilm development. The gene product of *algC* has been shown to be a bifunctional enzyme, having phosphomannomutase activity, converting mannose 6-phosphate to mannose 1-phosphate, as well as phosphoglucomutase activity, interconverting glucose 6-phosphate and glucose 1-phosphate (YE ET AL. 1994). In its role as phosphomannomutase, the *algC* gene product is responsible for the formation of alginate and also, possibly lipopolysaccharide O-antigen subunits. In its role as phosphoglucomutase, the *algC* gene product is responsible for the production of a complete LPS core (COYNE ET AL. 1994). Due to its role as a bifunctional enzyme, *algC* might be transcribed without producing alginate as a final product, although this has not been experimentally established. Using a continuous culture apparatus, it was demonstrated that *algC* reporter gene product specific activity in biofilm bacteria was more than 19 times higher than in planktonic bacteria. Uronic acids accumulation was also shown to be higher in continuous culture biofilms when compared with planktonic bacteria.

In order to study the temporal aspects of initiation of *algC* transcription during biofilm development by *P. aeruginosa* 8830, continuous culture studies were performed using a plasmid borne reporter construct in which the *algC* promoter was fused to a *lacZ* gene encoding β-galactosidase (ZIELINSKI ET AL. 1991). A flow-cell was designed to be mounted to a microscope with which to observe activity of the reporter gene during the initial events of biofilm development (DAVIES ET AL. 1993). The bacteria were cultured in minimal medium supplemented with the substrate methylumbelliferyl-β-D-galactoside which produces a fluorescent blue indicator when cleaved by β-galactosidase. Figure 5 shows the

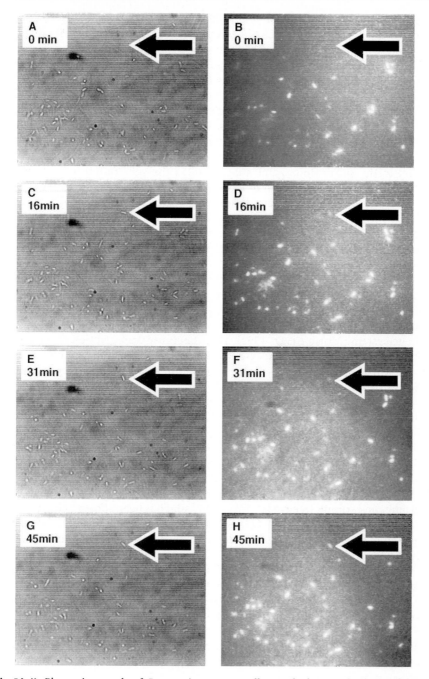

Fig. 5A–H. Photomicrograph of *P. aeruginosa* 8830 cells attached to a glass substratum. Magnification, ×1250. Phase contrast image of total cells (*left*) and fluorescent cells displaying β-galactosidase activity (*right*). The *arrow* denotes a cell attaching to the surface, with subsequent *algC* reporter gene up-expression

progression from attachment of a cell to the glass substratum to the initiation of *algC* transcription under conditions of continuous flow. In the left column of the figure all of the cells attached to the glass substratum are seen using transmitted light. In the right column only those bacteria expressing *algC* reporter activity are illuminated by β-galactosidase mediated fluorescence. In the top panel (A, B) the field is depicted prior to attachment of a bacterium (to the left of the arrow). At 16 min (C, D), the bacterium indicated by the arrow has left the bulk medium and attached to the glass substratum. At this point the bacterium does not demonstrate fluorescence, indicating that transcription of *algC* does not occur in the bulk liquid, and that *algC* activity is not required for attachment to the substratum. After attachment to the substratum for a period of 31 min, the bacterium is seen to have activated *algC* transcription and to have synthesized sufficient β-galactosidase to produce whole cell fluorescence (G, H). This work and a subsequent study (DAVIES AND GEESEY 1995) demonstrated that activation of the alginate biosynthetic operon *algC* was modulated by attachment to a substratum and indicated that the bacteria are capable of sensing the presence of surface.

The gene *algD* encodes the enzyme GDP-mannose dehydrogenase which catalyzes the oxidation of GDP-mannose into GDP-mannuronic acid, a direct precursor of alginate (DERETIC ET AL. 1987). The *algD* promoter has become a benchmark for monitoring molecular events which govern expression of the alginate system (MOHR ET AL. 1992). This is because transcription of *algD* is the first committed step in the biosynthesis of alginate, resulting in a functional enzyme which catalyzes a four-electron transfer in an apparently irreversible step (ROYCHOUDHURY ET AL. 1989).

The sequence of the *algD* gene and its promoter have been determined by DERETIC ET AL. (1987). The promoter has a novel secondary structure, possessing multiple direct and inverted repeats throughout the −50 to −110 region, as well as in the −35 and −10 regions (BERRY ET AL. 1989; MOHR ET AL. 1992). The *algD* gene has been shown to be transcriptionally activated by the DNA binding protein AlgR1 (DEVAULT ET AL. 1989). The AlgR1 protein has been shown to possess significant amino acid homology to a class of proteins responsive to environmental stimuli (DERETIC ET AL. 1989), and has been shown to be functionally interchangeable with the OmpR DNA binding protein of *E. coli* which can be used to activate *algD* transcription in *E. coli* in response to medium osmolarity (BERRY ET AL. 1989).

In tests to determine whether *algD* is itself an environmentally responsive gene, BERRY ET AL. (1989) and DEVAULT ET AL. (1990) investigated the influence of medium osmolarity and ethanol on alginate biosynthesis. These researchers showed that when either salt or ethanol was added to growing cultures of *P. aeruginosa* wild type 8821 or mucoid 8830, *algD* transcription and alginate biosynthesis were significantly increased as a result of enhanced *algD* promoter activity in liquid medium and increased alginate production after prolonged exposure on solid medium.

The activity of the *algD* promoter has been shown to follow the rule set by the *algC* operon. The number of bacteria that up-regulated *algC* transcription while attached to the surface of the flow cell coverslip as a percent of the total number of bacteria to attach to the surface over the course of the experiment was 57%. The number of bacteria that up-expressed *algD* transcription as a percent of the total bacteria that attached to the surface was only 31%. It may be presumed that there is a preference for attachment of bacteria that are up-regulated for alginate biosynthesis. This is not believed to be the case, since a strain not able to produce alginate (*P. aeruginosa* 8822) has been shown to colonize glass surfaces at a higher rate than the mucoid strain *P. aeruginosa* 8830 (observations made previously in our laboratory). The behavior of the *algD* promoter is, therefore, similar to that which was observed for the *algC* promoter. Both promoters showed a trend to become up-regulated following attachment to a glass surface. Both promoters were active in the majority of the attached populations and, when attaching or detaching, both promoters tend to be down expressed.

The results described in this section indicate that transcription of *algC* and *algD* by *P. aeruginosa* depends on whether the organism is free-living or growing as a biofilm. When in the bulk liquid the majority of cells are not up-regulated for the production of alginate. If these bacteria subsequently interact with a (glass) substratum, they may attach rather tenuously to that surface. The majority of bacteria that do so are washed from the surface and reenter the bulk medium. Of those bacteria that remain at the surface, most ultimately up-regulate *algD* and begin to produce alginate. This production of alginate is continued as long as the bacteria remain at the surface. If and when there is a change in the environment, the bacteria respond by down-regulating *algD* and ultimately detach from the surface and are carried further downstream where they may repeat this cycle.

It is important to note that the organism whose behavior is described in these studies is a clinical isolate that has been selected for alginate overproduction following treatment with a chemical mutagen. The behavior of strains of *P. aeruginosa* isolated from other environments may be different. Unfortunately, alginate production in environmental isolates of *P. aeruginosa* is not generally detectable (MAY ET AL. 1991). Furthermore, the number of genetic constructs and mutants of such bacteria do not compare with what is available of *P. aeruginosa* 8821 and its derivatives.

In nature, biofilms are generally composed of multiple species, unless they are found in extreme environments. Furthermore, there is little information on the variety of matrix polymers produced in nature by bacterial assemblages, or on the prevalence of alginate as a matrix polymer. One notable exception is found in work performed by WALLACE ET AL. (1994), in which 11 out of 120 bacterial isolates from biofilms growing in a nuclear power plant cooling water system were shown to have alginate biosynthesis genes.

It is hoped that in the future, more such studies will be performed to examine the distribution of exopolysaccharides in the environment, and that studies of the regulation of matrix polymer material in bacteria will extend to include environmental isolates in both pure and mixed cultures.

3
Control of Biofilm Dispersion

The control and removal of biofilm material from pipe and conduit surfaces has historically been carried out by the addition of corrosive chemicals such as chlorine or strong alkali solutions or through mechanical means. Such treatments are generally harsh to both the plumbing system and the environment, and have been necessary due to the recalcitrant nature of biofilms within those systems. The resistance to treatment by biocides has been due in large measure to the protective character of intact biofilm matrix polymers (SRINIVASAN ET AL. 1995; STEWART 1994). In medicine the use of elevated doses of antibiotics has been necessary in treatment when biofilms are believed to be involved. This is due at least in part to the enhanced protection of biofilm bacteria by the exopolysaccharide matrix material outside the cells (COSTERTON ET AL. 1987; NICHOLS ET AL. 1989; TACHIRO ET AL. 1991; ANWAR ET AL. 1992).

The control of biofilm development and persistence could be enhanced if the matrix polymers which protect biofilms could be degraded. The availability of a matrix degrading enzyme is therefore of considerable interest. Genetic studies of *P. aeruginosa* have shown that the *algD* operon encodes a gene for alginate lyase, an enzyme that is capable of cleaving the linkage between the uronic acid moieties which make up the alginate polymer. This enzyme is, therefore, capable of breaking down the alginate polymer into small subunits.

It has been shown by BOYD AND CHAKRABARTY (1994) that, when alginate lyase is induced artificially in these bacteria when grown on agar based medium, the bacteria growing on the surface of that medium are more easily removed by gentle rinsing when compared with uninduced bacteria. It is believed that cleavage of the chemical bonds between the uronic acid residues of the alginate polymers results in a breakdown of the alginate matrix allowing dispersal of the bacteria.

It is interesting that the gene *algL*, which encodes the information for production of alginate lyase, is nested within the *algD* gene cluster (BOYD ET AL. 1993; SCHILLER ET AL. 1993). This would suggest that *algL* is transcribed along with the biosynthesis genes required for alginate production. Such positioning is unexpected, since the activity of alginate lyase is known to result in the cleavage of the chemical linkage between uronic acid residues in alginate polymers (GACESA 1987). Such activity should lead to the complete or partial destruction of synthesized alginate polymer strands. The activity of alginate lyase has been characterized as an endolytic enzyme which cleaves alginate polymer strands by the β-elimination of the 4-O-glycosidic bond to yield two shorter saccharide polymers (GACESA 1987). An unsaturated sugar residue (4-deoxy-L-erythro-hex-4-ene-pyranosyluronate) is created on the non-reducing end of the 5′ product at the site of cleavage (PREISS AND ASHWELL 1962 A). Continued enzymatic activity results in the conversion of alginate to 2-keto-3-deoxy-D-gluconate (PREISS AND ASHWELL 1962 B).

It has been postulated that alginate lyase is used by the bacterium as a processing enzyme, necessary for production of alginate polymers in their final form prior to export from the cell envelope (BOYD AND CHAKRABARTY 1994).

There is, however, no evidence of lyase activity during alginate synthesis (BOYD ET AL. 1993). It has been shown previously, and in this study, that alginate can be produced and biofilm formation can occur in bacteria from which alginate lyase can be recovered and detected. It is postulated, therefore, that the lyase produced by these bacteria is somehow prevented from acting on alginate polymers outside of the cell. This could be achieved by regulating the transport of alginate lyase across the cell envelope or by processing of the enzyme after it is produced. When isolated from whole cell extracts, lyase has been shown to contain a leader sequence at the N-terminus (BOYD ET AL. 1993). This leader may prevent transport of the enzyme outside the cell envelope and/or it may affect the ability of the enzyme to act upon the substrate.

The above information suggests that alginate lyase is produced while alginate is being synthesized but, through some form of regulation, is prevented from interfering with polymerization of the acid polysaccharide. Previous work in our laboratory has shown that *P. aeruginosa* biofilms undergo unpredictable sloughing events with large sections of biofilm becoming detached from either the matrix or the substratum. It is believed that these events may be triggered by the release of active alginate lyase into the external milieu where alginate strands are broken apart releasing fragments of biofilm from the overall matrix.

In sloughing events induced by turning off flow within the continuous culture system, bacteria in biofilm cell clusters have been shown in our laboratories to break apart from one another and to disperse without the assistance of fluid shear. This demonstrated primarily that dispersion is not a strictly hydrodynamic process and secondarily that, following a sloughing event, bacteria actively disperse away from the site of the cell cluster. During the dispersion event, cells from the clusters were able to swim actively away from one another and could be imaged afterward as individual bacteria present in the bulk medium. These observations suggest that the polymer matrix responsible for maintaining the integrity of the cell clusters (biofilm) in a biofilm can be broken down to the extent that motility of individual bacteria does not appear to be impaired, with cells no longer connected to one another via polymer strand binding. The implication is that in a dispersion event induced by medium stagnation, the biofilm matrix polymer is significantly degraded.

In addition to the breaking apart of cell clusters, stagnation of the medium results in the detachment of bacteria directly from the substratum. It is believed that attachment to the substratum is not mediated by alginate; therefore, in addition to alginate lyase, other enzymes may be involved in a dispersion event induced by stagnation. When biofilm dispersion is artificially induced by the over-production of alginate lyase along with either salt or SDS addition, those bacteria directly attached to the substratum do not appear to be affected.

The extensive breakdown of the matrix polymer following medium stagnation may be due to the release of alginate lyase by the biofilm bacteria into the surrounding medium. During the period of alginate synthesis (at all times during development of the biofilm), one may assume from the configuration of the *algD* gene cluster that alginate lyase is continuously produced. Therefore, there should be a reservoir of the enzyme (in either active or inactive form) within the bacteria which, after significant accumulation within the cell, may be released into the surrounding environment. As a triggering mechanism mediating such a release, two possibilities are proposed. In the first, stagnant conditions within the flow-cell may result in a build-up of waste products. Waste products may themselves act as a trigger, such as through interaction of cell receptors with a specific accumulated compound in the medium, or they may result in secondary effects, such as may result from the accumulation of CO_2 and/or organic acids which may depress the local pH. Alternatively the bacteria may be able to respond to a depletion of nutrients from the environment, such as loss of the organic carbon source or the loss of O_2. The second possibility is that bacteria may chemically signal one another via specific molecules which can indicate cell density or the metabolic conditions of the cell. The involvement of chemical signaling will be discussed further in the following section.

4
Activation of Alginate Biosynthesis and Chemical Communication in Biofilms

One of the critical unanswered questions concerning the regulation of matrix polymer formation and breakdown has been the characterization of the signal transduction mechanism which is responsible for the initiation of these processes. The regulation of alginate biosynthesis in P. aeruginosa is presumed to respond to numerous environmental as well as physiological cues. In CF lung infections, a succession from nonmucoid to mucoid cells (DOGGETT ET AL. 1966) is thought to reflect an adaptation by P. aeruginosa to an environment deficient in water and high in K^+, Na^+, HCO_3^-, and Cl^- (KNOWLES ET AL. 1983). Laboratory studies have shown that high osmolarity activates P. aeruginosa transcription of algD (BERRY ET AL. 1989) and algC (ZIELINSKI ET AL. 1991) leading to enhanced alginate production. Ethanol has likewise been shown to enhance alginate synthesis by that organism on solid and in liquid media (DEVAULT ET AL. 1990). The effects of ethanol and salt are believed to result in a response leading to membrane perturbations similar to what is found in the environment of the CF lung. In a separate study, ROBERSON AND FIRESTONE (1992) showed that desiccation of a soil Pseudomonas sp. growing in a sand matrix resulted in more exopolymer than when cells were grown at high water potential. Activation of algC and algD in cells grown on a surface may be the result of decreasing water or nutrient levels at the point of contact with the substratum. Some believe that the switch to mucoidy is brought about by a mutational event in either mucA or mucB which relieves the negative regulatory effect on algU, inducing alginate production. Such a mutational event would enhance the survival of P. aeruginosa in environments with a low water activity.

One of the more interesting possibilities is that alginate production and degradation are modulated by cell-cell communication. In the early 1970s it was demonstrated that bacterial bioluminescence is controlled by intercellular communication in the marine bacteria Vibrio fischeri (now Photobacterium fischeri) and the related species Vibrio harveyi (EBERHARD 1972; NEALSON ET AL. 1970).

As planktonic bacteria in the marine environment, these cells are found at densities of 10^2 per ml and are non-luminescent. As symbionts in the light organs of certain marine fishes and squids, these bacteria reach cell densities in excess of 10^{10} to 10^{11} cells per ml and are luminescent. The trigger for light production in these bacteria has been shown to be an acyl homoserine lactone (HSL) which acts as an autoinducer for the production of luciferin and luciferase.

In all cases, HSL autoinducers are known to associate with a DNA binding protein homologous to LuxR in *Photobacterium fischeri*, causing a conformational change in the protein, initiating transcriptional activation. This process couples the expression of specific genes to bacterial cell density (LATIFI ET AL. 1996). Regulation of this type has been called "quorum sensing" because it suggests the requirement for a "quorate" population of bacterial cells prior to activation of the target genes (FUQUA ET AL. 1994).

For several years it was presumed that the autoinducer involved in bacterial luminescence was unique to the few bacteria that produce light in the marine environment. Then, in 1992, the terrestrial bacterium *Erwinia carotovora* was shown to use an autoinducer system to regulate the production of the β-lactam antibiotic carbapenem (BAINTON ET AL. 1992B). The molecule found to be responsible for autoinduction of carbapenem was identified as an acylated homoserine lactone (HSL), a member of the same class of molecule responsible for autoinduction of bioluminescence. This finding led to a general search for HSLs in a wide range of bacteria. To affect the search, a bioluminescence sensor system was developed and used to screen HSL production in the spent supernatant liquids of a number of bacterial cultures. Many different organisms were shown by the screening to produce HSLs. These included *P. aeruginosa, Serratia marcescens, Erwinia herbicola, Citrobacter freundii, Enterobacter agglomerans,* and *Proteus mirabilis* (BAINTON ET AL. 1992A; SWIFT ET AL. 1993). More recently, the list has grown to include *Erwinia stewartii* (BECK AND FARRAND 1995), *Yersinia enterocolitica* (THROUP ET AL. 1995), *Agrobacterium tumefaciens* (ZHANG ET AL. 1993; FUQUA AND WINANS 1994), *Rhizobium leguminosarium* (SCHRIPSEMA ET AL. 1996), and others. Today it is generally assumed that all enteric bacteria, and the Gram-negative bacteria generally, are capable of cell density regulation using HSL autoinducers.

GAMBELLO ET AL. (1993) showed that the HSL product of the *lasI* gene of *P. aeruginosa* controls the production of exotoxin A, and of other virulence factors, in a cell density dependent manner. Since that time, the production of a large number of *Pseudomonas* virulence factors have been shown to be controlled by HSL compounds produced by the LasI and RhlI regulatory systems (OCHSNER AND REISER 1995; WINSON ET AL. 1995; LATIFI ET AL. 1995) in a manner reminiscent of the Lux system. LATIFI ET AL. (1996) have also shown that many stationary phase properties of *P. aeruginosa*, including those controlled by the stationary phase sigma factor (RpoS), are under the hierarchical control of the LasI and RhlI cell-cell signaling systems. BROWN AND WILLIAMS (1985) have suggested that many of the properties of biofilm bacteria, including their remarkable resistance to antibiotics (NICKEL ET AL. 1985), may derive from the fact that some of their component cells exhibit characteristics of stationary phase planktonic cells.

In *P. aeruginosa*, quorum sensing has been shown to be involved in the regulation of a large number of exoproducts including elastase, alkaline protease, LasA protease, hemolysin, cyanide, pyocyanin, and rhamnolipid (GAMBELLO ET AL. 1993; LATIFI ET AL. 1995; WINSON ET AL. 1995; OCHSNER AND REISER 1995). Most of these exoproducts are synthesized and exported maximally as *P. aeruginosa* enters stationary phase.

It is also during stationary phase that Gram-negative bacteria have been shown to develop stress response resistance that is coordinately regulated through the induction of a stationary-phase sigma factor known as RpoS (HENGGE-ARONIS 1993). Biofilm bacteria are generally considered to show physiological similarity to stationary phase bacteria in batch cultures. Thus, it is presumed that the synthesis and export of stationary-phase autoinducer-mediated exoproducts occurs generally within biofilms. The stationary phase behavior of biofilm bacteria may be explained by the activity of accumulated HSL within cell clusters. The mechanism causing biofilm bacteria to demonstrate stationary-phase behavior is hinted at by the recent discovery that RpoS is produced in response to accumulation of $3OC_4$-HSL in *P. aeruginosa* cultures (LATIFI ET AL. 1996). MCLEAN ET AL. (1997) have shown that HSL autoinducers are detectable in naturally occurring biofilms. In their study, the authors sampled a stream in Texas, USA and demonstrated that the autoinduction system of *A. tumifaciens* could be activated on one side of an agar plate when biofilm containing rocks were placed on the other side of the plate. No activation was observed when the biofilm-coated rocks had been previously autoclaved or if the rocks did not contain biofilm.

In our laboratories we have shown that *P. aeruginosa* PAO1 requires the *lasI* gene product $3OC_{12}$-HSL in order to develop a normal differentiated biofilm (DAVIES ET AL. 1998). In this study we observed that bacteria knocked out in *lasI* produced biofilm cell clusters were 20% the thickness of the wild-type organism. In addition, these mutants grew as continuous sheets on the substratum, lacking differentiation and not demonstrating evidence of matrix polymer (Fig. 6B). By contrast, the wild-type organism formed characteristic microcolonies composed of groups of cells separated by intervening matrix polymer and separated from one another by water channels (Fig. 6A). When the autoinducer $3OC_{12}$-HSL was added to the medium of growing *lasI* mutant bacteria, these cells developed biofilms that were indistinguishable from the wild-type organism (Fig. 6C). Since biofilm architecture is presumed to be influenced by matrix polymer, we can hypothesize that autoinduction is at least partly responsible for regulation of *P. aeruginosa* EPS. In support of this hypothesis, we have demonstrated that the alginate genes *algC* and *algD* are induced in *P. aeruginosa* strain 8830 by $3OC_{12}$-HSL and by $3OC_4$-HSL. These data indicate the possible role of quorum sensing as a signal transduction system by which *P. aeruginosa* may initiate the production of alginate and possibly other types of EPS. It is possible that when *P. aeruginosa* attaches to a surface, the HSL which it produces accumulates between the bacterium and the substratum to a sufficient level to allow induction of the quorum sensing system, turning on alginate biosynthesis and perhaps other EPS biosynthetic machinery. As biofilm development progresses, bacterial density within the cell clusters is

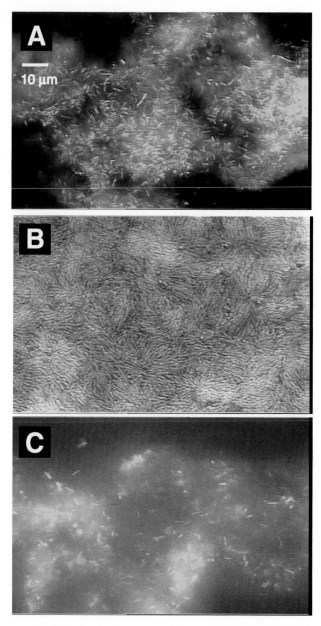

Fig. 6A–C. Demonstration of the influence of homoserine lactone $3OC_{12}$-HSL on biofilm development in *P. aeruginosa*: **A** normal biofilm cell cluster development by wild-type *P. aeruginosa* exhibiting void spaces, separated cells and matrix polymer (*fuzzy areas*); **B** cell cluster development by a mutant strain of *P. aeruginosa* deficient in the ability to produce $3OC_{12}$-HSL. The cell cluster is made of closely packed bacteria, lacking void spaces and without observable matrix polymer; **C** the same mutant strain as in B but grown in the presence of exogenous homoserine lactone (10 µmol l^{-1} $3OC_{12}$-HSL), demonstrating recovery of the wild-type phenotype

high enough to maintain activation of the quorum sensing system, resulting in continued production of EPS.

It is also possible that cell density dependent regulation is responsible for the release of enzymes which can degrade biofilm matrix polymers allowing bacteria to disperse from a biofilm. It has been observed in our laboratories and in the laboratories of Dr. Lapin-Scott at the University of Exeter, UK that when certain bacteria (including *P. aeruginosa*) reach high cell densities in biofilm cell clusters, the bacteria often undergo a dispersion event. Furthermore, biofilm dispersion has been shown to occur reproducibly under conditions of medium stagnation. Within both of these circumstances, HSL concentrations can build up to levels necessary for activation of cell density dependent genes.

The role of autoinduction in biofilm bacteria has been only superficially investigated. However, cell-cell communication as a biofilm phenomenon appears to be firmly established. Research on the degree to which quorum sensing and other forms of intercellular communication operate is needed to understand more perfectly the role of cell signaling in regulating biofilm development and persistence.

5
P. aeruginosa Biofilm Life Cycle

A working hypothesis has been developed which describes the life cycle of *P. aeruginosa* as it progresses through the stages of planktonic growth, biofilm development, and ultimately biofilm destruction following changes in the regulation of the promoters for the *algC* and *algD* operons, and regulation of AlgL. *P. aeruginosa* is able to respond to environmental stimuli which act as signals to the bacteria and to autoinducer molecules, influencing its mode of growth and physiology. Following attachment to a surface, at least two genes have been found to be turned on which enhance formation of a biofilm. Figure 7 shows the progression of regulation of *algC*, *algD*, and AlgL as biofilm formation, development, and breakdown progresses. Bacteria represented schematically are depicted as existing in a continuous or semi-continuous flowing system. Such environments are known to exist in streams (GEESEY ET AL. 1978), in industrial systems (GROBE ET AL. 1995), and in pulmonary infections involving *P. aeruginosa* (DEVAULT ET AL. 1989). The model presented here for a biofilm life cycle is, therefore, applicable to any of these diverse situations.

In the bulk phase (A) the bacteria are motile and are shown to swim in the menstruum. Occasionally a bacterium will interact with the substratum via one or more of its cell surface components. If the shear is not too high, the bacteria may attach to the surface. During this period, the *algC* and *algD* promoters are down-regulated in the majority of the population. Such a situation should result in production of a shortened LPS, lacking A-band polysaccharide (LIGHTFOOT AND LAM 1991). This condition has been shown to enhance the ability of *P. aeruginosa* PAO1 to attach to a glass substratum. In addition to depression of LPS A-band synthesis, alginate production has been shown to be down-regulated in these bacteria. The presence of alginate has been demonstrated by ALLISON AND SUTHERLAND (1987), and in our lab, to inhibit attachment of bacteria to a

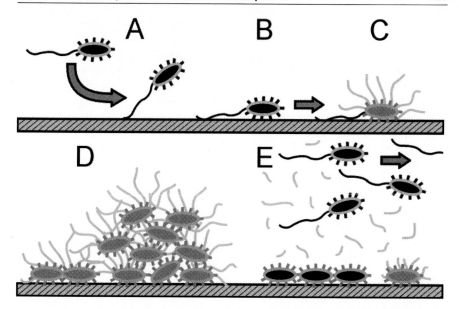

Fig. 7 A–E. Schematic representation of *P. aeruginosa* 8830 proposed life cycle as a biofilm: **A** planktonic bacteria in the bulk liquid interact with the substratum and initiate attachment; **B** cell surface molecules interact with the substratum resulting in reversible binding; **C** *algC* and *algD* regulation change, resulting in alteration of LPS and production of alginate; **D** cell cluster development progresses as alginate is produced; **E** changes in environmental conditions result in the down regulation of *algC* and *algD*, and the activation of alginate lyase resulting in destruction of cell cluster matrix polymer, releasing bacteria into the bulk liquid

glass substratum. In Fig. 7 B, bacteria are shown to lie down on the substratum and exhibit a more intimate association with the surface. The majority of bacteria that stick in this way remain at the surface only for a brief period. If the bacterium remains at the surface and is not washed downstream, it has been shown in the majority of cases to up-regulate both *algC* and *algD*. These bacteria then begin to produce alginate and presumably enhance their ability to remain attached to the substratum. It has been noted in our lab and in the laboratory of Dr. Gary Sayler, that *P. aeruginosa* 8822 (which does not produce alginate) sticks better to a glass surface than does *P. aeruginosa* 8830. However, strain 8822 is not capable of forming biofilm matrix polymer and only develops as a monolayer of cells at the glass surface. In Fig. 7 D, alginate production leads to the formation of cell clusters, with bacteria enmeshed in the alginate matrix. The condition of these cells continues until some change occurs in their environment. When this happens, both *algC* and *algD* promoter activity have been observed to down-regulate. In addition to the reduction in alginate biosynthesis, it appears that AlgL is activated and exported from the bacteria. The release of alginate lyase by biofilm bacteria results in the cleavage of alginate polymer strands and bacteria in cell clusters are released into the bulk liquid.

This complex regulation of surface attachment, biofilm formation, and dispersion is strongly suggested by the research results presented in this chapter.

The exact mechanisms of environmental signaling, the possibility of cell-cell communication, and the post-transcriptional regulation of genes described in this report have yet to be fully elucidated. The proposed model of biofilm development presented here is only presumed to apply to *P. aeruginosa* growing in axenic biofilms. In nature, bacterial biofilms are complex assemblages composed of multiple species and quite probably multiple types of matrix polymer. The ability of biofilms to slough in the environment has, however, been noted and implies that common response mechanisms to environmental stimuli may exist. The author hopes that this work will be of some use to those who in the future will try to further our understanding of the regulation of biofilms by bacteria in environmental, industrial, and medical ecosystems.

References

Aarons SJ, Sutherland IW, Chakrabarty AM, Gallagher MP (1997) A novel gene, *algK*, from the alginate biosynthetic cluster of *Pseudomonas aeruginosa*. Microbiology 143: 641–652

Abrahamson M, Lewandowski Z, Geesey GG, Skjåk-Bræk G, Strand W, Christensen BE (1996) Development of an artificial biofilm to study the effects of a single microcolony on mass transport. J Microbiol Methods 26:161–169

Allison DG, Sutherland IW (1984) A staining technique for attached bacteria and its correlation to extracellular carbohydrate production. J Microbiol Methods 2:93–99

Allison DG, Sutherland IW (1987) The role of exopolysaccharides in adhesion of freshwater bacteria. J Gen Microbiol 133:1319–1327

Anwar H, Strap JL, Chen K, Costerton JW (1992) Dynamic interactions of biofilms on mucoid *Pseudomonas aeruginosa* with tobramycin and piperacillin. Antimicrob Agents Chemother 36:1208–1214

Bainton NJ, Bycroft BW, Chhabra SR, Stead P, Gledhill L, Hill PJ, Rees CED, Winson MK, Salmond JPC, Stewart GSAB, Williams P (1992a) A general role for the *lux* autoinducer in bacterial cell signaling: control of antibiotic synthesis in *Erwinia*. Gene 116:87–91

Bainton NJ, Stead P, Chhabra SR, Bycroft BW, Salmond GPC, Stewart GSAB, Williams P (1992b) N-(3-oxohexanoyl)-l-homoserine lactone regulates carbepenem antibiotic production in *Erwinia carotovora*. Biochem J 288:997–1004

Banerjee PC, Vanags RI, Chakrabarty AM, Maitra PK (1983) Alginic acid synthesis in *Pseudomonas aeruginosa* mutants defective in carbohydrate metabolism. J Bacteriol 155:238–245

Beck von BS, Farrand SK (1995) Capsular polysaccharide biosynthesis and pathogenicity in *Erwinia carotovora*. J Bacteriol 177:5000–5008

Belas R, Simon M, Silverman M (1986) Regulation of lateral flagella gene transcription in *Vibrio parahaemolyticus*. J Bacteriol 167:210–218

Berry A, DeVault JD, Chakrabarty AM (1989) High osmolarity is a signal for enhanced *algD* transcription in mucoid and nonmucoid *Pseudomonas aeruginosa* strains. J Bacteriol 171:2312–2317

Boivin J, Costerton JW (1991) Biofilms and biodeterioration. In: Rossmore HW (ed) Biodeterioration and biodegradation 8. Elsevier Appl Sci, London, pp 53–62

Boyd A, Chakrabarty AM (1994) Role of alginate lyase in cell detachment of *Pseudomonas aeruginosa*. Appl Environ Microbiol 60:2355–2359

Boyd A, Ghosh M, May TB, Shinabarger D, Keogh R, Chakrabarty AM (1993) Sequence of the *algL* gene of *Pseudomonas aeruginosa* and purification of its alginate lyase product. Gene 131:1–8

Brown MRW, Williams P (1985) The influence of the environment of envelope properties affecting survival of bacteria in infections. Annu Rev Microbiol 39:527–556

Chartrand SA, Marks MI (1983) Pulmonary infections in cystic fibrosis: pathogenesis and therapy. In: Pennington JE (ed) Respiratory infections: diagnosis and management. Raven Press, New York, pp 201–216

Chitnis CE, Ohman DE (1990) Cloning of the *Pseudomonas aeruginosa algG*, which controls alginate structure. Mol Microbiol 8:583–590

Christensen BE, Characklis WG (1990) Physical and chemical properties of biofilms. In: Characklis WG, Marshall KC (eds) Biofilms. Wiley, New York, pp 93–130

Costerton JW, Cheng KJ, Geesey GG, Ladd TI, Michel JC, Dasgupta M, Marrie TJ (1987) Bacterial biofilms in nature and disease. Annu Rev Microbiol 41:435–464

Costerton JW, Lewandowski Z, DeBeer D, Caldwell DE, Korber DR, James GA (1994) Biofilms: the customized microniche. J Bacteriol 176:2137–2142

Coyne MJ Jr, Russel KS, Coyle CL, Goldberg JB (1994) The *Pseudomonas aeruginosa algC* gene encodes phosphoglucomutase, required for the synthesis of a complete lipopolysaccharide core. J Bacteriol 176:3500–3507

Dagostino L, Goodman AE, Marshall KC (1991) Physiological responses induced in bacteria adhering to surfaces. Biofouling 4:113–119

Darzins A, Chakrabarty AM (1984) Cloning of genes controlling alginate biosynthesis from a mucoid cystic fibrosis isolate of *Pseudomonas aeruginosa*. J Bacteriol 159:9–18

Darzins A, Nixon LL, Vanags RI, Chakrabarty AM (1985a) Cloning of *Escherichia coli* and *Pseudomonas aeruginosa* and phosphomannose isomerase genes and their expression in alginate-negative mutants of *Pseudomonas aeruginosa*. J Bacteriol 164:249–257

Darzins A, Wang SK, Vanags RI, Chakrabarty AM (1985b) Clustering of mutations affecting alginic acid biosynthesis in mucoid *Pseudomonas aeruginosa*. J Bacteriol 164:516–524

Davies, DG, Geesey GG (1995) Regulation of the alginate biosynthesis gene *algC* in *Pseudomonas aeruginosa* during biofilm development in continuous culture. Appl Environ Microbiol 61:860–867

Davies, DG, Chakrabarty AM, Geesey GG (1993) Exopolysaccharide production in biofilms: substratum activation of alginate gene expression by *Pseudomonas aeruginosa*. Appl Environ Microbiol 59:1181–1186

Davies, DG, Parsek MR, Pearson JP, Iglewski BH, Costerton JW, Greenberg EP (1998) The involvement of cell-to-cell signals in the development of a bacterial biofilm. Science 280:295–298

Dempsey MJ (1981) Marine bacterial fouling: a scanning electron microscope study. Mar Biol 61:305–315

Deretic V, Gill F, Chakrabarty AM (1987) *Pseudomonas aeruginosa* infection in cystic fibrosis:nucleotide sequence and transcriptional regulation of the *algD* gene. Nucleic Acids Res 15:4567–4581

Deretic V, Dikshit Konyecsni WM, Chakrabarty AM (1989) The *algR* gene, which regulates mucoidy in *Pseudomonas aeruginosa*, belongs to a class of environmentally responsive genes. J Bacteriol 169:351–358

Deretic V, Martin W, Schurr MJ, Mudd MH, Hibler NS, Curcic R, Boucher JC (1993) Conversion to mucoidy in *Pseudomonas aeruginosa*. Bio/Technology 11:1133–1136

DeVault JD, Berry A, Misra TK, Chakrabarty AM (1989) Environmental sensory signals and microbial pathogenesis: *Pseudomonas aeruginosa* infection in cystic fibrosis. Bio/Technology 7:352–357

DeVault JD, Kimbara K, Chakrabarty AM (1990) Pulmonary dehydration and infection in cystic fibrosis: evidence that ethanol activates alginate gene expression and induction of mucoidy in *Pseudomonas aeruginosa*. Mol Microbiol 4:737–745

DeVries CA, Ohman DE (1994) Mucoid-to-nonmucoid conversion in alginate-producing *Pseudomonas aeruginosa* often results from spontaneous mutations in *algT*, encoding a putative alternate sigma factor, and shows evidence for autoregulation. J Bacteriol 176:6677–6687

Doggett RG (1969) Incidence of mucoid *Pseudomonas aeruginosa* from clinical sources. Appl Microbiol 18:936–937

Doggett RG, Harrison GM, Stillwell RN, Wallis ES (1966) An atypical *Pseudomonas aeruginosa* associated with cystic fibrosis of the pancreas. J Pediatr 68:215–221

Doggett RG, Harrison GM, Carter RE (1977) Mucoid *Pseudomonas aeruginosa* in patients with chronic illnesses. Lancet I:236–237

Eberhard A (1972) Inhibition and activation of bacterial luciferase synthesis. J Bacteriol 109:1101–1105

Fialho AM, Zielinski NA, Fett WF, Chakrabarty AM, Berry A (1990) Distribution of alginate gene sequences in the *Pseudomonas* rRNA homology group1-*Azomonas-Azotobacter* lineage of superfamily B procaryotes. Appl Environ Microbiol 56:436–443

Fletcher M (1980) Adherence of marine microorganisms to smooth surfaces. In: Beachey EH (ed) Bacterial adherence (receptors and recognition, series B, vol 6). Chapman & Hall, London, pp 347–371

Floodgate GD (1972) The mechanism of bacterial attachment to detritus in aquatic systems. Memorie dell'Istituto italiano di idrobiologica Dott Carco di Marchi 29[Suppl]:309–323

Franklin M, Ohman DE (1993) Identification of *algF* in the alginate biosynthetic gene-cluster of *Pseudomonas aeruginosa* which is required for alginate acetylation. J Bacteriol 175:5057–5065

Franklin M, Ohman DE (1996) Identification of *algI* and *algJ* in the *Pseudomonas aeruginosa* alginate biosynthetic gene cluster which are required for alginate O-acetylation. J Bacteriol 178:2186–2195

Franklin M, Chitnis CE, Gacesa P, Sonesson A, White DC, Ohman DE (1994) *Pseudomonas aeruginosa algG* is a polymer level alginate C5-mannuronan epimerase. J Bacteriol 176:1821–1830

Fuqua WC, Winans SC (1994) A LuxR-LuxI type regulatory system activates *Agrobacterium* Ti plasmid conjugal transfer in the presence of a plant tumor metabolite. J Bacteriol 176:2796–2806

Fuqua WC, Winans SC, Greenberg EP (1994) Quorum sensing in bacteria: the LuxR-LuxI family of cell density-responsive transcriptional regulators. J Bacteriol 176:269–275

Gacesa P (1987) Alginate-modifying-enzymes: a proposed unified mechanism of action for the lyases and epimerases. FEBS Lett 212:199–202

Gacesa P (1992) Enzymatic degradation of alginates. Int J Biochem 24:545–552

Gambello MJ, Kaye S, Iglewski BH (1993) LasR of *Pseudomonas aeruginosa* is a transcriptional activator of the alkaline protease gene (*apr*) and an enhancer of exotoxin A expression. Infect Immun 61:1180–1184

Geesey GG, Richardson WT, Yeomans HG, Irvin RT, Costerton JW (1977) Microscopic examination of natural sessile bacterial populations from Alpine streams. Can J Microbiol 23:1733–1736

Geesey GG, Mutch R, Costerton JW, Green RB (1978) Sessile bacteria: an important microbial population in small mountain streams. Limnol Oceanogr 23:1214–1222

Gorin PAJ, Spencer JFT (1966) Exocellular alginic acid from *Azotobacter vinelandii*. Can J Chem 44:993–998

Govan JRW (1975) Mucoid strains of *Pseudomonas aeruginosa*: the influence of culture medium on the stability of mucus production. J Med Microbiol 8:513–522

Govan JRW (1990) Characteristics of mucoid *Pseudomonas aeruginosa* in vitro and in vivo. In: Gacesa P, Russell NJ (eds) *Pseudomonas* infection and alginates. Biochemistry, genetics and pathology. Chapman & Hall, London, pp 50–75

Govan JRW, Fyfe JAM, McMillan C (1979) The instability of mucoid *Pseudomonas aeruginosa*-fluctuation test and improved stability of the mucoid form in shaker culture. J Gen Microbiol 110:229–232

Grobe S, Wingender J, Trüper HG (1995) Characterization of mucoid *Pseudomonas aeruginosa* strains isolated from technical water systems. J Appl Bacteriol 79:94–102

Hengge-Aronis R (1993) Survival of hunger and stress: the role of *rpoS* in early stationary phase regulation in *Escherichia coli*. Cell 72:165–168

Hulton CSJ, Seirafi A, Hinton JCD, Sidebotham JM, Waddel L, Pavitt GD, Owen-Hughes T, Spassky A, Buc H, Higgins CF (1990) Histone-like protein H1 (H-NS), DNA supercoiling, and gene expression in bacteria. Cell 63:631–642

Jones HC, Roth IL, Sanders WM (1969) Electron microscopic study of a slime layer. J Bacteriol 99:316–325

Kato J, Chu L, Kitano K, DeVault JD, Kimbara K, Chakrabarty AM, Misra TK (1989) Nucleotide sequence of a regulatory region controlling alginate synthesis in *Pseudomonas aeruginosa*:characterization of the *algR2* gene. Gene 84:31–38

Kato J, Misra TK, Chakrabarty AM (1990) AlgR3, a protein resembling eukaryotic histone H1, regulates alginate synthesis in *Pseudomonas aeruginosa*. Proc Natl Acad Sci USA 87:2887–2891

Khoury AE, Lam K, Ellis BD, Costerton JW (1992) Prevention and control of bacterial infections associated with medical devices. ASAIO J 38:M174–178

Knowles MR, Stutts MJ, Spock A, Fischer N, Gutzy JJ, Boucher RC (1983) Abnormal ion permeation through cystic fibrosis respiratory epithelium. Science 221:1067–1070

Latifi A, Winson KM, Foglino M, Bycroft BS, Stewart GSAB, Lazdunski A, Williams P (1995) Multiple homologues of LuxR and LuxI control expression of virulence determinants and secondary metabolites through quorum sensing in *Pseudomonas aeruginosa* PAO1. Mol Microbiol 17:333–344

Latifi A, Foglino M, Tanaka K, Williams P, Lazdunski A (1996) A hierarchical quorum-sensing cascade in *Pseudomonas aeruginosa* links the transcriptional activators LasR and RhlR (VsmR) to expression of the stationary-phase sigma factor RpoS. Mol Microbiol 21:1137–1146

Lightfoot J, Lam JS (1991) Molecular cloning of genes involved with expression of A-band lipopolysaccharide, an antigenically conserved form in *Pseudomonas aeruginosa*. J Bacteriol 173:5624–5630

Lin TY, Hassid WZ (1966a) Isolation of guanosine diphosphate uronic acids from a marine brown alga, *Fucus gardneri* Silva. J Biol Chem 241:3283–3293

Lin TY, Hassid WZ (1966b) Pathway of alginic acid synthesis in the marine brown alga, *Fucus gardneri* Silva. J Biol Chem 241:5284–5297

Linker A, Jones RS (1964) A polysaccharide resembling alginic acid from a *Pseudomonas* micro-organism. Nature 204:187–188

Linker A, Jones RS (1966) A new polysaccharide resembling alginic acid isolated from pseudomonads. J Biol Chem 241:3845–3851

Maharaj R, May TB, Wang SK, Chakrabarty AM (1993) Sequence of the *alg8* and *alg44* genes involved in the synthesis of alginate by *Pseudomonas aeruginosa*. Gene 136:267–269

Martin DW, Holloway BW, Deretic V (1993a) Characterization of a locus determining the mucoid status of *Pseudomonas aeruginosa*: AlgU shows sequence similarities with a *Bacillus* sigma factor. J Bacteriol 175:1153–1164

Martin DW, Schurr MJ, Mudd MH, Deretic V (1993b) Differentiation of *Pseudomonas aeruginosa* into the alginate-producing form: inactivation of *mucB* causes conversion to mucoidy. Mol Microbiol 9:495–506

Martin DW, Schurr MJ, Mudd MH, Govan JRW, Holloway BW, Deretic V (1993c) Mechanism of conversion to mucoidy in *Pseudomonas aeruginosa* infecting cystic fibrosis patients. Proc Natl Acad Sci USA 90:8377–8381

Martin DW, Schurr MJ, Yu H, Deretic V (1994) Analysis of promoters controlled by the putative sigma factor AlgU regulating conversion to mucoidy in *Pseudomonas aeruginosa*:relationship to σ^E and stress response. J Bacteriol 176:6688–6696

May TB, Shinabarger D, Maharaj R, Kato J, Chu L, DeVault JD, Roychoudhury S, Zielinski NA, Berry A, Rothmel RK, Misra TK, Chakrabarty AM (1991) Alginate synthesis by *Pseudomonas aeruginosa*: a key pathogenic factor in chronic pulmonary infections of cystic fibrosis patients. Clin Microbiol Rev 4:191–206

McLean RJ, Whiteley M, Stickler DJ, Fuqua WC (1997) Evidence of autoinducer activity in naturally occurring biofilms. FEMS Microbiol Lett 154:259–263

McPherson MA, Goodchild MC (1988) The biochemical defect in cystic fibrosis. Clin Sci 74:337–345

Mohr CD, Leveau JHJ, Krieg DP, Hibler NS, Deretic V (1992) AlgR-binding sites within the *algD* promoter make up a set of inverted repeats separated by a large intervening segment of DNA. J Bacteriol 174:6624–6633

Nealson KH, Platt T, Hastings JW (1970) Cellular control of the synthesis and activity of the bacterial luminescent system. J Bacteriol 104:313–322

Nichols WW, Evans MJ, Slack MPE, Walmsley HL (1989) The penetration of antibiotics into aggregates of mucoid and nonmucoid *Pseudomonas aeruginosa*. J Gen Microbiol 135:1291–1303

Nickel J, Ruseska CK, Wright JB, Costerton JW (1985) Tobramycin resistance of cells of *Pseudomonas aeruginosa* growing as a biofilm on urinary catheter material. Antimicrob Agents Chemother 27:619–624

Ochsner UA, Reiser J (1995) Autoinducer-mediated regulation of rhamnolipid biosurfactant synthesis in *Pseudomonas aeruginosa*. Proc Natl Acad Sci USA 92:6424–6428

Padgett PJ, Phibbs PV Jr (1986) Phosphomannomutase activity in wild-type and alginate-producing strains of *Pseudomonas aeruginosa*. Curr Microbiol 14:187–192

Piggott NH, Sutherland IW, Jarman TR (1981) Enzymes involved in the biosynthesis of alginate by *Pseudomonas aeruginosa*. Eur J Appl Microbiol Biotechnol 13:179–183

Pindar DF, Bucke C (1975) The biosynthesis of alginic acid by *Azotobacter vinelandii*. Biochem J 152:617–622

Preiss J, Ashwell G (1962a) Alginic acid metabolism in bacteria I. Enzymatic formation of unsaturated oligosaccharides and 4-deoxy-1-erythro-5-hexoseulose uronic acid. J Biol Chem 237:309–316

Preiss J, Ashwell G (1962b) Alginic acid metabolism in bacteria II. The enzymatic reduction of 4-deoxy-1-erythro-5-hexoseulose uronic acid to 2-keto-3-deoxy-D-gluconic acid. J Biol Chem 237:317–321

Rehm BHA, Valla S (1997) Bacterial alginates:biosynthesis and applications. Appl Microbiol Biotechnol 48:281–288

Rehm BHA, Boheim G, Tommassen J, Winkler UK (1994) Overexpression of AlgE in *Escherichia coli*:subcellular localization, purification, and ion channel properties. J Bacteriol 176:5639–5647

Roberson EB, Firestone MK (1992) Relationship between desiccation and exopolysaccharide production in a soil *Pseudomonas* sp. Appl Environ Microbiol 58:1284–1291

Roychoudhury S, May TB, Gill JF, Singh SK, Feingold DS, Chakrabarty AM (1989) Purification and characterization of guanosine diphospho-D-mannose dehydrogenase: a key enzyme in the biosynthesis of alginate by *Pseudomonas aeruginosa*. J Biol Chem 264:9380–9385

Sá-Correia I, Darzins A, Wang SK, Berry A, Chakrabarty AM (1987) Alginate biosynthetic enzymes in mucoid and nonmucoid *Pseudomonas aeruginosa*: overproduction of phosphomannose isomerase, phosphomannomutase, and GDP-mannose pyrophosphorylase by overexpression of the phosphomannose isomerase (*pmi*) gene. J Bacteriol 169:3224–3231

Schiller NL, MondaySR, Boyd CM, Keen NT, Ohman DE (1993) Characterization of the *Pseudomonas aeruginosa* alginate lyase gene (*algL*): cloning, sequencing and expression in *Escherichia coli*. J Bacteriol 175:4780–4789

Schripsema J, de Rudder KEE, van Vleit TG, Lankhorst PP, de Vroom E, Kijne JW, van Brussel AAN (1996) Bacteriocin *small* of *Rhizobium leguminosarum* belongs to the class of N-acyl-L-homoserine lactone molecules, known as autoinducers and as quorum sensing cotranscription factors. J Bacteriol 178:366–371

Shinabarger D, May TB, Boyd A, Ghosh M, Chakrabarty AM (1993) Nucleotide sequence and expression of the *Pseudomonas aeruginosa algF* gene controlling acetylation of alginate. Mol Microbiol 9:1027–1035

Srinivasan R, Stewart PS, Griebe T, Chen CI, Xu X (1995) Biofilm parameters influencing biocide efficacy. Biotech Bioeng 46:553–560

Stewart PS (1994) Biofilm accumulation model that predicts antibiotic resistance of *Pseudomonas aeruginosa* biofilms. Antimicrob Agents Chemother 38:1052–1058

Sutherland IW (1980) Polysaccharides in the adhesion of marine and freshwater bacteria. In: Berkeley RCW, Lynch JM, Melling J, Rutter PR, Vincent B (eds) Microbial adhesion to surfaces. Ellis Horwood, London, pp 329–338

Swift S, Winson MK, Chan PF, Bainton NJ, Birstall M, Reeves PJ, Rees CEC, Chhabra SR, Hill PJ, Stewart GSAB (1993) A novel strategy for the isolation of *luxI* homologues: evidence for the widespread distribution of a LuxR:LuxI superfamily in enteric bacteria. Mol Microbiol 10:511–520

Tachiro H, Numakura T, Nishikawa S, Miyaji Y (1991) Penetration of biocides into biofilms. Wat Sci Technol 23:1395–1403

Throup J, Camara M, Briggs G, Winson MK, Chhabra SR, Bycroft BW, Williams P, Stewart GSAB (1995) Characterization of the *yenI/yenR* locus from *Yersinia enterocolitica* mediating the synthesis of the *N*-acylhomoserine lactone signal molecules. Mol Microbiol 17:345–356

Wallace WH, Fleming JT, White DC, Sayler GS (1994) An *algD*-bioluminescent reporter plasmid to monitor alginate production in biofilms. Microb Ecol 27:225–239

Wardell JN, Brown CM, Flannigan B (1983) In: Microbes and surfaces. Symposia of the Society for General Microbiology, 34:351–378

Winson MK, Camara M, Latifi A, Foglino M, Chhabra SR, Daykin M, Bally M, Chapon V, Salmond GPC, Bycroft BW, Lazdunski A, Stewart GSAB, Williams P (1995) Multiple *N*-acyl-L-homoserine lactone signal molecules regulate production of virulence determinants and secondary metabolites in *Pseudomonas aeruginosa*. Proc Natl Acad Sci USA 92:9427–9431

Ye RW, Zielinski NA, Chakrabarty AM (1994) Purification and characterization of phosphomannomutase/phosphoglucomutase from *Pseudomonas aeruginosa* involved in biosynthesis of both alginate and lipopolysaccharide. J Bacteriol 176:4851–4857

Zhang L, Murphy PJ, Max IT (1993) Agrobacterium conjugation and gene regulation by *N*-acyl-L-homoserine lactones. Nature 362:446–448

Zielinski NA, Chakrabarty AM, Berry A (1991) Characterization and regulation of the *Pseudomonas aeruginosa algC* gene encoding phosphomannomutase. J Biol Chem 266:9754–9763

Zobell CE (1943) The effect of solid surfaces upon bacterial activity. J Bacteriol 46:39–56

CHAPTER 6

Exopolymers of Sulphate-Reducing Bacteria

Iwona B. Beech[1] · Rudi C. Tapper[2]

University of Portsmouth, Microbiology Research Laboratory, St. Michaels' Building, White Swan Road, Portsmouth, PO1 2DT, UK,
[1] E-mail: Iwona.Beech@port.ac.uk
[2] E-mail: Rudi.Tapper@port.ac.uk

Keywords. Sulphate-reducing bacteria, SRB, Exopolymers, EPS, Atomic force microscopy, Environmental scanning electron microscopy, Metal binding

Sulphate-reducing bacteria (SRB) are a group of phylogenetically diverse anaerobic microorganisms of several genera which carry out dissimilatory reduction of sulphur compounds such as sulphate, sulphite, thiosulphate and even sulphur itself to sulphide (BAK AND CYPIONKA 1987; LOVLEY AND PHILIPS 1994). SRB are often present in biofilms developed on inanimate surfaces in terrestrial, freshwater and marine habitats (VON WOLZOGEN KUHR AND VAN DER VLUGT 1934; BEECH ET AL. 1995; GUBNER AND BEECH 1996). These microorganisms of varying nutritional needs are capable of utilising a wide range of carbon sources including alcohols, CO_2, fatty acids and hydrocarbons. Although strictly anaerobic, some SRB genera tolerate oxygen (HARDY AND HAMILTON 1981). Excellent reviews on sulphate-reducers have been published during the past decade (POSTGATE 1984; WIDDEL 1988; BARTON 1995).

Unlike aerobic bacteria, EPS production by environmental strains of anaerobes including SRB has not been a subject of extensive study. Secretion of extracellular polymeric substances (EPS) by SRB was first reported by SENEZ (1953) on growing *Desulfovibrio desulfuricans* spp. as batch cultures in a yeast extract based medium. Sessile proliferation along the surface of the vessel was associated with an amorphous viscous substance that was thought to be polysaccharide. Abundant slime formation was observed by GROSSMAN AND POSTGATE (1955) in old cultures of *Desulfovibrio desulfuricans* spp. and in continuous cultures of *Desulfovibrio vulgaris* spp. Paper chromatography (PC) analysis of this material produced in SRB batch cultures, conducted by OCHYNSKI AND POSTGATE (1963) indicated the presence of both protein and polysaccharide compounds that were not chemically related to the cell wall. The only type of carbohydrate detected using PC was an aldol-hexose (mannose), and hence the EPS material recovered from the cultures was referred to as mucopolymannoside. The synthesis of this polymer was more pronounced in marine strains of *Desulfovibrio* spp. or fresh water strains adapted to saline environments. Almost thirty years later, scanning electron microscopy (SEM), environmental scanning electron microscopy (ESEM) and atomic force microscopy (AFM) techniques (Figs. 1–3, respectively) have been used to image exopolymers in SRB biofilms formed on mild and stainless steel surfaces (BEECH ET AL. 1991;

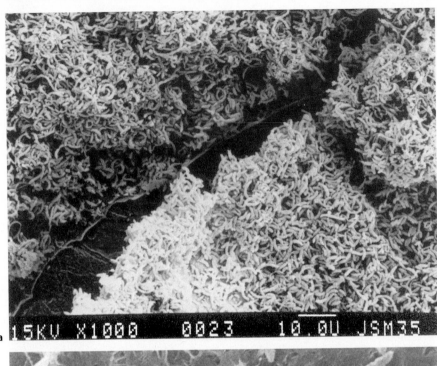

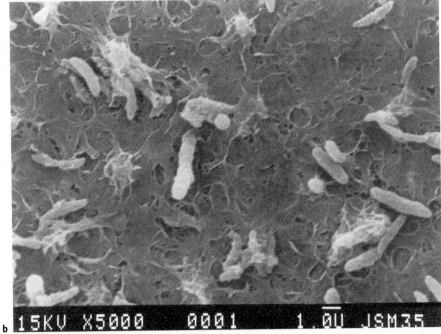

Fig. 1a, b. SEM micrographs of a seven day old biofilm developed on the surface of mild steel exposed to pure cultures of *Desulfovibrio indonensis* demonstrating: **a** dense coverage of the surface; **b** the presence of exopolymeric matrix (CHEUNG 1995)

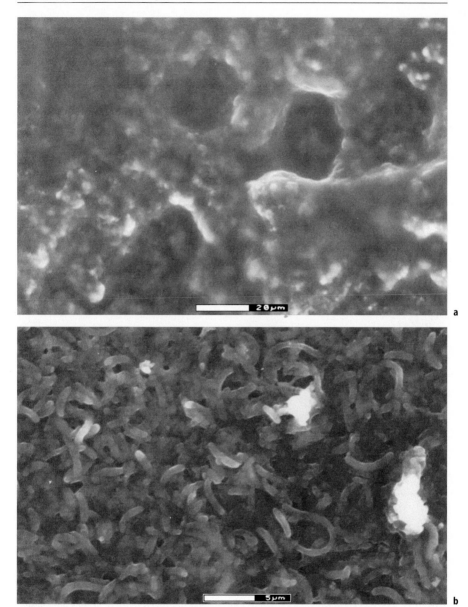

Fig. 2a, b. ESEM micrographs of a seven day old biofilm of *Desulfovibrio indonensis* formed on a mild steel surface: **a** in a fully hydrated state showing characteristic stacks and pores (voids); **b** following 40% dehydration to reveal individual cells previously invisible due to the presence of exopolymer (BEECH ET AL. 1996 with permission from Gordon and Breach Publishers, Lausanne, Switzerland)

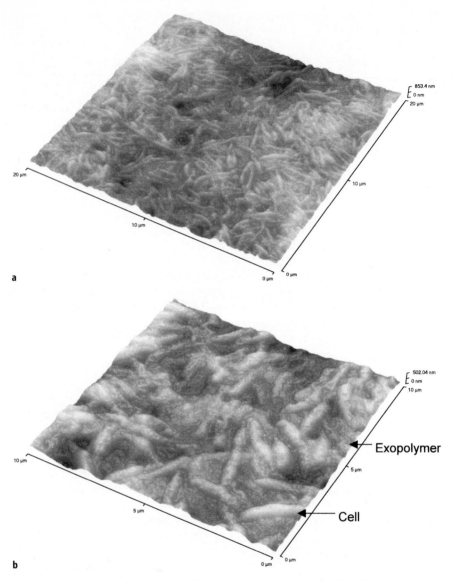

Fig. 3a, b. AFM images of a three day old *Desulfovibrio indonensis* biofilm developed on the surface of AISI 316 stainless steel showing: **a** monolayer of cells; **b** individual cells embedded in exopolymer (adapted from KANG 1998)

COUTINHO ET AL. 1993; CHEUNG 1995; BEECH ET AL. 1996A; KANG 1998). A variety of analytical techniques such as colorimetric assays, gas chromatography-mass spectrometry (GC-MS), protein (SDS-PAGE) and nucleic acid gel electrophoresis as well as Fourier Transform Infrared Spectroscopy (FTIR) have aided the analysis of the EPS released by different strains of freshwater and marine

SRB into the bulk phase of culture media and present in biofilms formed on mild and stainless steel surfaces (BEECH AND GAYLARDE 1991; ZINKEVICH ET AL. 1996; KANG 1998). These studies demonstrated that polysaccharides produced by different SRB species all contained neutral hexose, amino sugars and uronic acids in varying concentrations. In addition to polysaccharides, the presence of proteins and nucleic acid as part of exopolymer released either to the bulk phase (free EPS) or secreted by sessile cells (biofilm EPS) has also been documented. Several neutral carbohydrates such as glucose, mannose, galactose, ribose, rhamnose, xylose and allose have been identified using GC-MS in free and biofilm EPS of *Desulfovibrio desulfuricans* subspecies *desulfuricans* (BEECH ET AL. 1991), later reclassified by PEREIRA ET AL. (1996) as *Desulfomicrobium baculatus,* and in the new species of marine SRB *Desulfovibrio alaskensis* (KANG 1998) and *Desulfovibrio indonensis* (FEIO ET AL. 1998). Amino sugars, although detected with colorimetric methods, were not identified in *D. baculatus* exopolymers. The presence of mannosamine and galactosamine was, however, confirmed in EPS from *D. alaskensis* and *D. indonensis* using GC-MS analysis (KANG 1998). Additional studies conducted employing FTIR, SDS-PAGE and GC-MS confirmed considerable differences in protein profiles and carbohydrate concentration between free and biofilm exopolymers of SRB as well as between free exopolymers released in the presence and the absence of mild steel surfaces, indicating that the type of SRB and their cellular mode of growth (sessile vs planktonic) and the presence of a metal surface, possibly providing a source of iron ions, can influence the type of exopolymer synthesised (BEECH AND GAYLARDE 1991; ZINKEVICH ET AL. 1996; KANG 1998).

It is widely accepted that the yield and composition of bacterial EPS depend on the species and growth conditions (SUTHERLAND 1985). SRB are no exception to this rule. Marked differences were detected in the amount and chemical signature of free EPS between different SRB isolates belonging to the same genus grown under identical conditions (CHEUNG 1995; BEECH AND CHEUNG 1995). Contrary to some reports, there was no difference in the growth rates between these isolates (BEECH ET AL. 1994; CHEUNG AND BEECH 1996) that would account for the difference in EPS chemistry and abundance. Exopolymers secreted by the SRB *D. indonensis* varied with the type of carbon source used, metal ion species added as salts and antimicrobial compounds present in the growth media. In the media supplemented with lactate, acetate or propionate, the highest yield of free EPS harvested from *D. indonensis* batch cultures was recorded while using propionate as carbon source. Furthermore, the polymer synthesised when growing on propionate was considerably enriched with uronic acid (CHAN 1993). EPS yield was decreased compared to that in metal salt-free growth media when *D. indonensis* was cultivated in the presence of Cr, Mo and Ni ions (BEECH AND CHEUNG 1995). Treatment of *D. indonensis* with the monovalent carboxylic ionophore Monensin, reported to inhibit exopolymer synthesis in fungi (KOULALI ET AL. 1996), resulted in a significant decrease of EPS yield as compared to untreated cultures (TAPPER 1998).

The involvement of exopolymers secreted by different genera of aerobic bacteria in the process of metal deterioration (FORD ET AL. 1987 and references therein), is univocally accepted (GEESEY AND MITTELMAN 1985; GEESEY ET AL.

1986; WHITE ET AL. 1985; GEESEY ET AL. 1998). Biofilm and free EPS may selectively bind metal ions, thus creating metal concentration cells, promoting galvanic coupling and therefore influencing the electrochemical behaviour of steel. The role of SRB in pitting corrosion of various metals and their alloys in both aquatic and terrestrial environments under anoxic as well as oxygenated conditions is well documented (HAMILTON 1985). Several models have been proposed to explain the mechanisms by which these bacteria can influence the corrosion process of mild steel. Of these, cell metabolism such as activity of hydrogenase enzymes, production of hydrogen sulphide leading to the formation of metal sulphides, and liberation of volatile phosphorous compounds, are thought to play a key role in SRB-influenced corrosion (IVERSON 1987; FORD ET AL. 1990; ODOM AND SINGLETON 1992). Recent reviews by LEE ET AL. (1995) and HAMILTON (1998) clearly state that one predominant mechanism may not exist and that a number of factors are involved. Although binding of metal ions such as Fe, Cr, Mo and Ni by SRB exopolymers has been demonstrated using techniques of atomic absorption spectrometry (AA), X-ray photoelectron spectroscopy (XPS) and Time-of-Flight Secondary Ion Mass Spectrometry (TOF-SIMS) (BEECH AND CHEUNG 1995; BEECH ET AL. 1996B; BEECH ET AL. 1998A), no study elucidating the role of SRB exopolymers in steel deterioration has been reported until recently. The ability of SRB exopolymers to influence corrosion of mild steel and determination of the type of EPS macromolecules involved in the deterioration process has been demonstrated in vitro (BEECH ET AL. 1998B). A thermostable polysaccharide-protein complex of molecular mass greater than 200 kDa (estimated using gel filtration) was isolated from free EPS secreted by D. indonensis by applying low-pressure chromatography. This complex was capable of accelerating the deterioration of mild steel, observed as grain boundary and intercrystalline attack. It is hypothesised that the exceptional affinity of the exopolymer complex for Fe ions, as determined using TOF-SIMS and XPS, could account for its corrosive nature (BEECH ET AL. 1996B).

The implication of EPS chemistry in the biocorrosion of carbon steel in the presence of D. alaskensis and D. indonensis, known to promote different corrosion rates of steel under identical growth conditions (BEECH ET AL. 1993), has recently been investigated. It appears that there is not only a marked difference in the capacity of EPS to bind Fe depending on the SRB isolate but also in the type of macromolecules involved. Combined XPS and TOF-SIMS investigation revealed that EPS produced by D. indonensis, which causes greater corrosion rates of steel compared to D. alaskensis, accumulates 20 times more Fe originating from the surface of mild steel than D. alaskensis exopolymer (Beech, unpublished). It appears that the likely importance of EPS released by SRB in the process of corrosion has been largely ignored as the attention has been drawn to inorganic deposits formed on steel surfaces in the presence of these bacteria. Furthermore, it seems that species specificity ought to be considered when discussing the possible involvement of SRB exopolymers in the biocorrosion process.

References

Bak F, Cypionka H (1987) A novel type of energy metabolism involving fermentation of inorganic sulfur compounds. Nature 326:891–892

Barton LL (ed) (1995) Sulfate-reducing bacteria. Plenum Press, New York

Beech IB, Cheung CWS (1995) Interactions of exopolymers produced by sulphate reducing bacteria with metal ions. Int Biodet Biodegr 35:59–72

Beech IB, Gaylarde CC (1991) Microbial polysaccharides and corrosion. Int Biodeterior 27:95–107

Beech IB, Gaylarde CC, Smith JJ, Geessey GG (1991) Extracelluar polysaccharides from *Desulfovibrio desulfuricans* and *Pseudomonas fluorescens* in the presence of mild and stainless steel. Appl Microbiol Biotechnol 25:65–71

Beech IB, Campbell SA, Walsh FC (1993) Microbiological aspects of the low water corrosion of carbon steel. In: Proceedings of the 12th International Corrosion Congress, NACE, Houston, Texas, pp 3735–3746

Beech IB, Cheung CWS, Hill MA, Franco R, Lino AR (1994) Study of parameters implicated in biodeterioration of steel in the presence of different species of sulphate-reducing bacteria. Int Biodet Biodegr 34:289–303

Beech IB, Edyvean RGJ, Cheung CWS, Turner A (1995) Bacteria and corrosion in potable water mains. In: Tiller AK, Sequeira CAC (eds) Microbial corrosion. Proceedings of the 3rd International EFC Workshop. European Federation of Corrosion Publication 15. Institute of Materials, London, pp 328–337

Beech IB, Cheung CWS, Johnson DB, Smith JR (1996a) Comparative studies of bacterial biofilms on steel surfaces using techniques of atomic microscopy and environmental scanning electron microscopy. Biofouling 10:65–77

Beech IB, Zinkevich V, Tapper RC, Gubner R, Avci R (1996b) The interaction of exopolymers produced by marine sulphate-reducing bacteria with iron. In: Sand W (ed) Biodeterioration and biodegradation. Dechema Monographs vol 133, Dechema, Frankfurt, pp 333–338

Beech IB, Zinkevich V, Tapper RC, Avci R (1998a) The interaction of exopolymers produced by a marine sulphate-reducing bacterium with metal ions using X-ray photoelectron spectroscopy and time-of-flight secondary ionisation mass spectrometry. In: Gaylarde C, Barbosa T (eds) Proceedings of the Third Latin American Biodegradation and Biodeterioration Symposium (Labs 3), Paper P027

Beech IB, Zinkevich V, Tapper RC, Gubner R (1998b) The direct involvement of extracellular compounds from a marine sulphate-reducing bacterium in deterioration of steel. Geomicrobiol J 15:119–132

Chan CP (1993) The effect of Fe, Cr and Mo on growth and exopolymer production of sulphate-reducing bacteria. Diploma thesis. University of Portsmouth, UK

Cheung CWS (1995) Biofilms of marine sulphate-reducing bacteria on mild steel. PhD thesis, University of Portsmouth, UK

Cheung CWS, Beech IB (1996) The use of biocides to control sulphate-reducing bacteria in biofilms on mild steel surfaces. Biofouling 9:231–249

Coutinho CMLM, Maghalaes FCM, Araujo-Jorge TC (1993) Scanning electron microscopy study of biofilm formation at different flow rates over metal surfaces with sulphate-reducing bacteria. Biofouling 7:19–27

Feio MJ, Beech IB, Carepo M, Lopes J, Cheung CWS, Franco R, Guezennec J, Smith J, Mitchell J, Moura JG, Lino AR (1998) Isolation and characterisation of a novel sulphate-reducing bacterium of the *Desulfovibrio* genus. Anaerobe 4:117–130

Ford TE, Maki JS, Mitchell R (1987) Metal binding bacterial exopolymers and corrosion process. In: Corrosion 87, NACE, Houston, Texas, Paper 380

Ford TE, Black JP, Mitchell R (1990) Relationship between bacterial exopolymers and corroding metal surfaces. In: Corrosion 90, NACE, Houston, Texas, Paper 110

Geesey GG, Mittelman MW (1985) The role of high-affinity, metal-binding exopolymers of adherent bacteria in microbial-enhanced corrosion. In: Corrosion 85, NACE, Houston, Texas, Paper 297

Geesey GG, Mittelman MW, Iwaoka T, Griffiths PR (1986) Role of bacterial exopolymers in the deterioration of metallic copper surfaces. Mat Perform 25:37–40

Geesey GG, Jang L, Jolley JG, Hankins MR, Iwaoka T, Griffiths PR (1998) Binding of metal ions by extracellular polymers of biofilm bacteria. Wat Sci Technol 20:161–165

Grossman JP, Postgate JR (1955) The metabolism of malate and other compounds by *Desulfovibrio desulfuricans*. J Gen Microbiol 12:429–434

Gubner R, Beech IB (1996) The advantages of performing microbial corrosion studies using continuous culture systems. In: Sand W (ed) Biodeterioration and biodegradation. Dechema Monographs, vol 133, Dechema, Frankfurt, pp 185–191

Hamilton WA (1985) Sulphate-reducing bacteria and anaerobic corrosion. Ann Rev Microbiol 39:195–217

Hamilton WA (1998) Sulphate-reducing bacteria: physiology determines their environmental impact. Geomicrobiol J 15:19–28

Hardy JA, Hamilton WA (1981) The oxygen tolerance of sulfate-reducing bacteria isolated from the North Sea waters. Curr Microbiol 6:259–262

Iverson WP (1987) Microbial corrosion of metals. Adv Appl Microbiol 32:1–13

Kang H (1998) Characterisation of bacterial exopolymers using analytical techniques. PhD thesis. University of Portsmouth, UK

Koulali Y, Talouizte A, Fonvielle J-L, Dargent R (1996) Influence de la Monensine sur la croissance et la sécrétion des exopolysaccharides chez le *Botrytis cinerea pers.* et le *Sclerotium rolfsii sacc.* Can J Microbiol 42:965–972

Lee W, Lewandowski Z, Nielsen PH, Hamilton WA (1995) Role of sulphate-reducing bacteria in corrosion of mild steel: a review. Biofouling 8:165–194

Lovley DR, Philips EJP (1994) Novel processes for anaerobic sulfate production from elemental sulfur by sulfate-reducing bacteria. Appl Environ Microbiol 60:2394–2399

Ochynski FW, Postgate JR (1963) Some biological differences between fresh water and salt water strains of sulphate-reducing bacteria. In: Oppenheimer CH (ed) Marine microbiology III. Thomas, Springfield, pp 426–441

Odom JM, Singleton R (eds) (1992) The sulphate-reducing bacteria: contemporary perspectives. Springer, Berlin Heidelberg New York

Pereira A, Franco R, Feio MJ, Pinto C, Lampreia L, Reis MA, Calvete J, Moura I, Beech IB, Lino AR, Moura JG (1996) Characterisation of enzymes from a sulphate-reducing bacterium implicated in the corrosion of mild steel. Biophys Biochem Res Comm 221:414–421

Postgate JR (1984) The sulphate-reducing bacteria, 2nd edn. Cambridge University Press, London.

Senez JC (1953) The activity of anaerobic sulphate-reducing bacteria in semi autotrophic cultures. Ann Inst Pasteur 84:595–603

Sutherland IW (1985) Biosynthesis and composition of Gram-negative bacterial extracellular and cell wall polysaccharides. Ann Rev Microbiol 39:243–262

Tapper RC (1998) The use of biocides for the control of marine biofilms on stainless steel surfaces. PhD thesis. University of Portsmouth, UK

Von Wolzogen Kuhr CAV, Van der Vlugt SS (1934) Graphitisation of cast iron as an electrochemical process in anaerobic soils. Water (den Haag) 18:147–165

White DC, Nivens PD, Nichols AT, Kerger BD, Henson JM, Geesey GG, Clarke CK (1985) Corrosion of steel induced by aerobic bacteria and their extracellular polymers. In: Proceedings of International Workshop on Biodeterioration held at the University of La Plata, Argentina, March 1985. Aquatec Quimica, Sao Paulo, pp 73–86

Widdel F (1988) Microbiology and ecology of sulfate and sulfur-reducing bacteria. In: Zehnder AJB (ed) Biology of anaerobic microorganisms. Wiley-Liss, Wiley, New York, pp 469–586

Zinkevich V, Kang H, Bogdarina I, Hill MA, Beech IB (1996) The characterisation of exopolymers produced by different species of marine sulphate-reducing bacteria. Int Biodet Biodegr 37:163–172

Analysis and Function of the EPS from the Strong Acidophile Thiobacillus ferrooxidans

Wolfgang Sand[1] · Tilman Gehrke

Mikrobiologie, Universität Hamburg, Ohnhorststrasse 18, D-22609 Hamburg, Germany
[1] Mailing address: Institut für Allgemeine Botanik, Abteilung für Mikrobiologie, Universität Hamburg, Ohnhorststrasse 18, D-22609 Hamburg, Germany, *E-mail: FB6a042@mikrobiologie.uni-hamburg.de*

Keywords. Bacterial attachment, Bioleaching, Extracellular polymeric substances, Iron ions, Leaching mechanism, *Leptospirillum ferrooxidans*, Lipopolysaccharides, *Thiobacillus ferrooxidans*

1
Introduction

Extracellular polymeric substances (EPS) play a pivotal role in many processes (Costerton 1985). Often their importance is not fully known. One of these processes, in which we recently started to appreciate the role of EPS for its function, is the biological leaching of precious metals and its detrimental effect on acid mine/rock drainage.

Highly specialized bacteria oxidize insoluble metal sulfides to gain metabolic energy. The end products of this oxidation are sulfuric acid and solubilized metal ions. This process called bioleaching allows to recover (solubilized) precious metals like Cu^{2+}, Zn^{2+}, Co^{2+}, U^{6+}, or Au from low-grade sulfide ores (Fig. 1). In most cases, there is a concomitant negative effect on the environment termed acid mine/rock drainage (ARD; Fig. 2). It results from the pollution of water and soil by these end products. Because the leaching process does not stop when the precious metals are recovered, it continues until all metal sulfides are decomposed. Thus, low-value compounds like iron sulfides are oxidized in waste materials from mines and give rise to strongly acidic waters containing dissolved

Fig. 1. Mine waste heap as habitat for leaching bacteria

Fig. 2. Water and soil pollution in an abandoned sulfide mine area. Red water color results from ferric ions dissolved in sulfuric acid

heavy metals. Since these are liberated often without any control, they pose a serious threat to the environment. To mitigate ARD, the basic process of dissolution needs to be known.

2
Leaching Bacteria

The bacteria responsible for this dissolution process are mainly members of two species: *Thiobacillus ferrooxidans* and/or *Leptospirillum ferrooxidans* (Fig. 3). Many other species may contribute too, but the main processes to be discussed here are effected by these two species. They oxidize Fe^{2+} to Fe^{3+} ions, which are known to be strong oxidizing agents for metal sulfides. The function of the two bacteria is to keep the iron ions mainly in the oxidized state to allow the attack to proceed. *Leptospirillum ferrooxidans* is one of the few bacteria restricted to only one energy source – the oxidation of Fe^{2+} ions. *Thiobacillus ferrooxidans* may, in addition, oxidize sulfur compounds, allowing it to grow on a much larger variety of nutrients. In order to keep the iron ions in a soluble state (to be effective) the pH has to be low. Thus, these bacteria live in a range between pH 1 and 3, and sometimes they may be detected at even lower values. Their cell carbon is derived from the fixation of CO_2 by the Calvin cycle. Carboxysomes have only been detected in strains from *Thiobacillus ferrooxidans* (Fig. 3b). Summarizing, the metabolism is an example for a fully chemolithoautotrophic type requiring strongly acidic conditions. Most work has been focused on *Thiobacillus ferrooxidans*, as this bacterium has been known since 1950 (COLMER ET AL. 1950). *Leptospirillum ferrooxidans* was detected only in

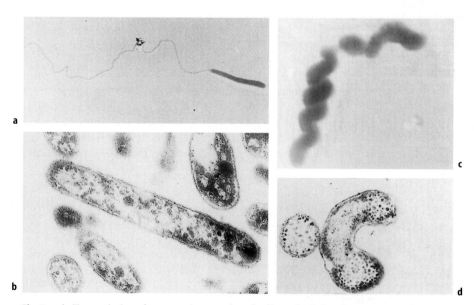

Fig. 3a–d. Transmission electron micrographs of cells and of ultrathin sections of cells of: **a, b** *Thiobacillus ferrooxidans*; **c, d** *Leptospirillum ferrooxidans*

1972 (MARKOSYAN 1972) and, due to its growth requirements, is much more difficult to obtain in pure culture (SAND ET AL. 1992). Consequently, most work on the metabolism of these specialized organisms has been performed using strains of *Thiobacillus ferrooxidans*. One strain of the latter has been selected for the chemical and functional analysis of its EPS.

3
Metal Sulfide Dissolution

In the course of (bio)leaching most metal sulfides are dissolved by the combined attack of protons and/or Fe^{3+} ions. The mechanisms, as described by SCHIPPERS (1998), comprise in one case an acid hydrolysis of the metal sulfide combined with an intermittent stage of H_2S/polysulfide formation which is followed by sulfur formation. The last step, the sulfur formation, is coupled to Fe^{3+} reduction. Consequently, this type of metal sulfides like ZnS or PbS yields mainly elemental sulfur as first degradation product. The other (few) metal sulfides (pyrite, molybdenite, and wolframite – FeS_2, MoS_2, WS_2) are only degradable by the oxidizing action of Fe^{3+} ions. Their first sulfur intermediate is thiosulfate, which in consecutive reactions via tetrathionate and trithionate gives rise to sulfate (SCHIPPERS ET AL. 1996). This is not the place to revive the old discussion about direct/indirect leaching mechanisms. Following the outlined mechanisms above involving protons and/or Fe^{3+} ions, the indirect leaching mechanism is the only one explaining appropriately the phenomena connected with bioleaching (SAND ET AL. 1995). The theoretical assumption of an enzyme/enzyme complex oxidizing without any detectable intermediates, only by incorporating oxygen, metal sulfides to sulfate and metal ions, the so-called direct leaching mechanism, has lost any evidence of existence. Besides, no proof has been presented up to now. Only ambiguous results based on indirect techniques, like deduction of reactions from data obtained by oxygen monitor experiments, were used for supporting this theory.

In particular, the often repeated argument of attachment to the substratum metal sulfide, which should be required and, thus, be an indication of the direct mechanism (GROUDEV 1979; LUNDGREN AND SILVER 1980; EHRLICH 1996), is invalid, since *Leptospirillum ferrooxidans* attaches rapidly and firmly to metal sulfide surfaces. However, this bacterium is known to be able only to act via the oxidation of Fe^{2+} to Fe^{3+} ions. The latter comprises essentially the indirect mechanism.

The following equations (Eqs. 1–5) summarize the current state of knowledge concerning the leaching reactions by the indirect mechanism. The metal sulfides need to be divided into two groups. For one group the main degradation intermediate is thiosulfate, for the other it is polysulfide. Pyrite (FeS_2) belongs to the first group, sphalerite (ZnS) to the second (SCHIPPERS 1998).

Thiosulfate mechanism

$$FeS_2 + 6Fe^{3+} + 3H_2O \rightarrow S_2O_3^{2-} + 7Fe^{2+} + 6H^+ \qquad (1)$$

$$S_2O_3^{2-} + 8Fe^{3+} + 5H_2O \rightarrow 2SO_4^{2-} + 8Fe^{2+} + 10H^+ \qquad (2)$$

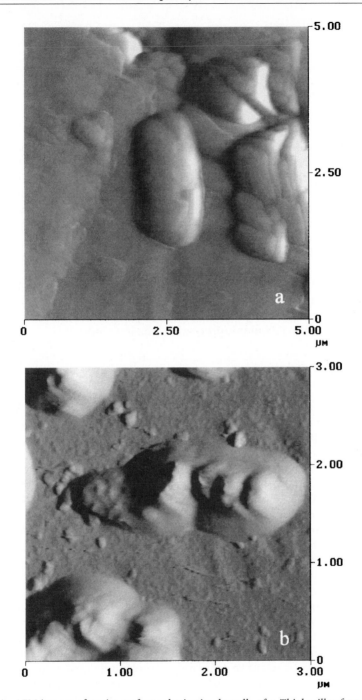

Fig. 4a, b. AFM images of pyrite surface colonization by cells of **a** *Thiobacillus ferrooxidans* or **b** *Leptospirillum ferrooxidans* after two days incubation

Polysulfide mechanism

$$M_{etal}S + Fe^{3+} \rightarrow + H^+ \rightarrow M_{etal}^{2+} + Fe^{2+} + 0.5H_2S_n \quad (n \geq 2) \tag{3}$$

$$0.5H_2S_n + Fe^{3+} \rightarrow 0.125S_8 + Fe^{2+} + H^+ \tag{4}$$

$$0.125S_8 + 1.5O_2 + H_2O \rightarrow SO_4^{2-} + 2H^+ \tag{5}$$

For further details about the historic discussion of direct/indirect mechanism the reader is referred to EHRLICH (1996).

It becomes evident from these equations that Fe^{3+} ions are of crucial importance for the process. Somehow the cells must bring these ions near to the crystal surface to start the degradation process. This can only be effected by a complexation followed by a transport to the metal sulfide surface. Consequently, the need for attachment is evident. EPS are generally involved in attachment to surfaces. The EPS are known to be involved in processes like microbially influenced corrosion (MIC) (GEESEY 1991; SAND 1995), sometimes exerting an effect even without a living cell. In Fig. 4 images are shown of cells of *Thiobacillus ferrooxidans* and *Leptospirillum ferrooxidans*, respectively, which have been obtained by atomic force microscopy. The cells are attached to a pyrite surface; defects of the latter are also visible.

Based on this obvious need for attachment, questions about the composition and function of the EPS in this process became immanent.

4
EPS Analysis

For studying the EPS composition with regard to attachment to substrates which are differently metabolized, strain R1 of *Thiobacillus ferrooxidans* was selected. This strain originates from a mine habitat in Romania (SAND ET AL. 1992). For EPS isolation, first the strain was cultured on media either containing ferrous iron (as sulfate) or solids like (finely ground) pyrite or sulfur. The EPS from the planktonic (iron(II) sulfate-grown) cells were harvested by centrifugation of the culture suspension. EPS from (pyrite- or sulfur-attached) sessile cells were obtained in the same way, except that a homogenization step of the substratum slurry in the presence of a weak detergent and a metal-complexing agent (CAMPER ET AL. 1985) was additionally included (GEHRKE ET AL. 1998). The resultant EPS-containing supernatants were extensively dialyzed and freeze-dried (crude EPS). For subsequent chemical characterization, the crude EPS extracts were analyzed by gas-liquid chromatography (detection of organic constituents) and spectrophotometric assays (detection of elements like phosphorus, nitrogen, or iron species). The results of these analyses are given in Table 1. Generally, the EPS consisted of sugars and lipids. Because of the high lipid content, the EPS occur as distinct accumulations of slimy material on the surface of pyrite (in contrast to mainly carbohydrate-containing EPS which would occur as diffuse accumulations; Fig. 5). Besides, some nitrogen, phosphorus, and free (extractable) fatty acids (FFA) were detectable. Whereas there was no general difference in the chemical composition of the EPS from iron(II) sulfate- or pyrite-grown cells detectable, a higher content of bound fatty acids,

Table 1. Amount and chemical composition of EPS from 10^{10} cells of *Thiobacillus ferrooxidans* grown with iron(II) sulfate, pyrite, or sulfur

Substrate	Total ($\mu g/10^{10}$ cells)	Wt% of total EPS				
		Sugars	Lipids	FFA [a]	N [b]	P [c]
Iron(II) sulfate	215 ± 30	52.2	36.9	5.5	0.5	0.7
Pyrite	2760 ± 301	48.5	39.4	5.8	0.5	0.8
Sulfur	1155 ± 94	40.9	53.8	8.0	0.6	2.8

[a] Free fatty acids.
[b] Nitrogen.
[c] Phosphorus.

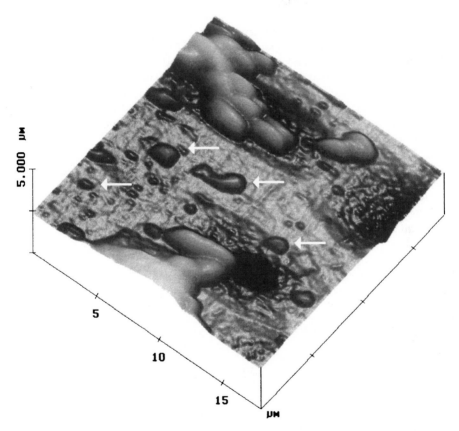

Fig. 5. AFM image of cells of *Thiobacillus ferrooxidans* adjacent to and in corrosion pits on a pyrite surface. Some distinct slime accumulations, probably bacterial "footprints" consisting of EPS, are also visible (indicated by *arrows*)

FFA, and phosphorus (indicating exopolymeric phospholipids) was detectable for the cells grown with sulfur. Protein and hexosamine were not detectable regardless of the growth substrate (data not shown). The generally low nitrogen content indicates that only negligible amounts of other, unidentified N-compounds (e.g., nucleic acids) may have been present.

The individual constituents of the sugar and lipid fraction for all three types of cells are indicated in Table 2. Included are data for the monomers of FFA. The data for the sugar fraction from iron(II) sulfate-grown cells revealed the presence of neutral sugars, glucuronic acid, and iron exclusively as iron(III) ions. The latter were presumably bound by glucuronic acid molecules as part of the carbohydrate moiety. The molar ratio amounted to 2 moles glucuronic acid to 1 mole iron(III) ions suggesting the formation of iron ion-glucuronic acid complexes. The composition of the polysaccharides from pyrite-grown cells varied only slightly from the previous one, whereas sulfur-grown cells revealed a considerably different composition. In the latter case, glucose was the only monosaccharide, and the content of glucuronic acid was reduced by about 85%. Consequently, iron species were also no longer detectable. The lipids of all EPS preparations were not significantly different in their chemical composition and consisted of saturated long chain fatty acids with equivalent chain lengths from 12 to 20 carbon atoms, except that the C_{13} and C_{15} acids did not occur. Stearic acid ($C_{18:0}$) was the most abundant compound, contributing to about 55 wt% to the total lipid fraction. Generally, small amounts of free fatty acids were detectable. These were palmitic ($C_{16:0}$) and stearic acids. Unsaturated fatty acids were not detectable.

Table 2. Chemical constituents of the sugar, lipid, and FFA (free fatty acids) fraction of EPS from cells of *Thiobacillus ferrooxidans* grown with iron(II) sulfate, pyrite, or sulfur

EPS fraction	Constituent	Wt% of total EPS from cells grown on		
		Iron(II) sulfate	Pyrite	Sulfur
Sugar	Rhamnose	13.9	10.8	nd[a]
	Fucose	20.5	17.1	nd[a]
	Xylose	0.9	0.8	nd[a]
	Mannose	0.4	0.7	nd[a]
	Glucose	11.4	15.2	40.4
	Glucuronic acid	4.4	3.3	0.6
	Iron(III) ions	0.7	0.5	nd[a]
Lipid	$C_{12:0}$[b]	1.9	2.0	2.7
	$C_{14:0}$[b]	0.4	0.4	0.6
	$C_{16:0}$[b]	8.8	9.4	12.9
	$C_{17:0}$[b]	0.9	1.0	1.3
	$C_{18:0}$[b]	20.2	21.6	29.5
	$C_{19:0}$[b]	3.9	4.2	5.7
	$C_{20:0}$[b]	0.8	0.8	1.1
FFA	$C_{16:0}$[b]	1.7	1.8	2.4
	$C_{18:0}$[b]	3.8	4.0	5.6

[a] Not detected ($< 0.08\%$).
[b] Equivalent chain length of fatty acids.

5
Function of EPS

The data as shown in Table 1 indicate that the cells of this bacterium possess different amounts of EPS depending on the substrate. The cells grown with the dissolved substrate $FeSO_4$ contained the lowest amount of EPS, whereas cells grown on solid substrates like sulfur or pyrite contained a medium or high amount of EPS. Obviously, the cells adapt their EPS production according to the need for attachment.

The second finding is concerned with the adaptation of the components of the EPS to the type of substrate. In case of the dissolved $FeSO_4$, a variety of sugars, uronic acids, and fatty acids was present in the EPS. For the substrate pyrite the composition of the EPS only changed slightly. The amount of sugars decreased a bit, while the amount of fatty acids increased to the same amount. In both types of EPS, Fe^{3+} ions were detectable in a stoichiometry of 1:2 to uronic acid. Considering the net charge of Fe^{3+} ions as 3+ and the charge of the carboxyl group of uronic acid as 1–, a net positive charge of 1+ remains. This net

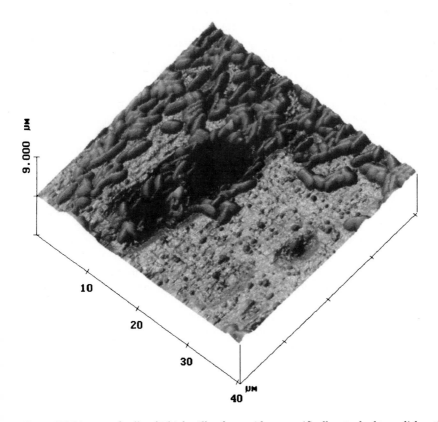

Fig. 6. AFM image of cells of *Thiobacillus ferrooxidans* specifically attached to a dislocation area (surface fault) of pyrite after 4 h incubation

positive charge renders the EPS of *Thiobacillus ferrooxidans* electropositive. As a consequence in case of the negatively charged pyrite surface, an attraction of the cells/EPS results, allowing for a primary attachment due to electrostatic interactions. BLAKE ET AL. (1994) produced similar results using zeta potential measurements. This may explain why these bacteria attach primarily at sites on the pyrite surface which exhibit surface defects. These defects and the bacterial attachment are shown in Fig. 6 in an atomic-force-microscopic image (*Thiobacillus ferrooxidans* on pyrite after 4 h incubation).

If the cells of *Thiobacillus ferrooxidans* R1 had been cultured with elemental sulfur, the composition of the EPS would have differed considerably from the previous type. The amount of sugars substantially decreased, finally only 41 % of the total EPS being polysaccharides. In addition, the variety of sugars decreased too. Only glucose remained detectable. Also the uronic acids fell below the limit of detectability. Instead, the amount of fatty acids increased, becoming the major component of the EPS. Quite obviously, the cells adapted to the hydrophobicity of the substrate sulfur by regulating the excretion of sugars, uronic acids, and fatty acids accordingly. The regulation of uronic acid excretion is of special interest, because these are a prerequisite for an attachment of the cells to, for example, pyrite. Since the uronic acids are able to complex Fe^{3+} ions, attachment becomes possible and, at the same time, the degradation mechanism using these ions as oxidizing agents is started. In case of elemental sulfur, the degradation mechanism is not dependent on Fe^{3+} ions and, thus, uronic acids are not excreted nor incorporated into the EPS. In contrast to the sugar composition of the EPS, which changed depending on the substrate, the fatty acid composition did not vary. Only the amount of fatty acids in the EPS was adapted to the substrate, not the composition of the various acids.

The extracellular lipid components, especially those from the pyrite-grown cells, may in part be considered as surface active compounds (biosurfactants). They could increase the solubility of hydrophobic compounds and/or create a conditioning film at the interface or might serve as an anchor for the hydrophilic compounds of the EPS in the hydrophobic part of the outer membrane (NEU 1996). Such compounds have already been detected at surfaces of cells of *Thiobacillus* sp. and may be involved in the initial stages of adhesion to hydrophobic surfaces (JONES AND STARKEY 1961; BEEBE AND UMBREIT 1971; BRYANT ET AL. 1984).

Summarizing, cells of *Thiobacillus ferrooxidans* adapt to their substrate by a change in the chemical composition of their EPS according to the hydrophobicity of the substrate and to the necessary degradation mechanism.

As indicated above, the Fe^{3+} ions complexed by uronic acids play a crucial role in the degradation of pyrite and also of other metal sulfides. It has recently been shown that Fe^{3+} ions dissolve pyrite by an oxidizing attack. The products of this reaction are Fe^{2+} ions and thiosulfate, which will consecutively be oxidized by peri- and/or cytoplasmic reactions (SAND ET AL. 1995; SCHIPPERS ET AL. 1996). The dissolution process does not start unless living cells containing EPS with complexed Fe^{3+} ions are present. If these ions were removed from the cells, e.g., by centrifugation, even cells pregrown with ferrous sulfate or pyrite were not able to degrade pyrite (GEHRKE ET AL. 1995). Only after supplying Fe^{3+} ions

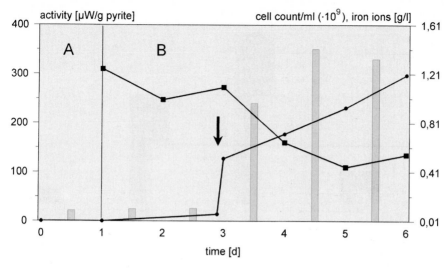

Fig. 7. Importance of iron(III) ion concentration for the onset of pyrite dissolution (bioleaching) by cells of *Thiobacillus ferrooxidans*. Pyrite dissolution was measured before (*A*) and after (*B*) inoculation with cells as an increase of total iron concentration in the medium (●) and activity on the substratum (*shaded bars*). ■ is the cell count of planktonic bacteria. *Arrow* indicates the addition of 0.5 g/l iron(III) ions

to those assays did the dissolution process start. The correlation between pyrite dissolution, metabolic activity, and Fe^{3+} ion concentration is shown in Fig. 7.

Dead cells are not able to oxidize pyrite, as experiments with a Kelvin probe unequivocally demonstrated (GEHRKE ET AL. 1998). The corrosion potential above a pyrite surface increased only in the case of living bacteria containing EPS and/or a sufficient amount of iron ions, as demonstrated in Fig. 8. In case of other metal sulfides like sphalerite (ZnS) or hauerite (MnS_2), Fe^{3+} ions considerably enhance the dissolution rate, although the primary attack on these sulfides seems to be mediated by protons (SCHIPPERS 1998). Consequently, the dissolution mechanism is different from the above-mentioned one. Elemental sulfur instead of thiosulfate is the main intermediate for the latter sulfur compounds.

However, in the case of elemental sulfur, EPS-complexed Fe^{3+} ions are probably not involved in the process of degradation. Other mechanisms need to be discussed like exoenzyme excretion (SAND 1996) possibly combined with an excretion of emulsifying agents (biosurfactants) like phospholipids (SHIVELY AND BENSON 1967; SAND 1996).

There exists a report describing that Fe^{3+} ions may be involved in the degradation of elemental sulfur (SUGIO ET AL. 1989). The report assumes a reduction of elemental sulfur by glutathione to sulfide, which will then be oxidized by the enzyme sulfur/sulfide:ferric-iron oxidoreductase to sulfite. At the sulfide stage an interaction with polysulfides might also occur leading to storage globules (STEUDEL 1989, 1996).

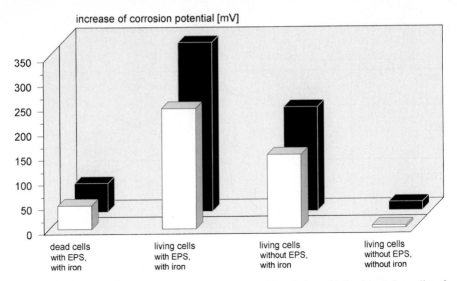

increase of corrosion potential [mV]

Fig. 8. Importance of EPS for the onset of pyrite degradation (bioleaching) by cells of *Thiobacillus ferrooxidans*. Pyrite degradation was measured 4 h (□) and 18 h (■) after inoculation with EPS-containing or -deficient cells of living or dead iron(II) sulfate-grown bacteria as an increase of the surface potential by the Kelvin electrode

However, no further evidence is available elucidating the role of Fe^{3+} ions in sulfur degradation.

6
Conclusions

Summarizing the findings it becomes obvious that the EPS of the acidophile *Thiobacillus ferrooxidans* have two prime functions: first, to mediate the attachment to the substrate/substratum either via mainly hydrophobic or mainly electrostatic interactions and, second, to supply the oxidizing agent, Fe^{3+} ions, for metal sulfide degradation in the direct vicinity of the substrate. In Fig. 9 this is summarized.

Based on these functions, several, partly speculative conclusions may be drawn concerning the role of EPS in the metabolism of this microorganism which, finally, may also turn out to be valid in general for the function of EPS of other bacteria.

First, the finding of the importance of the EPS-complexed Fe^{3+} ions, which are localized outside of the cell of *Thiobacillus ferrooxidans*, because the highly insoluble substrate metal sulfide cannot be taken up via the cytoplasmic membrane to be degraded in the bacterial metabolism, indicates that the bacteria extend their reactive space by these EPS. All areas on the pyrite surface to be covered by EPS are subject to the degradative attack of the complexed Fe^{3+} ions. Thus, the bacterium has access to a much larger area of substrate surface and is

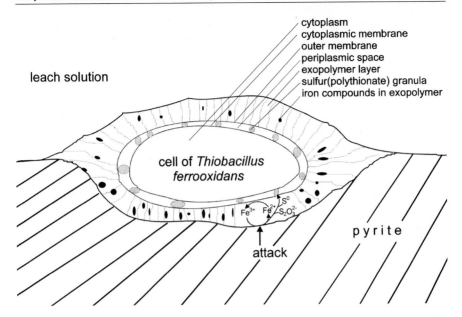

Fig. 9. A model for the mechanism of the indirect leaching attack as catalyzed by a pyrite-attached cell of *Thiobacillus ferrooxidans*

not restricted only to its cell area. As a consequence, sessile bacteria are provided with much more substrate than planktonic cells. The dissolution products, which cannot be used by the sessile cells, are the nutrients for the planktonic ones (besides those which result from chemical dissolution reactions these are at least ten times slower than the biological ones). It can further be speculated that the planktonic and the sessile populations may differ in their K_M-values for ferrous iron and sulfur compounds as a result of the different availability of these to the cells.

Second, in order to be able to use the improved access to the surface, substrate, and metabolizable compounds, the cells must necessarily exert some extracellular metabolic reactions. Most probably these will be the first degradative steps in the process of metal sulfide or sulfur dissolution. Similar evidence has been presented by Scotto and coworkers (SCOTTO ET AL. 1985; SCOTTO AND LAI 1998) for a totally different system (the ennoblement of steel in a marine environment caused by a biofilm reversibly collapsed after azide addition indicating the biological, extracellular nature of the phenomenon).

Third, the surface coverage by EPS may considerably interfere with the recognition systems of these bacteria for their substrates. Since the surface is covered by an organic film, other leaching bacteria are not able to recognize the substrate/substratum anymore. This may explain why in bioleaching no multilayered biofilm has ever been detected or visualized. All photos exhibit single cell layers. Furthermore, calculations of the surface coverage demonstrate that, even if the planktonic part of the population is ten times more abundant than the sessile (biofilm) one (5×10^8 to 5×10^9 cells/g pyrite, sessile to planktonic cells), still only 40% of the surface is maximally covered by these bacteria (SAND ET AL. 1996). It may be speculated that the remaining 60% (at least) is covered by EPS. Some evidence for this hypothesis may be derived from AFM-images showing for *Leptospirillum ferrooxidans* a total surface coverage by EPS after three days of growth on pyrite crystals (TELEGDI ET AL. 1998).

Fourth, the real concentrations of the reactive compounds in the EPS-layer between bacterial outer membrane and substrate/substratum interface are unknown. It is highly probable that this concentration is considerably higher than known up to now. Based on the analyses of the Fe^{3+} ions complexed in the EPS from *Thiobacillus* or *Leptospirillum ferrooxidans*, a concentration of about 44 g/l Fe^{3+} ions was estimated for the EPS-filled space. Even at a pH of approximately 2, the usual value for these bacteria, precipitation reactions would almost immediately occur at such a concentration. Because the bacteria are obviously able to keep these cations dissolved, it becomes explainable why they are able to enhance the biological metal sulfide dissolution at least by a factor of ten over the chemical one. Considerable research will be necessary to elucidate and verify the hypotheses presented above.

Acknowledgements. We appreciate the excellent help of J. Telegdi, Z. Keresztes, D. Thierry, A. Nazarov, and F. Zou with atomic force microscopy and potential measurements, respectively.

References

Beebe JL, Umbreit WW (1971) Extracellular lipid of *Thiobacillus thiooxidans*. J Bacteriol 108:612–614

Blake RC, Shute EA, Howard GT (1994) Solubilization of minerals by bacteria: electrophoretic mobility of *Thiobacillus ferrooxidans* in the presence of iron, pyrite, and sulfur. Appl Environ Microbiol 60:3349–3357

Bryant RD, Costerton JW, Laishley EJ (1984) The role of *Thiobacillus albertis* glycocalix in the adhesion of cells to elemental sulfur. Can J Microbiol 30:81–90

Camper AK, LeChevallier MW, Broadaway SC, McFeters GA (1985) Evaluation of procedures to desorb bacteria from granular activated carbon. J Microbiol Methods 3:187–198

Colmer AR, Temple KT, Hinkle ME (1950) An iron-oxidizing bacterium from the acid mine drainage of some bituminous coal mines. J Bacteriol 59:317–328

Costerton JW (1985) The role of bacterial exopolysaccharides in nature and disease. Dev Ind Microbiol 26:249–261

Ehrlich HL (1996) Geomicrobiology. Marcel Dekker, New York

Geesey GG (1991) What is biocorrosion? In: Flemming HC, Geesey GG (eds) Biofouling and biocorrosion in industrial water systems. Springer, Berlin Heidelberg New York, pp 155–165

Gehrke T, Telegdi J, Thierry D, Sand W (1998) Importance of extracellular polymeric substances from *Thiobacillus ferrooxidans* for bioleaching. Appl Environ Microbiol 64: 2743–2747

Gehrke T, Hallmann R, Sand W (1995) Importance of exopolymers from *Thiobacillus ferrooxidans* and *Leptospirillum ferrooxidans* for bioleaching. In: Vargas T, Jerez CA, Wiertz JV, Toledo H (eds) Biohydrometallurgical processing, vol I. University of Chile, Santiago, Chile, pp 1–11

Groudev SN (1979) Mechanism of bacterial oxidation of pyrite. Mikrobiologiya 16:75–87

Jones GE, Starkey RL (1961) Surface-active substances produced by *Thiobacillus thiooxidans*. J Bacteriol 82:788–789

Lundgren DG, Silver M (1980) Ore leaching by bacteria. Ann Rev Microbiol 34:263–283

Markosyan GE (1972) A new acidophilic iron bacterium *Leptospirillum ferrooxidans*. Biol Zh Armenii 25:26

Neu TR (1996) Significance of bacterial surface-active compounds in interaction of bacteria with interfaces. Microbiol Rev 60:151–166

Sand W (1995) Mineralische Werkstoffe. In: Brill H (ed) Mikrobielle Materialzerstörung und Materialschutz: Schädigungsmechanismen und Schutzmaßnahmen. Gustav Fischer, Jena, 780–110

Sand W (1996) Microbial mechanisms. In: Heitz E, Flemming HC, Sand W (eds) Microbially influenced corrosion of materials. Springer, Berlin Heidelberg New York, pp 15–25

Sand W, Rohde K, Sobotke B, Zenneck C (1992) Evaluation of *Leptospirillum ferrooxidans* for leaching. Appl Environ Microbiol 58:85–92

Sand W, Gehrke T, Hallmann R, Schippers A (1995) Sulfur chemistry, biofilm, and the (in) direct attack mechanism – a critical evaluation of bacterial leaching. Appl Microbiol Biotechnol 43:961–966

Sand W, Gehrke T, Hallmann R, Schippers A (1996) Towards a novel bioleaching mechanism. In: Kuyucak N, Costerton JW (eds) Minerals bioprocessing and biorecovery III. Engineering Foundation, New York, pp 80–85

Schippers A (1998) Untersuchungen zur Schwefelchemie der biologischen Laugung von Metallsulfiden. PhD thesis, University of Hamburg, Hamburg, Germany

Schippers A, Jozsa PG, Sand W (1996) Sulfur chemistry of bacterial leaching of pyrite. Appl Environ Microbiol 62:3424–3431

Scotto V, Lai ME (1998) The ennoblement of stainless steels in seawater: a likely explanation coming from the field. Corros Sci 38:1–12

Scotto V, Di Cintio R, Marcenaro G (1985) The influence of marine aerobic microbial film on stainless steel corrosion behaviour. Corros Sci 25:185–194

Shively JM, Benson AA (1967) Phospholipids of *Thiobacillus thiooxidans*. J Bacteriol 94:1679–1683

Steudel R (1989) On the nature of the "elemental sulfur" (S^0) produced by sulfur-oxidizing bacteria – a model for S^0 globules. In: Schlegel HG, Bowien B (eds) Autotrophic bacteria. Springer, Berlin Heidelberg New York, pp 289–303

Steudel R (1996) Mechanism for the formation of elemental sulfur from aqueous sulfide in chemical and microbiological desulfurization processes. Ind Eng Chem Res 35: 1417–1423

Sugio T, Katagiri T, Inagaki K, Tano T (1989) Actual substrate for elemental sulfur: ferric ion oxidoreductase purified from *Thiobacillus ferrooxidans*. Biochim Biophys Acta 973: 250–256

Telegdi J, Keresztes Z, Palinkas G, Kalman E, Sand W (1998) Microbially influenced corrosion visualized by atomic force microscopy. Appl Phys A 66:639–642

Physical and Chemical Properties of Extracellular Polysaccharides Associated with Biofilms and Related Systems

Bjørn E. Christensen

Norwegian Biopolymer Laboratory (NOBIPOL), Department of Biotechnology, Norwegian University of Science and Technology, N-7034 Trondheim, Norway, *E-mail: Bjoern.E.Christensen@chembio.ntnu.no*

Keywords. Extracellular polysaccharides Gel, network, Gelation, Solubilisation, Depolymerisation, Rheological characterization, Bacterial alginate

List of Symbols and Abbreviations

ω	frequency (Hz)
G'	elastic modulus (Nm^{-2})
G''	viscous modulus (Nm^{-2})
KDO	2-keto-3-deoxy-octulosonic acid
ORD	oxidative-reductive depolymerization

1
Introduction

The presence of gel-like matrices surrounding the microbial cells in biofilms is well established. The thickness and general appearance may vary considerably, from essentially monolayers of microbial cells to thick algal mats. The matrices themselves are easily visualized by microscopic techniques, especially in combination with various stainings. Upon dehydration, for instance during the preparation of samples for electron microscopy, the gels tend to collapse to form fiber-like structures which reflect the precipitation of high molecular weight polymers. When touched, biofilms usually appear as slimy or sticky, and a certain elastic behavior may be observed as well. All in all, biofilms resemble in many ways classical biopolymer gels such as those prepared from gelatin, agarose or some calcium alginates, and it is assumed that the microorganisms produce extracellular polymers to form gel-like structures and thereby form biofilms.

Many microorganisms produce polysaccharides that are excreted into the medium or form a capsule around the cells. The chemical structure and physical properties are in many cases well known, mainly because some microbial polysaccharides have established industrial applications as viscosifyers or gelling agents, for instance in the food industry. Xanthan is one example of a bacterial polysaccharide used in this way. Other microbial polysaccharides are used in the biomedical area. Examples include dextran and hyaluronan, while certain bacterial alginates show great promise as immunostimulants (SKJÅK-BRÆK AND ESPEVIK 1996). More recently, information about the genetics and biosynthesis of microbial polysaccharides has significantly expanded our knowledge in the field. In any case, a wide range of methods for chemical, biochemical, and physical characterization of such biopolymers have been developed, and may in principle be applied to all types of biopolymers.

Biopolymers other than polysaccharides may in principle be involved in the formation of biofilm matrices, for instance proteins. Nevertheless, in this chapter I will restrict myself to discuss gelation aspects of polysaccharides alone.

Most methods – in particular methods for physical characterization – require quite large amounts of purified polymers. For naturally occurring biofilms, bioaggregates, and similar structures, the situation is often very different from that of the pure (liquid) cultures used to produces large amounts of industrial polysaccharides. Although the microbial species involved may be known, and the presence of a gel-like glycocalyx may be observed, small amounts of biopolymers are usually present. For example, a homogeneous biofilm of thickness 10 μm containing 1% (w/v) of polymer corresponds to only 10 μg of pure polymer per square cm. Furthermore, the gel-like state often prevents a facile purification of the individual biopolymers constituting the biofilm. Thus, methods are needed to isolate each component in a pure state, both for their quantification as well as for structural analyses.

There seem to be few cases where biopolymers comprising the major structural components of biofilm or bioaggregates actually have been identified. Although the gel-like state itself may be crucial for the function of biopolymers in biofilms, it is the author's opinion that it is highly beneficial to study the production of polysaccharide in pure culture, preferentially suspended culture, to obtain background information for subsequent work. Another useful and supplementary approach involves the preparation of artificial biofilms from purified polysaccharides. Because of the complexity and variability of biofilms, it is further crucial to work on simple systems, preferentially pure cultures and a single gel-forming polymer. Until we have a fairly good understanding of simple systems there is little hope that we may obtain detailed and useful understanding of mixed systems such as natural biofilms.

In the following, the nature of polysaccharide gels, with particular emphasis on microbial polysaccharides, will be discussed in more detail. We need to define a "gel" and to discuss the major factors which influence the gel properties. Finally, the disintegration of gels will be discussed, because this aspect is closely related to biofilm removal. Focus will be placed on bacterial alginates, a family of polysaccharides which displays a wide variety of gelling properties, and which in the case of *Pseudomonas aeruginosa* has been shown to be involved in the formation (DAVIES ET AL. 1993) and detachment of biofilms (BOYD AND CHAKRABARTY 1995).

2
Biopolymer Gels

According to ALMDAL ET AL. (1993) "the term 'gel' is used so indiscriminately that it has been become ambiguous". For the present purpose, however, a gel may be defined as a polymeric, three-dimensional *network* which is swollen with a large excess of solvent. Aqueous gels are swollen with water, and are considered to be in equilibrium with the water and co-solutes such as salts, H^+, etc. A characteristic feature of the network is *cross-links*, which link the polymer chains together. These may be *permanent*, as in covalently cross-linked gels such as poly(acrylamide) gels, which are cross-linked by means of bifunctional monomers (*bis*-acrylamide). Cation-mediated cross-linking of polysaccharide chains (for instance in calcium alginate gels) can also be considered as permanent. In contrast, cross-links may be *temporary*, implying that weaker and reversible, non-covalent forces are involved. Examples of the latter include hydrogen bonds, hydrophobic interactions, etc. Whether these in practice are categorized as temporary or permanent depends also on the time-scale of the observation as well as external forces acting on the system. The well known gels formed upon cooling hot solutions of agarose are cross-linked by means of non-covalent linkages. The gels are nevertheless permanent on a practical time-scale.

Gelation may also occur as a result of *entanglement* of long chains. This occurs at quite high concentrations and does not depend much on the chemical nature of the polymers, whereas the chain stiffness is important. Such gels are clearly reversible and dissolve upon dilution. Nevertheless, the time-scale is important because entanglement gels appear to be stable for long periods due to the slow process of disentanglement. A classical example of entanglement gels are the mucus layers of the gastrointestinal tract or the respiratory system. The mucus forms a protective layer covering the epithelial cells, and provides in the respiratory tract a slowly flowing medium for transport of particles, etc. The mucus layer of the stomach also contributes to maintaining a steep pH gradient between the epithelium and the lumen, and prevents the transport of microorganisms, viruses, and proteases (e.g., pepsin) from the lumen to the cell surfaces. The major components of mucus gels are the mucins, which are highly glycosylated (50–80% sugar) glycoproteins of very high molecular weight (around 10^6). The mucin peptide chains are further interconnected (end-to-end) through disulfide bridges to form even longer structures. Mucin gels dissolve slowly upon dilution, and may be reconstituted from purified mucin (ALLEN AND PEARSON 1993). Mucin gels also dissolve by breaking the disulfide bridges that connect the subunits (individual mucins).

As biofilms (in most cases) are not composed on the basis of mucin-like glycoproteins, but rather of polysaccharides, the function and physical characteristics of biofilms appear to be quite similar to that of mammalian mucus. In fact, entanglement gels are readily formed by almost any water soluble polymer provided that the molecular weight and the concentration is high enough. This implies that the microorganisms may – in a strain specific manner –"choose" an EPS which in addition to the gelling properties give rise to other specific ecological advantages, for instance resistance towards degradation by other microorganisms. It is well established that bacterial EPS often is specific even for individual strains.

Strong gels such as those formed by agarose (for cultivation of microorganisms) or calcium alginate are probably not representative for most biofilms. Neither are biofilms chemically or physically homogeneous. Voids and channels have commonly been observed (COSTERTON ET AL. 1994).

In an attempt to reconstitute a biofilm which enabled penetration of microelectrodes for the study of mass transport of oxygen ABRAHAMSON ET AL. (1996) used the purified EPS produced by *P. aeruginosa*. The EPS, which upon analysis turned out to be a classical, non-gelling, bacterial alginate with a very low content of L-guluronic acid (see below), formed a stable entanglement gel simply by concentration of a dilute aqueous solution of EPS. The gel (containing microcolonies added by means of a micropipet) was placed in a biofilm reactor where subsequent development of the artificial biofilm could be studied by microelectrodes and microscopy. This example illustrates in many ways the basic physical requirements for forming a biofilm.

3
Physical Characterization of Biofilms

Any progress in the characterization of a specific biofilm depends on a successful dissolution of the involved biopolymers in order to purify and quantify them. Dissolution may be obtained only if we understand the nature of the cross-linking mechanisms. This includes both the chemical characteristics (type of linkage) and the life-time of the cross-links. Of particular interest are the entanglement gels, which can be distinguished from gels with more permanent cross-links by rheological methods.

The major rheological method which is much used in the study of biopolymer gels is mechanical spectroscopy. A wide variety of advanced instrumentation is commercially available for such studies. The material is subjected to small and reversible oscillatory deformations over a wide range of frequencies (ω). The frequency dependence of the elastic ($G'(\omega)$) and viscous ($G''(\omega)$) moduli (often termed mechanical spectra) are the major diagnostic tools in classifying the gels. The frequency range studied may typically be 10^{-2} to 10^2 s^{-1}. Strong gels with permanent cross-links are relatively solid-like, i.e., $G' \gg G''$ for all frequencies, and G' is more or less independent of the frequency except at low frequencies. In contrast, entanglement gels may be 'solid-like' only at very high frequencies. At lower frequencies the viscous modulus (G'') may dominate ($G'' \gg G'$), reflecting a more liquid-like behavior.

There seems to be few – if any – instances where biofilms (native or reconstituted) have been subjected to detailed rheological analyses. One reason is ob-

vious: biofilms cannot easily be transferred from the substratum to the rheo-
meters, which usually require large amounts of material. Alternative approaches
are therefore needed, particularly for in situ measurements. One future possibility could be to use sub-micron, magnetic particles, which by micromani-
pulation could be placed at various locations in the biofilm. An oscillating magnetic field would then be needed to induce an oscillatory stress (force), and the
resulting movement of the particle (oscillatory strain) could be monitored
through the microscope using a video-based system. Computerized image analysis would of course be required to obtain fundamental parameters such as
$G'(\omega)$ and $G''(\omega)$. The viscoelastic character of the biofilm could in this way be
determined at many different locations, which would give useful information
about rheological and structural inhomogeneities (channels, voids etc.).

4
Chemical and Biochemical Characterization of Biofilms

A multitude of experimental approaches are in principle available to characterize the biofilms and their biopolymer components. However, they all require,
at least to some extent, the solubilization of the biofilm to obtain pure components. There are, however, certain approaches where solubilization can be
avoided.

First of all, if the biofilm-forming microorganisms may be identified and isolated, they should be studied in suspended culture, for instance in simple batch
cultures. The production of extracellular polymers (EPS) is most readily studied in this way, and their purification can easily be obtained. We may also study
how a variety of factors, i.e., growth conditions, influence the production of
EPS.

A second and quite obvious approach is to use model organisms where the
EPS production is already well understood, and study their behavior in a biofilm situation. Such organisms might include the producers of industrial gums
such as xanthan, gellan, dextran, or bacterial alginate (see below).

When the genes involved in the biosynthesis and export of biopolymers are
known then genetic tools may be used to study where and when these genes are
activated during the formation of a biofilm (DAVIES ET AL. 1993; BOYD AND
CHAKRABARTY 1995).

Direct chemical analysis EPS is usually not possible, particularly not for
in situ studies. Generally, the biofilm has to be dissolved and EPS must be
isolated in a pure form. Many bacterial EPS contain unusual sugars (KENNE
AND LINDBERG 1983) that may be analyzed and serve as probes for a particular
EPS.

If dissolution procedures are required, the identification of an efficient method may also throw light on the nature of the cross-links. This is easily demonstrated in calcium alginate gels. The removal of calcium by chelating agents
such as EDTA dissolves the gels because the calcium ions are involved in the formation of ionic cross-links. If carboxylate groups ($-COO^-$) are involved in cation mediated cross-linking, such as in alginate, protonation of the carboxyl
group by lowering pH well below pK_a prevents cation binding, and thus de-

stroys the gel structure. It may be noted that such neutralization often renders the polysaccharides insoluble in water.

An important method for solubilization of gels, including biofilms, is to degrade the polymers that are responsible for the formation of gels. This approach does not aim towards the disintegration of cross-links, but breaks up the chains (elastic segments) between the junction zones. The solubilization is thus irreversible. The susceptibility towards degradation depends both on the agent which leads to the breaking of chains, and the chemical nature of the polymer. Polysaccharides are generally labile in strong acids, which leads to acid hydrolysis of the glycosidic linkages. Some linkages are very resistant towards hydrolysis, for instance the glycosidic linkages of uronic acids and 2-amino sugars such as glucosamine. In other cases very acid labile bonds are found, for instance in certain 2- and 6-deoxysugars, ketosugars (e.g., KDO or sialic acids) or in furanoses. In these cases mild acid treatment (although at elevated temperatures) may bring about dissolution.

Polysaccharides are also quite stable towards degradation in alkali. Exceptions are sugars where the glycosidic linkage is located in a β-position relative to a carbonylic group ($> C=O$). In this case alkaline β-elimination can bring about depolymerization. An example is again alginate, where all sugar residues are 4-linked, which corresponds to the β-position relative to the carboxylate group at C-6. Alginates are therefore unstable in alkali, especially at high temperature.

Strong alkali may often solubilize gels not only because of chemical degradation, but also because of the ionization of the hydroxyl groups ($-OH \rightarrow -O^-$) which leads to extensive swelling and subsequent solubilisation.

Alkaline solutions also favor another type of degradation, called oxidative-reductive depolymerization (ORD). All polymers are susceptible to ORD, and the degradation of alginate by ORD has been studied in detail (SMIDSRØD ET AL. 1963 A, B, 1965). ORD involves a series of free radical reactions which ultimately lead to chain scission. Autooxidable compounds like ascorbates, sulfites, or phenols may initiate ORD, whereas molecular oxygen and transition metals ions (e.g., Fe^{2+}/Fe^{3+}) are efficient catalysts. The Fenton reagent (H_2O_2 and Fe^{2+} salt) effectively degrades polymers, and this approach has been applied to remove marine biofilms (CHRISTENSEN ET AL. 1990). ORD has the advantage that relatively high depolymerization rates may be obtained at ambient temperature, whereas acid or alkaline hydrolysis require high temperatures.

Methods such as ORD are very nonspecific and degrade all polysaccharides, whereas enzymes, on the other hand, are often very specific. As all naturally occurring polysaccharides are biodegradable, enzymes that are capable of degrading them must necessarily exist. For microbial EPS, few enzymes are readily available, but it should in principle be possible to isolate EPS degrading enzymes from other microorganisms or bacteriophages. Such isolation of EPS specific enzymes is much used in the structure elucidation of microbial polysaccharides. To date, this approach seems to be much neglected in the biofilm area, despite its obvious potential. One exception is alginate lyases (see below). Because of their specificities, enzymes can act on a single EPS in, for instance, a mixed biofilm and thus clarify the role of that EPS in the biofilm. Specific enzymes, once identified and produced in larger quantities, have in this author's opinion a great potential as biofilm controlling agents in almost all areas where biofilm removal is desired.

5
Bacterial Alginates

Bacterial alginates are a family of polysaccharides that display a wide variety in structure and properties. Because of this variety they serve as useful models in the study of EPS in relation to biofilms. Most of our knowledge of bacterial alginates is derived from studies of *Azotobacter vinelandii*, where the production of alginate is associated with the formation of cysts, and from *P. aeruginosa*, a human pathogen commonly associated with lung infections. *P. aeruginosa* has also been used extensively in the study of biofilm formation (BAKKE ET AL. 1984; STEWART ET AL. 1993). Bacterial alginates seem to be more widespread than previously believed, and have been found in several other species of *Pseudomonas*, especially plant pathogenic strains. In a recent study of bacterial isolates taken from corroded metal surfaces, genes associated with alginate biosynthesis were found in 11 of 120 isolates, 9 of which were identified as *Pseudomonas* spp. (WALLACE ET AL. 1994).

The chemical composition of bacterial alginates is summarized in Fig. 1. Mannuronan, the homopolymer containing only M-residues (i.e., 1,4-linked residues of β-D-mannuronic acid) is the 'starting polymer' which is produced prior to subsequent modification and diversification. Pure mannuronan can be produced by certain genetically modified bacteria, but has not been found in natural isolates. Mannuronan is a non-gelling alginate. However, the molecular weight is high (typically 10^6 Da), enabling the formation of soft entanglement gels at high concentrations.

Bacterial alginates are, in contrast to algal alginates, usually O-acetylated (in position 2 or 3) of the M-residues (SKJÅK-BRÆK ET AL. 1986B). Acetylation occurs prior to the major enzymatic modification of alginates, namely the C-5-epimerisation of the M-residues to form the C-5-epimer of D-mannuronic acid which according to standard nomenclature is named L-guluronic acid (G) (Fig. 1). The epimerisation occurs at the polymer level, i.e., after the formation of mannuronan from precursor sugar nucleotides (ERTESVÅG ET AL. 1996). It may be noted that epimerisation does only occur in M-residues *without* O-acetyl groups (SKJÅK-BRÆK ET AL. 1986 A, B).

The chemical and physical properties of the resulting (epimerised) alginate are determined by the extent of epimerisation (expressed in terms of the fraction of residues that are converted to G, denoted F_G, see Table 1) and the distribution of G-residues, as well as the chain length (i.e., molecular weight). The distribution can in practice only be described in terms of statistical parameters, preferentially parameters which can be obtained experimentally. The major parameters such as F_G, F_{GG}, F_{MGM}, and $N_{G>1}$ are defined in Table 1. These parameters may be readily obtained by high-field proton NMR (GRASDALEN ET AL. 1979; GRASDALEN 1983).

D-fructose-6-phosphate
$$\downarrow$$
$$\downarrow$$
GDP-D-mannuronic acid
$\quad\downarrow$ (polymerization)
Mannuronan (β-1,4-linked polymannuronic acid) (--M-M-M-M-M-M---)

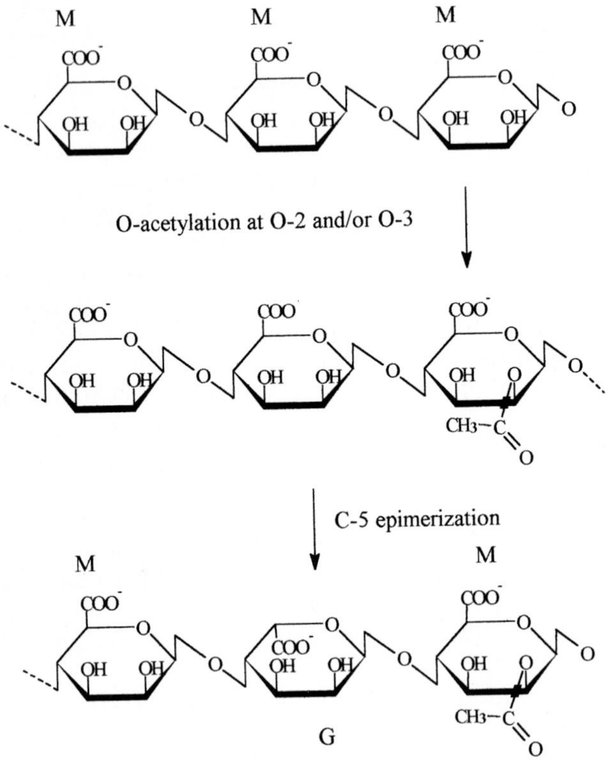

O-acetylation at O-2 and/or O-3

C-5 epimerization

G = L-guluronic acid (C-5 epimer of D-mannuronic acid)

Further epimerization (of unacetylated M-residues) results in different types of sequences and blocks:

.........M-M-M-M-M-G-M-M-M........ → ......M-G-M-G-M-G-M-G-M-........
$\qquad\qquad\downarrow$ $\qquad\qquad\qquad\qquad\qquad\qquad$ (alternating sequence)
.........M-G-G-G-G-G-G-G-M........
$\qquad$ (G-block)

Fig. 1. Key steps in the biosynthesis of bacterial alginates

Table. 1. Parameters describing the chemical composition of alginates

---

DA (degree of acetylation): Average number of O-acetyl groups per sugar residue (0–2.0)

Monad frequencies (G, M):

$F_G = n_G/(n_M + n_G)$ $\quad\quad$ n_G: the number of G-residues

$F_M = 1 - F_G = n_M/(n_M + n_G)$ $\quad$ n_M: the number of M-residues

Diad frequencies (GG, GM, MG, MM):

$F_{GG} = n_{GG}/(n_M + n_G)$ $\quad\quad$ n_{GG}: the number of G-residues containing an adjacent G-residue

$F_{GM} = n_{GM}/(n_M + n_G)$ $\quad\quad$ n_{GM}: the number of G-residues containing an adjacent M-residue

etc.

$F_{GG} + F_{GM} = F_G$

$F_{MM} + F_{MG} = F_M$

etc.

Triad frequencies (GGG, GGM, GMM, GMG, MMM, MMG, MGG, MGM):

$F_{GGG} = n_{GGG}/(n_M + n_G)$ $\quad$ n_{GGG}: the number of G-residues containing two adjacent G-residues

etc. $\quad\quad\quad\quad\quad\quad\quad\quad$ n_{GGG}: the number of G-residues containing two adjacent G-residues

Average length of G-blocks:

$N_{G>1} = (F_G - F_{MGM})/F_{GGM}$

Table 2 summarizes the chemical composition of a range of bacterial alginates. Some algal alginates used commercially as gelling agents and viscosifyers are included for comparison. In spite of the variability in chemical composition we may divide the bacterial alginates into three classes:

1. Alginates with $F_G < 0.10$. Such alginates are produced by several *Pseudomonas* strains. The G-residues are invariably single (non-consecutive) residues, and G-blocks (consecutive G-residues) are totally absent. These alginates cannot form the characteristic stiff gels in the presence of Ca^{2+}, but may form entanglement gels at high polymer concentrations. They may also form alginic acid gels (in the absence of Ca^{2+}) at pH values below 2.5.
2. Alginates with $F_G > 0.1$, but with $F_{GG} \approx 0$. Alginates with 40–50% G, but with no or very few G-blocks consist primarily of alternating sequences (…–M–G–M–G–M–G–M–G– …). They resemble certain algal alginates (e.g., alginate from *Ascophyllum nodosum*), and are also poor gelling agents in combination with Ca^{2+} or H^+ (Smidsrød and Draget 1996).
3. Alginates with $F_G > 0.5$ containing G-blocks. Such alginates are produced by *A. vinelandii*. Following de-O-acetylation these alginates resemble the gelling alginates produced by brown algae such as *Laminaria* sp. The ability to form stiff gels in the presence of Ca^{2+} is linked to the G-blocks, and the elastic modulus of Ca-alginate gels increases with the average G-block length ($N_{G>1}$). It remains to be demonstrated that bacteria other than *A. vinelandii* can produce such alginates.

The variability in the relative content and average length of G-blocks is caused by the different mannuronan C-5-epimerases. Different enzymes give rise to very different structures. Six genes encoding for mannuronan C-5-epimerases in *A. vinelandii* have so far been identified and sequenced (Ertesvåg et al. 1996; Rehm and Valla 1997), and several of these have been cloned and ex-

Table. 2. Chemical composition of some alginates (literature data)

Source	F_G	F_M	F_{GG}	F_{MM}	F_{GM} (= F_{MG})	$N_{G>1}$
Brown algae[a]						
Laminaria hyperborea, stipe[b]	0.68	0.32	0.56	0.20	0.12	11
Laminaria hyperborea	0.72	0.28	0.60			16
L. hyperborea, leaf[b]	0.55	0.45	0.38	0.28	0.17	7
Ascophyllum nodosum[c]	0.36	0.64	0.16			4
Bacteria						
Azotobacter vinelandii TL[c]	0.45	0.55	0.42	0.52	0.03	
Azotobacter vinelandii IV[c]	0.94	0.06	0.93	0.04	0.01	0
Pseudomonas aeruginosa[c,d]	0–0.45	1.00[d]	0	1.00[d]	0	0
Pseudomonas mendocina 10541[c]	0.26	0.74	0	0.48	0.26	0
Pseudomonas putida 1007[c]	0.37	0.63	0	0.26	0.37	0
Pseudomonas fluorescens 10255[c]	0.40	0.60	0	0.20	0.40	0
Pseudomonas corrugata	0–0.1[d]		0			0
Pseudomonas syringae	0–0.2[d]		0			0

[a] A more comprehensive list of algal alginates is provided by SKJÅK-BRÆK AND MARTINSEN (1991).
[b] SKJÅK-BRÆK ET AL. (1986a).
[c] SKJÅK-BRÆK ET AL. (1986b).
[d] Produces virtually homopolymeric mannuronan when grown at low temperature (G. Skjåk-Bræk, personal communication). Pure mannuronan is only produced by AlgG-negative mutants of Pseudomonas (FRANKLIN ET AL. 1994).

pressed in *Escherichia coli*. The purified enzymes have indeed been shown to produce very different alginates in vitro, from the alternating sequences produced by an epimerase denoted AlgE4, to the G-blocks introduced by AlgE2 (ERTESVÅG ET AL. 1996). The epimerisation by *A. vinelandii* epimerases takes place outside the cell (LARSEN AND HAUG 1971). *P. aeruginosa*, on the other hand, synthesizes a single (periplasmic) epimerase (AlgG) which is unable to introduce G-blocks (REHM AND VALLA 1997).

The knowledge about the genes encoding for mannuronan C-5 epimerases and the type of alginate they give rise to is a powerful tool in studying the role of bacterial alginates in relation to biofilms. By monitoring the expression of specific genes, for instance by using appropriate mutants, we may possibly relate biofilm structure and properties to the production and modification of alginate (BOYD AND CHAKRABARTY 1995).

The structure and properties of EPS gels are also influenced by the length of the polymer chains or in other words, the molecular weight. Bacterial alginates seem to have quite high molecular weights, values up to 1.8×10^6 g mol^{-1} for the weight average molecular weight (M_w) have been reported (ABRAHAMSON ET AL. 1996) as compared to commercial algal alginates with values in the range $2–4 \times 10^5$ g mol^{-1}. However, alginate producing bacteria may also produce a class of alginate degrading enzymes called alginate lyases. The lyases cleave glycosidic linkages in the alginate and reduce the molecular weight. Such depolymerization will weaken alginate gels and will eventually lead to complete dissolution. This applies both to entanglement gels and Ca-alginate gels. Alginate lyases may play important roles in the detachment and sloughing of biofilms containing alginate producing bacteria (BOYD AND CHAKRABARTY 1995). This important observation demonstrates that EPS-degrading enzymes in general play a role in controlling biofilm removal. This aspect clearly deserves attention in future biofilm research.

This survey on bacterial alginates intends to demonstrate that strong structure-function relationships exist which has obvious relevance in biofilm research. A thorough understanding of the role of EPS in biofilms requires detailed knowledge of the EPS structure as well as the genes and enzymes involved in biosynthesis, export and degradation. Work with pure cultures should be encouraged, especially in cases where EPS also can be produced in larger quantities for separate studies of gel formation.

Acknowledgement Professor Gudmund Skjåk-Bræk is thanked for helpful discussions during the preparation of the manuscript.

References

Abrahamson M, Lewandowski Z, Geesey G, Skjåk-Bræk G, Strand W, Christensen BE (1996) Development of an artificial biofilm to study the effects of a single microcolony on mass transport. J Microbiol Methods 26:161–169

Allen A, Pearson JP (1993) Mucus glycoproteins of the normal gastrointestinal tract. Eur J Gastroenter Hepatol 5:193–199

Almdal K, Dyre J, Hvidt S, Kramer O (1993) Towards a phenomenological definition of the term 'gel'. Polym Gels Networks 1:5–17

Bakke R, Trulear MG, Robinson JA, Characklis WG (1984) Activity of Pseudomonas aeruginosa biofilms:steady state. Biotechnol Bioeng 26:1418–1424

Boyd A, Chakrabarty AM (1995) Pseudomonas aeruginosa biofilms – role of the alginate exopolysaccharide. J Ind Microbiol 15:162–168

Christensen BE, Naper Trønnes H, Vollan K, Bakke R (1990) Biofilm removal by low concentrations of hydrogen peroxide. Biofouling 2:165–175

Costerton JW, Lewandowski Z, DeBeer D, Caldwell D, Korber D, James G (1994) Biofilms, the customized microniche. J Bacteriol 176:2137–2142

Davies DG, Chakrabarty AM, Geesey GG (1993) Exopolysaccharide production in biofilms – substratum activation of alginate gene-expression by Pseudomonas aeruginosa. Appl Environ Microbiol 59:1181–1186

Ertesvåg H, Valla S, Skjåk-Bræk G (1996) Genetics and biosynthesis of alginates. Carbohydr Eur 14:14–18

Franklin MJ, Chitnis CE, Gacesa P, Sonesson A, White DC, Ohman DE (1994) Pseudomonas aeruginosa AlgG is a polymer level alginate C5-mannuronan epimerase. J Bacteriol 176:1821–1830

Grasdalen H (1983) High-field ^{1}H NMR. spectroscopy of alginate: Sequential structure and linkage conformations. Carbohydr Res 118:255–260

Grasdalen H, Larsen B, Smidsrød O (1979) A p.m.r. study of the composition and sequence of uronate residues in alginates. Carbohydr Res 68:23–31

Kenne L, Lindberg B (1983) Bacterial polysaccharides. In: Aspinall GO (ed) The polysaccharides, vol 2. Academic Press, Orlando, pp 287–363

Larsen B, Haug A (1971) Biosynthesis of alginate, pt I. Compostion and structure of alginate produced by Azotobacter vinelandii (Lipman). Carbohydr Res 17:287–296

Rehm BHA, Valla S (1997) Bacterial alginates:biosynthesis and applications. Appl Microbiol Biotechnol 48:281–288

Skjåk-Bræk G, Espevik T (1996) Application of alginate gels in biotechnology and biomedicine. Carbohydr Eur 14:19–25

Skjåk-Bræk G, Martinsen A (1991) Applications of some algal polysaccharides in biotechnology. In: Guiry MD, Blunden G (eds) Seaweed resources in Europe:uses and potential. Wiley, Chichester, pp 219–257

Skjåk-Bræk G, Smidsrød O, Larsen B (1986a) Tailoring of alginates by enzymatic modification in vitro. Int J Biol Macromol 8:330–336

Skjåk-Bræk G, Grasdalen H, Larsen B (1986b) Monomer sequence and acetylation pattern in some bacterial alginates. Carbohydr Res 154:239–250

Smidsrød O, Draget KI (1996) Chemistry and physical properties of alginates. Carbohydr Eur 14:6–13

Smidsrød O, Haug A, Larsen B (1963a) The influence of reducing substances on the rate of degradation of alginates. Acta Chem Scand 17:1473–1474

Smidsrød O, Haug A, Larsen B (1963b) Degradation of alginate in the presence of reducing compounds. Acta Chem Scand 17:2628–2637

Smidsrød O, Haug A, Larsen B (1965) Kinetic studies on the degradation of alginic acid by hydrogen peroxide in the presence of iron salts. Acta Chem Scand 19:143–152

Stewart PS, Peyton BM, Drury WJ, Murga R (1993) Quantitative observations of heterogeneities in *Pseudomonas aeruginosa* biofilms. Appl Environ Microbiol 59:327–329

Wallace WH, Rice JF, White DC, Sayler GS (1994) Distribution of alginate genes in bacterial isolates from corroded metal-surfaces. Microb Ecol 27:213–223

Chemical Communication Within Microbial Biofilms: Chemotaxis and Quorum Sensing in Bacterial Cells

Alan W. Decho

Department of Environmental Health Sciences, School of Public Health, University of South Carolina, Columbia, SC. 29208, USA, *E-mail: adecho@sph.sc.edu*

Keywords. Chemical signals, Quorum sensing, Biofilms, Extracellular polymers

1
Introduction

Bacteria obtain a wide range of information from their surrounding environment. This includes physical cues such as temperature and viscosity, inorganic cues such as pH, oxygen and ionic concentration, and various organic chemical cues (KARP-BOSS ET AL. 1996). The successful adaptation of bacteria to changing natural conditions requires that the organism is able to sense and respond to its external environment and modulate gene expression accordingly (PIRHONEN ET AL. 1993).

A recent series of studies has examined the roles of specific organic molecules which act as signals for relaying information, and facilitating communication between bacterial cells, or between bacteria and other organisms (see RUBY 1996; FUQUA ET AL. 1996 for reviews). Both free-living cells and attached bacteria have been shown to utilize chemical communication pathways. Chemical sensing and communication may occur in several major forms: chemotaxis, autoregulation, and autoinduction (i.e., quorum sensing). General "chemotaxis" involves the directed movement of a bacterial cell toward or away from a chemical concentration gradient (BLAIR 1995). The second form, autoregulation, will not be covered here (see PARKINSON 1993 for review). A third, and more specific form of chemical communication is "quorum sensing."

Quorum sensing is based on the process of autoinduction (NEALSON 1977; FUQUA ET AL. 1994). It involves an environmental sensing system that allows bacteria to monitor and respond to their own (and perhaps other bacterial) population densities. The bacteria produce a diffusible organic "signal," originally called an autoinducer molecule, which accumulates in the surrounding environment during growth. High cell densities result in high concentrations of signal, and induce expression of certain genes and/or physiological change in neighboring cells (FUQUA ET AL. 1996). By definition, responses to chemical signals in both chemotaxis and quorum sensing are concentration-dependent processes. However, in quorum sensing a critical threshold concentration of the signal molecule must be reached before a physiological (or genetic) response will be elicited.

This review addresses the more-general process of chemical communication, and the specific process of "quorum sensing" within bacterial biofilm systems. Some of the clearest examples of quorum sensing involve bacterial cells which in nature are associated with biofilms on surfaces. Its potential role is examined in affecting active migration, structured spatial arrangements, and controlled activities of cells which have been observed to occur within specific areas of biofilms.

2
Bacterial Chemotaxis and Initial Aggregation of Cells

The chemical taxis literature for bacteria is vast (ARMITAGE 1992; see BLAIR 1995 for reviews). Chemical taxis is defined as the active movement of bacteria in response to a chemical gradient. Flagellated bacteria often swim in stretches of smooth runs interrupted by intervals of chaotic motion called tumbling. These movements correspond to opposite directions of flagellar rotation (SILVERMAN AND SIMON 1977). Movement away from an attractant source or toward increasing concentrations of a repellent produces more frequent tumbling (ADLER 1983). A wide range of molecules may stimulate chemotaxis, often utilizing specific taxis systems (HEDBLOM AND ADLER 1983). Bacteria exhibit taxis toward a range of molecules, which include amino acids, particularly basic amino acids (CRAVEN AND MONTIE 1985), peptides (KELLY-WINTENBERG AND MONTIE 1994), small aromatic compounds (LOPEZ-DE-VICTORIA ET AL. 1994) and carbohydrates (MCNAB 1987; YU ET AL. 1993).

Initial attachment of bacterial cells to a surface is generally described as a two-step process which includes a "reversible attachment phase," followed by either irreversible cell attachment or detachment (see LAWRENCE ET AL. 1995 for review). Originally it was thought that once irreversible cell attachment had occurred and a biofilm matrix had been secreted, cells did not move (POWELL AND SLATER 1983). However, a number of studies involving natural biofilms and pure cultures have shown that irreversible attachment is less frequent in occurrence than previously thought (see LAWRENCE ET AL. 1987; MARSHALL 1988; KORBER ET AL. 1990 and others). The *"motile attachment theory,"* posited by LAWRENCE ET AL. (1987), suggests that bacteria may move along surfaces, independent of flow, while remaining in a semi-attached state. Cells will periodically reposition themselves, perhaps in response to chemosensory signals. This modification provides a more encompassing explanation for the chemically-directed movement of cells to a surface, and once on a surface (LAWRENCE ET AL. 1995).

3
Microspatial Patterns and Movement of Bacteria Within Biofilms

Recent studies of microbial biofilms demonstrate that cells are often organized into highly-structured units or assemblages (see LAWRENCE ET AL. 1991; COSTERTON ET AL. 1994; MOLLER ET AL. 1996; NEU AND LAWRENCE 1997). The extracellular polymer matrix of a biofilm is composed of diverse microenvironments, some of which may be more suitable for cells than other areas. Frequently, tightly-packed microcolonies of cells are observed in biofilms. Structured assemblages of cells may arise from selective colonization of specific parts of the biofilm, from reproductive accumulations of cells in specific areas, or from active migration of cells to specific areas. Packing behavior has been studied as a colony-forming mechanism in a number of bacterial strains (LAWRENCE ET AL. 1987, 1992; LAWRENCE AND KORBER 1993; DELAQUIS 1990). This mechanism begins with a few attached cells which grow and divide, forming a tightly packed monolayer of cells on a surface. Complex patterns of cells have been shown to occur during growth of *Escherichia coli* (BUDRENE AND BERG 1991).

Locations of areas having high cell abundances appear to change over time and are affected by the presence of other cells. Modeling studies of microbial food webs (BLACKBURN ET AL. 1997) show that microorganisms in diffusion transport systems can benefit significantly in the uptake of nutrients and energy sources by gathering in patches. Three-dimensional scenarios predict that patch-generating events (i.e., nutrient releases due to lysis, predation, etc.) probably occur several times per hour. Chemotactic responses by cells quickly follow. While these studies do not consider the presence of an extracellular polymer matrix, the resulting models illustrate the dynamically changing microenvironments which microbial cells encounter over rather frequent time scales.

The biofilm matrix is not an easy place for a cell to move. Entwined within extracellular polymers, cells are often constrained in their ability to move. Also, upon attachment and secretion of polymer, cells often loose their flagella, further restricting their ability to move. Despite this, the movement of cells within a biofilm has been correlated with the formation of small pits and channels (KORBER ET AL. 1993) suggesting that bacterial movement induces channeling within biofilms.

In some bacteria, such as *Vibrio parahaemolyticus*, cell movement influences packing behavior. There are two major types of motility, swimming-only motility and swarming motility (SAR ET AL. 1990). Changes from swimming to swarming often involves cellular differentiation (i.e., phenotypic changes in cells). Swimming cells have a single polar flagella, while swarmer cells possess numerous lateral flagella (MCCARTER ET AL. 1992). Cells of *V. parahaemolyticus* actually exclude other bacterial cells during swarming (BELAS AND COLWELL 1982). The multicellular swarming response is thought to allow bacteria to forage more efficiently for nutrients on solid surfaces, especially under low nutrient conditions (BELAS 1996).

Biofilm structure also develops in response to flow (GJALTEMA ET AL. 1994; NEU AND LAWRENCE 1997). Ridges develop oriented parallel to flow direction. Highest bacterial biomass develops in a "shell-type growth," located near the surface of the biofilm (NEU AND LAWRENCE 1997).

4
Bacterial Quorum Sensing: Induction and Regulation of Activities

The autoinduction process potentially allows cells having the same or similar physiological capabilities to regulate expression of these activities. It assumes that these microbial activities should be conducted only when cells are at high densities, and that during high cell abundances, cells are better served when "acting as a coordinated unit" rather than a group of opportunistic individuals.

In quorum sensing, activation of a sensory mechanism in a bacterium leads to alterations in gene expression which facilitate an adaptive response (BAINTON ET AL. 1992). The activation is mediated by various response-regulator proteins, which are triggered by small diffusible signal molecules. The "signal" molecules are therefore key elements for intercellular communication in bacteria. They are produced by bacteria and freely diffuse to other cells where they will enter into the cells (via diffusion) in response to local concentration gradients. Therefore, high densities of similar (signal-producing) cells will result in a locally high concentration of the autoinducer signal. When the concentration of the signal exceeds a threshold, activation of the response-regulator proteins occurs and this is followed by alterations in gene expression. This represents a unique but highly efficient system of communication (and regulation) for bacteria in proximity to each other.

Quorum sensing systems are now known to be involved in a range of important microbial activities. These include extracellular enzyme biosynthesis (JONES ET AL. 1993; PIRHONEN ET AL. 1993; CHATTERJEE ET AL. 1995; CUI ET AL. 1995), biofilm development (DAVIES ET AL. 1998) antibiotic biosynthesis (BAINTON ET AL. 1992; PIERSON ET AL. 1994; MCGOWAN ET AL. 1995; SCHRIPSEMA ET AL. 1996), biosurfactant production (BRINT AND OHMAN 1995; LATIFI ET AL. 1995; OCHSNER ET AL. 1994; OCHSNER AND REISER 1995; PEARSON ET AL. 1995;

WINSON ET AL. 1995), exopolysaccharide synthesis (TORRES-CABASSA ET AL. 1987; BECK VON BODMAN AND FARRAND 1995), conjugal transfer of plasmids in *Agrobacterium tumefaciens* Ti plasmids (PIPER ET AL. 1993; FUQUA AND WINANS 1994; HWANG ET AL. 1995; ZHANG ET AL. 1993), and the extracellular virulence factors of *Pseudomonas aeruginosa, Erwinia carotovora, Rhizobium leguminosarum, E. coli,* and other Gram-negative bacteria (JONES ET AL. 1993; PASSADOR ET AL. 1993; PIRHONEN ET AL. 1993; PEARSON ET AL. 1994).

4.1
Model Systems

4.1.1
Homoserine Lactone (Autoinducer) Signals

A large family of similar molecules which act as signals in bacteria are the homoserine lactones (HSL) (EBERHARD ET AL. 1981). These signal molecules are secreted by bacteria into the surrounding medium. At threshold concentrations, ranging from low nmol/l to mmol/l, the signal molecules are proposed to bind to transcriptional activator proteins (R proteins), and the R protein-Signal complex stimulates expression of a target gene (SEED ET AL. 1995; STEVENS AND GREENBERG 1997).

In most systems characterized thus far, the HSL signals show similarities in structure with each other. The molecules contain various features critical for interaction with their target proteins (PASSADOR ET AL. 1996). They consist of a homoserine lactone base with additional acyl side chains of varying length and hydrophobicity (Fig. 1). The length and structure of the acyl side chains provides a unique signature for different signals. Structural homologs can act as both agonists and antagonists for activity. A very detailed study by Passador and colleagues (PASSADOR ET AL. 1996) examined various structural homologs of known signals, and indicated that proper chain length appears to be critical factor for binding and induction of expression. They found that even rotation of the acyl side chain may play a role in its ability to activate the target gene.

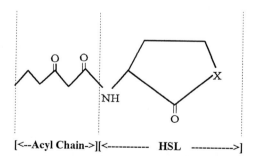

[<--Acyl Chain->][<----------- HSL ----------->]

Fig. 1. Basic structure of an *N*-acyl L-homoserine lactone autoinducer signal. The molecule consists of an L-homoserine lactone (HSL) base with an acyl chain which varies in length and structure; X=O, S, or N

Two trends have become apparent during recent years after examinations of many systems (RUBY 1996). First, more than one autoinduction system may exist within a given bacterium (BASSLER ET AL. 1994; PEARSON ET AL. 1995; THROUP ET AL. 1995; GILSON ET AL. 1996). Second, a single autoinduction system may control the expression of multiple, and apparently functionally unrelated genes (GRAY ET AL. 1994; FUQUA ET AL. 1996).

4.1.2
The Vibrio fischeri System

The bacterium *Vibrio fischeri* is the specific symbiont in the light organs of certain marine fishes and squids (DUNLAP AND GREENBERG 1991; HAYGOOD 1994; RUBY 1996 for reviews). At high cell densities of *V. fischeri*, signal accumulates in the light organ to stimulate light production by bacterial cells. Light production in *V. fischeri* is controlled by the Vibrio Auto-Inducer (VAI) signal, also referred to as *N*-(3-oxohexanoyl) homoserine lactone (EBERHARD ET AL. 1981, 1986; SCHAEFER ET AL. 1996A). Cell membrane is permeable to VAI. At low cell densities, VAI passively diffuses across the membrane and out of the cells down a concentrations gradient (FUQUA ET AL. 1994). At sufficiently high concentrations of VAI, the signal interacts with the LuxR, which then activates transcription for light production. The autoinduction system appears to be designed to allow *V. fischeri* to discriminate between the free-living low densities which occur in marine waters, and the higher densities which occur in the host, and to induce the luminescence system only when the bacterium host is associated. Within the light organ of the squid *Euprymna scolopes*, cells of *V. fischeri* are specifically associated with the host tissue (RUBY 1996). Recently, it has become apparent that several different bacterial genera possess regulatory systems which are homologous to the LuxR and LuxI proteins involved in symbiotic light production. The signal of another bacterium *Vibrio harveyi* is a closely related compound, *B*-hydroxybutryl homoserine lactone (CAO AND MEIGHEN 1989, 1993). The signal is produced at a low constitutive rate during the early stages of growth (ENGEBRECHT ET AL. 1983). *B*-Ketocaproyl homoserine lactone is produced by a wide range of bacteria, in addition to *V. harveyi*. These include the terrestrial bacteria *P. aeruginosa*, *E. carotovora*, *Erwinia herbicola*, and *Seratia marcescens* (BAINTON ET AL. 1992). The "A Factor" (isocapryl-δ butyryl lactone) in the bacterium *Streptomyces griseus* is a structurally similar compound and induces sporulation, streptomycin synthesis, and streptomycin resistance (HORINOUCHI AND BEPPU 1992).

4.2
Other Chemical Signals

4.2.1
Conjugation and Pheromone-Inducible Gene Expression in Enterococcus faecalis

Conjugation involves the attachment of two bacteria, and the exchange of plasmids between a donor cell (carrying a plasmid) and a recipient cell (initially plasmid-free). *Enterococcus faecalis* is a nonmotile Gram-positive bacterium. Prior to conjugation, recipient cells secrete small heat- and nuclease-stable, but trypsin- and chymotrypsin-sensitive, pheromones (DUNNY ET AL. 1978). This peptide pheromone signals donor cells to synthesize adhesin, which causes them to clump with recipient cells and other donor cells to form aggregates. Then conjugation takes place. Pheromones are plasmid specific, and over 18 plasmids that encode a pheromone response have been identified already (GALLI ET AL. 1992). The pheromones are structurally different from the HSL autoinducers described above. They have been shown to be hydrophobic octapeptides or heptapeptides (CLEWELL AND WEAVER 1989; JI ET AL. 1995). Pheromones are extremely sensitive and highly specific molecules (CLEWELL 1993). At least five different pheromones may be produced by a single recipient cell. These peptides are typically active at very low concentrations (less than 5×10^{-11} mol/l); as few as two molecules per cell may be sufficient to activate gene transcription (GALLI ET AL. 1992). Pheromone-induced surface-bound adhesins in *E. faecalis* are large closely-related proteins that form dense hair-like structures on the cell wall of induced bacteria. The ligand for the adhesin on *E. feacalis* cells is a surface constituent which appears to be present on all cells (GALLI ET AL. 1989).

4.2.2
Cyanobacteria Systems

The filamentous cyanobacterium *Anabaena* forms a linear array of cells. Certain cells undergo differentiation from vegetative cells into thicker-walled cells called heterocysts, which are capable of fixing atmospheric N_2 (BUIKEMA AND HASELKORN 1993). The developing cell or proheterocyst undergoes a ariety of distinctive morphological, physiological and genetic changes that adapt it to the task of fixing atmospheric N_2. Heterocyst differentiation does not occur uniformly in all cells along the filament. Early work under nitrogen-limited conditions by WOLK (1967) and WILCOX ET AL. (1973) suggested that the development and maintenance of heterocysts was determined by interactions between neighboring cells. Breakage of a filament into short chains of cells resulted in an overall increase in the number of heterocysts formed. It was later postulated that heterocysts release an inhibitor to differentiation that diffuses to neighboring cells in the filament and prevents them from undergoing heterocyst formation (WOLK 1991). Diffusion sets up a gradient of inhibitor around each heterocyst and that gradient determines the spacing of heterocyst formation. The cells are contained within a polysaccharide sheath which would provide a localization matrix for the inhibitor molecule, slowing diffusion and enhancing the gradient. While this model is extremely attractive in explaining the spacing of heterocyst along filaments, direct evidence for the existence of an inhibitor molecule has yet to appear. It appears that intercellular communication between cells in the filament determines the spatial patterns of heterocyst formation (LIANG ET AL. 1992).

4.3
Functional Roles of Quorum Sensing

Information to date, indicates that intercellular communication between bacteria serves several functions. First, and perhaps foremost, it aids in monitoring cell densities. For example, the autoinduction system in *V. fischeri* appears to be designed to allow cells to discriminate between the free-living low densities which occur in marine waters, and the higher densities which occur in the host, and induce the luminescence system only when the bacterium host-associated (RUBY 1996). Second, intercellular communication is involved in coordinating gene expression among cells. In *Bacillus subtilis* a signal transduction pathway coordinates the timing of gene expression in the mother cell with the development of forespores (KAISER AND LOSICK 1993). The very close packing of cells and forespore development are derived from signal transduction. Finally, morphological differentiation is a third process mediated by intercellular signals. In the soil bacterium *Streptomyces coelicolor* a cascade of intercellular signals induce a large number of cells coordinately produce a morphogenetic protein that facilitates the formation of aerial hyphae (CHATER 1993).

4.4
Specificity of Autoinduction Process

The process of quorum sensing provides a mechanism for self-organization and self-regulation by microbial cells. Many of the initial studies on quorum sensing suggested that autoinducer molecules were highly specific, and would induce only individuals of the same species or even strain. These studies, however, were based on tests examining a limited number of laboratory strains of bacteria. Other observations of autoinduction showed that cell-free extracellular material from one bacterium (*V. fischeri*) was able to stimulate luminescence gene expression in a different species (*V. harveyi*). Certain autoinducers were somewhat general in their ability to induce bacterial cells of different species or genera (GREENBERG ET AL. 1979). More recently, a wider range of structural homologs for known autoinducers have been examined for autoinduction activity (MCKENNEY ET AL. 1995; PASSADOR ET AL. 1996). Many transcriptional activator proteins will accept autoinducer molecules with shorter or longer acyl chain lengths. For example, autoinducers isolated from *P. aeruginosa* are capable of stimulating virulence factor production in cultures of a different bacterium *Burkholderia cepacia* (MCKENNEY ET AL. 1995). These studies have suggested that, for a given bacterium, a range of molecules may act as effective agonists, and have raised the intriguing possibility of cross-species communication.

5
Enhancement of Chemotaxis and Quorum Sensing Processes by the Biofilm Matrix

Some of the clearest examples of chemical communication and quorum sensing have been observed to occur within biofilm systems. These include the bacteria

V. fischeri, Erwinia stewartii, and *P. aeruginosa.* Most autoinducer signals which have been characterized to date appear to be readily diffusible through cell membranes and in aqueous medium (i.e., culture medium). Therefore, it should not be expected that such molecules will bind with high affinities to the extracellular polymer polysaccharides of a biofilms matrix. Current information suggests that in order for an autoinducer to be active it must be in an unbound state until it enters the cells via passive diffusion (PIRHONEN ET AL. 1993).

It has been proposed by FUQUA ET AL. (1996) that autoinducer-receptor complexes form only when intracellular concentrations of autoinducer signals approach the K_d for receptor binding. Because signals are membrane permeable, sufficiently high intracellular concentrations require relatively high extracellular concentration. High extracellular concentrations would require a high density of cells possibly enclosed within some sort of diffusion barrier.

The extracellular polymer matrix of a biofilm (DECHO 1990), owing to its inherent diffusion-slowing and sorptive properties, may represent a suitable diffusion barrier (compared with the surrounding water) to localize chemical signals released by microbial cells. In open water, rapid diffusion, convection, and turbulent flow will quickly disperse inducer molecules, preventing their build-up to high concentrations. The biofilm represents "immobilized water," interlaced with an extracellular polymer matrix. Even though convective flow may occur within portions of the biofilm to enhance mass exchange (LEWANDOSKI ET AL. 1993), areas enclosed in dense extracellular polymer gels may exhibit considerably slower diffusion rates than surrounding areas, and thus permit the accumulation of autoinducer molecules. Studies examining the diffusional movement of autoinducers indicate that these molecules readily diffuse through cells and other potential barriers (KAPLAN AND GREENBERG 1985). Owing to the highly diffusible nature of most autoinducers, it is unlikely that these molecules will be bound tightly (if at all) by extracellular polymer ligands.

Further, specific areas of biofilms called microdomains, owing to the specific physical/chemical properties of their extracellular polymers (DECHO 1998) may more efficiently concentrate inducer signals than adjacent neighboring areas, which can even lead cells to migrate to those locations. A hydrated region of an extracellular polymer matrix surrounded by a hydrophobic region may provide an ideal localization matrix for autoinducer molecules. Dense extracellular polymer gels are able to concentrate small organic molecules, such as amino acids and peptides, more efficiently than the same extracellular polymers in less dense gel states (Decho et al., in preparation). Studies directly examining the binding constants of autoinducers within extracellular polymer gels must be conducted to address this important process. The highly variable nature of the extracellular polymer matrix within biofilms may provide a continuum of microenvironments, some of which, owing to their chemical properties and gel densities, may be able to localize autoinducer signal molecules more efficiently than adjacent regions.

6
The Biofilm as a Microbial Organism

A new concept is emerging concerning the development and evolution of bacterial communities. The concept addresses bacteria, not as individuals, but rather as an "organism." The concept promotes "self-organization" and cooperativity among cells as a driving force in community development, rather than the classical natural selection of individuals. This concept becomes especially apparent when examining bacterial biofilm communities. Bacteria are considered to be far from solitary organisms, rather being colonial by nature and exploiting elaborate systems of intercellular interactions and communications to facilitate their adaptation to changing environments (KAISER AND LOSICK 1993; ANDREWS 1995; LOSICK AND KAISER 1997). The above-mentioned discussion of quorum sensing and chemical communication demonstrates examples of autoinducers which operate largely between individuals of the same species or strain. GREENBERG ET AL. (1979), and later MCKENNEY ET AL. (1995), have proposed that in natural environments, which are crowded with a diverse array of bacteria, production of heterologous N-acyl homoserine-producing bacteria may facilitate intergeneric communication as well as between members of a single species (SCHAEFER ET AL. 1996B). Many bacterial isolates are capable of producing several autoinducer molecules (KUO ET AL. 1994). Further, it has now been shown that a single autoinducer from one bacterial species may elicit a transcriptional response in a number of other bacteria (BASSLER AND SILVERMAN 1995; MCKENNEY ET AL. 1995).

Many natural bacterial communities appear to exist as groups of "metabolic units" or consortia (WIMPENNY 1992). By definition, a microbial consortium implies a high degree of intercellular interaction, coordination, and communication between bacteria of different species and genera. The signal molecules involved in autoinduction have been demonstrated to facilitate intergeneric communication as well as between members of a single species (FUQUA ET AL. 1996), and provide a plausible means of maintaining microbial consortial activities.

In order for complex bacterial biofilm communities to exist and proliferate, the bacterial communities (and not the individual cells) must exist as units of ecological activity, similar to multicellular organisms (SONEA 1991; ANDREWS 1995; CALDWELL AND COSTERTON 1996). Our previously constrained view of community development and succession has developed largely from ecological studies of macroorganisms. Driving determinants in ecological succession are either biological (e.g., competition, predation) or physical (MAYR 1963). In these communities, selection is proposed to operate at the level of the individual. Examinations of the unique aspects of bacterial community development has even led us to re-examine selection processes and even the utility of the classical species concept when examining bacteria (SONEA 1991; CALDWELL ET AL. 1997). CALDWELL AND COSTERTON (1996) have stressed that microbial biofilm communities may develop and evolve through proliferation and association, rather than classical selection and competition. Especially important is the ability of bacteria to exist for prolonged periods of time in relatively inactive states (KJELLEBERG ET AL. 1993; HUISMAN AND KOLTER 1994). Such abilities tend to allow a high diversity of bacteria to exist, with abundant community members reflecting those phenotypes (or genotypes) which can proliferate and work together, in a metabolic sense. It is probable that the complex array of chemical signals which are used in natural microbial communities will affect interactions between many species of bacteria, and represent a regulating parameter in microbial community development.

References

Adler J (1983) Bacterial chemotaxis and molecular neurobiology. Cold Spring Harbor Symp Quant Biol 2:803–804

Andrews JH (1995) What if bacteria are modular organisms? ASM News 61:627–632

Armitage JP (1992) Behavioral responses in bacteria. Annu Rev Physiol 54:682–714

Bainton NJ, Bycroft BW, Chhabra SR, Stead P, Gledhill L, Hill PJ, Rees CED, Winson MK, Salmond GPC, Stewart GSAB, Williams P (1992) A general role for the *lux* autoinducer in bacterial cell signalling: control of antibiotic synthesis in *Erwinia*. Gene 116:87–91

Bassler BL, Silverman MR (1995) Intercellular communication in marine *Vibrio* species: density-dependent regulation of the expression of bioluminescence. In: Hoch JA, Silhavy TJ (eds) Two-component signal transduction. ASM Press, Washingtion, pp 431–445

Bassler BL, Wright M, Silverman MT (1994) Multiple signalling systems controlling expression of luminescence in *Vibrio harveyi*: sequence and function of genes encoding a second sensory pathway. Mol Microbiol 13:273–286

Beck von Bodman S, Farrand SK (1995) Capsular polysaccharide biosynthesis and pathogenicity in *Erwinia stewartii* require induction by an *N*-acylhomoserine lactone autoinducer. J Bacteriol 177:5000–5008

Belas R (1996) Sensing, response and adaptation to surfaces: swarmer cell differentiation and behaviour. In: Fletcher M (ed) Bacterial adhesion: molecular and ecological diversity. Wiley, New York, pp 281–332

Belas R, Colwell RR (1982) Adsorption kinetics of laterally and polarly flagellated *Vibrio*. J Bacteriol 151:1569–1580

Blackburn N, Azam F, Hagstrom A (1997) Spatially explicit simulations of a microbial food web. Limnol Oceanogr 42:613–622

Blair DF (1995) How bacteria sense and swim. Annu Rev Microbiol 49:489–522

Brint JM, Ohman DE (1995) Synthesis of multiple exoproducts in *Pseudomonas aeruginosa* is under the control of RhlR-RhlI, another set of regulators in strain PAO1 with homology to the autoinducer responsive LuxR-LuxI family. J Bacteriol 177:7155–7163

Budrene EO, Berg HC (1991) Complex patterns formed by motile cells of *Escherichia coli*. Nature 349:630–636

Buikema WJ, Haselkorn R (1993) Molecular genetics of cyanobacterial developement. Annu Rev Plant Physiol Plant Mol Biol 44:33–52

Caldwell DE, Costerton JW (1996) Are bacterial biofilms constrained to Darwin's concept of evolution through natural selection? Microbiologia SEM 12:347–358

Caldwell DE, Wolfaardt GE, Korber DR, Lawrence JR (1997) Do bacterial communities transcend contemporary theories of ecology and evolution? Adv Microbial Ecol 15:1–86

Cao JG, Meighan EA (1989) Purification and structural identification of an autoinducer for the luminescence system of *V. harveyi*. J Biol Chem 264:21,670–21,676

Cao JG, Meighan EA (1993) Biosynthesis and stereochemistry of the autoinducer controlling bioluminescence in *Vibrio harveyi*. J Bacteriol 175:3856–3862

Chater KF (1993) Genetic differentiation in *Streptomyces*. Annu Rev Microbiol 47:685–713

Chatterjee A, Cui Y, Lui Y, Dumenyo CK, Chatterjee AK (1995) Inactivation of *rsmA* leads to overproduction of extracellular pectinases, cellulases, and proteases in *Erwinia caroto-*

vora subsp. *carotovora* in the absence of the starvation/cell density-sensing signal, *N*-(3-oxohexanoyl)-L-homoserine lactone. Appl Environ Microbiol 61:1959–1967

Clewell DB (1993) Bacterial sex pheromone-induced plasmid transfer. Cell 73:9–121

Clewell DB, Weaver KE (1989) Sex pheromone and plasmid transfer in *Enterococcus faecalis*. Plasmid 21:175–184

Costerton JW, Lewandoski Z, DeBeer D, Caldwell DE, Korber DR, James G (1994) Biofilms: the customized microniche. J Bacteriol 176:2137–2142

Craven R, Montie TC (1985) Regulation of *Pseudomonas aeruginosa* chemotaxis by the nitrogen source. J Bacteriol 164:544–549

Cui Y, Chatterjee A, Liu Y, Dumenyo CK, Chatterjee AK (1995) Identification of a global repressor gene *rsmA* of *Erwinia carotovora* subsp. *carotovora* that controls extracellular enzymes, *N*-(3-oxohexanoyl)-L-homoserine lactone, and pathogenicity in soft-rotting *Erwinia* spp. J Bacteriol 177:5108–5115

Davies DG, Parsek MR, Pearson JP, Iglewski BH, Costerton JW, Greenberg EP (1998) The involvement of cell-to-cell signals in the development of a bacterial biofilm. Science 280:295–298

Decho AW (1990) Microbial exopolymers in ocean environments: their role(s) in food webs and marine processes. Oceanogr Mar Biol Annu Rev 28:73–153

Decho AW (1998) Exopolymer microdomains as a structuring agent for heterogeneity within microbial biofilms. In: Riding R (ed) Microbial sediments. Springer, Berlin Heidelberg New York (in press)

Delaquis PJ (1990) Colonization of model and meat surfaces by *Pseudomonas fragi* and *Pseudomonas fluorescens*, PhD thesis. University of Saskatchewan, Saskatoon, Canada

Dunlap PV, Greenberg EP (1991) Role of intercellular chemical communication in the *Vibrio fischeri* -monocentrid fish symbiosis. In: Dworkin M (ed) Microbial cell-cell interactions. ASM Press, pp 219–253

Dunny GM, Brown BL, Clewell DB (1978) Induced cell aggregation and mating in *Streptococcus faecalis*:evidence for bacterial sex pheromone. Proc Natl Acad Sci USA 75: 3479–3483

Eberhard A, Burlingame AL, Eberhard C, Kenyon GL, Nealson KH, Oppenheimer NJ (1981) Structural identification of autoinducer of *Photobacterium fischeri* luciferase. Biochemistry 20:2444–2449

Eberhard A, Widrig CA, McBath P, Schineller JB (1986) Analogs of the autoinducer of bioluminescence in *Vibrio fischeri*. Arch Microbiol 146:35–40

Engebrecht J, Nealson K, Silverman M (1983) Bacterial bioluminescence: isolation and genetic analysis of functions from *Vibrio fischeri*. Cell 32:773–781

Fuqua WC, Winans SC (1994) A LuxR-LuxI type regulatory system activates *Agrobacterium* Ti plasmid conjugal transfer in the presence of a plant tumor metabolite. J Bacteriol 176: 2796–2806

Fuqua WC, Winans SC, Greenberg EP (1994) Quorum sensing in bacteria: the LuxR/LuxI family of cell density responsive transcriptional regulators. J Bacteriol 176:269–275

Fuqua C, Winans SC, Greenberg EP (1996) Census and consensus in bacterial ecosystems: the LuxR-LuxI family of quorum sensing transcriptional regulators. Annu Rev Microbiol 50:727–751

Galli D, Wirth R, Wanner G (1989) Identification of aggregation substances of *Enterococcus faecalis* cells after induction by sex pheromones. Arch Microbiol 151:486–490

Galli D, Friesenegger A, Wirth R (1992) Transcriptional control of sex-pheromone-inducible genes on plasmid pADI of *Enterococcus faecalis* and sequence analysis of a third structural gene for (pPD1-encoded) aggregation substance. Mol Microbiol 6:1297–1308

Gilson L, Kuo A, Dunlap PV (1996) AinS and a new family of autoinducer synthesis proteins. J Bacteriol 177:6946–6951

Gjaltema A, Arts PAM, Van Loosdrecht MCM, Kuenen JG, Heijnen JJ (1994) Heterogeneity in biofilms in rotating annular reactors: occurrence, structure and consequences. Biotechnol Bioeng 44:194–204

Gray KM, Passador L, Iglewski BH (1994) Interchangability and specificity of components from the quorum sensing regulatory systems of *Vibrio fischeri* and *Pseudomonas aeruginosa*. J Bacteriol 176:3076–3080

Greenberg EP, Hastings JW, Ulitzur S (1979) Induction of luciferase synthesis in *Beneckea harveyi* by other marine bacteria. Arch Microbiol 120:87–91

Haygood MG (1994) Light organ symbioses in fishes. Crit Rev Microbiol 19:191–216

Hedblom ML, Adler J (1983) Chemotactic response of *Escherichia coli* to chemically synthesized amino acids. J Bacteriol 155:1463–1466

Horinouchi S, Beppu T (1992) Regulation of secondary metabolism and cell differentiation in *Streptomyces*: A-factor as a microbial hormone and the AfsR protein as a component of two-component regulatory system. Gene 115:167–172

Huisman GW, Kolter R (1994) Sensing starvation: a homoserine lactone-dependent signalling pathway in *Escherichia coli*. Science 265:537–539

Hwang I, Cook DM, Farrand SK (1995) A new regulatory element modulates homoserine lactone-mediated autoinduction of Ti plasmid conjugal transfer. J Bacteriol 177:449–458

Ji G, Beavis RC, Novick RP (1995) Cell density control of staphylococcal virulence mediated by an octapeptide pheromone. Proc Natl Acad Sci USA 92:12,055–12,059

Jones S, Yu B, Bainton NJ, Birdsall M, Bycroft BW, Chhabra SR, Cox AJR, Golby P, Reeves PJ, Stephens S, Winson MK, Salmond GSAB, Williams P (1993) The lux autoinducer regulates the production of exoenzyme virulence determinants in *Erwinia carotovora* and *Pseudomonas aeruginosa*. EMBO J 12:2477–2482

Kaiser D, Losick R (1993) How and why bacteria talk to each other. Cell 73:873–885

Kaplan HB, Greenberg EP (1985) Diffusion of autoinducer is involved in regulation of the *Vibrio fischeri* luminescence systems. J Bacteriol 163:1210–1214

Karp-Boss L, Boss E, Jumars PA (1996) Nutrient fluxes to planktonic osmotrophs in the presence of fluid motion. Oceanogr Mar Biol Annu Rev 34:71–107

Kelly-Wintenberg K, Montie TC (1994) Chemotaxis to oligopeptides by *Pseudomonas aeruginosa*. Appl Environ Microbiol 60:363–367

Kjelleberg S, Ostling J, Holmquist L, Flardh K, Svenblad B, Jouper-Jann A, Weichart D, Albertson N (1993) Starvation and recovery of *Vibrio*. In: Guerrero R, Pedros-Alio C (eds) Trends in microbial ecology. Spanish Society for Microbiology, Barcelona, Spain, pp 169–175

Korber DR, Lawrence JR, Zhang L, Caldwell DE (1990) Effect of gravity on bacterial deposition and orientation in laminar flow environments. Biofouling 2:335–350

Korber DR, Lawrence JR, Hendry MJ, Caldwell DE (1993) Analysis of spatial variability within mot+ and mot– *Pseudomonas fluorescens* biofilms using representative elements. Biofouling 7:339–358

Kuo A, Blough NV, Dunlap PV (1994) Multiple *N*-acyl-L-homoserine lactone autoinducers of luminescence genes in the marine symbiotic bacterium *Vibrio fischeri*. J Bacteriol 176:7558–7565

Latifi A, Winson MK, Foglino M, Bycroft BW, Stewart GSAB (1995) Multiple homologues of LuxR and LuxI control expression of virulence determinants and secondary metabolites through quorum sensing in *Pseudomonas aeruginosa* PAO1. Mol Microbiol 17:333–343

Lawrence JR, Korber DR (1993) Aspects of microbial surface colonization behavior. In: Guerrero R, Pedros-Alio C (eds) Trends in microbial ecology. Spanish Society for Microbiology, Barcelona, Spain, pp 113–118

Lawrence JR, Delaquis PJ, Korber DR, Caldwell DE (1987) Behaviour of *Pseudomonas fluorescens* within the hydrodynamic boundary layers of surface microenvironments. Microbiol Ecol 14:1–14

Lawrence JR, Korber DR, Hoyle BD, Costerton JW, Caldwell DE (1991) Optical sectioning of microbial biofilms. J Bacteriol 173:6558–6567

Lawrence JR, Korber DR, Caldwell DE (1992) Behavioural analysis of *Vibrio parahaemolyticus* variants in high- and low viscosity microenvironments by use of digital image processing. J Bacteriol 174:5732–5739

Lawrence JR, Korber DR, Wolfaardt GM, Caldwell DE (1995) Behavioral strategies of surface colonizing bacteria. Adv Microbiol Ecol 14:1–75

Lewandoski Z, Altobelli SA, Fukushima E (1993) NMR and microelectrode studies of hydrodynamics and kinetics in biofilms. Biotechnol Prog 9:40–45

Liang J, Scappino L, Haselkorn R (1992) The patA gene product, which contains a region similar to CheY of *Escherichia coli*, controls heterocyst pattern formation in the cyanobacterium *Anabaena* 7120. Proc Natl Acad Sci USA 89:5655–5659

Lopez-de-Victoria G, Fielder DR, Zimmer-Faust RK, Lovell CR (1994) Motility behaviour of *Azospirillum* species in response to aromatic compounds. Can J Microbiol 40:704–711

Losick R, Kaiser D (1997) Why and how bacteria communicate. Sci Amer 68–73

Marshall KC (1988) Adhesion and growth of bacteria at surfaces in oligotrophic habitats. Can J Microbiol 34:503–506

Mayr E (1963) Animal species and evolution. Harvard University Press, Cambridge MA

McCarter LL, Showalter RE, Silverman MR (1992) Genetic analysis of surface sensing in *Vibrio parahaemolyticus*. Biofouling 5:163–175

McGowan S, Sebaihia M, Jones S, Yu B, Bainton N (1995) Carbapenem antibiotic production in *Erwinia carotovora* is regulated by CarR, a homologue of the LuxR transcriptional activator. Microbiology 141:541–550

McKenney D, Brown KE, Allision DG (1995) Influence of *Pseudomonas aeruginosa* exoproducts on virulence factor production in *Burkolderia cepacia*: evidence of interspecies communication. J Bacteriol 177:6989–6992

McNab RM (1987) Motility and chemotaxis. In: Neidhardt IC, Ingraham JL, Low KB, Magasanik B, Schaechter M, Umbarger HE (eds) *Escherichia coli* and *Salmonella tymphimurium*: cellular and molecular biology. ASM Press, Washington, D.C., pp 732–759

Moller S, Pedersen AR, Poulsen LK, Arvin E, Molin S (1996) Activity of three-dimensional distribution of toluene-degrading *Pseudomonas putida* in a multispecies biofilm assessed by quantitative in situ hybridization and scanning confocal laser microscopy. Appl Environ Microbiol 62:4632–4640

Nealson KH (1977) Autoinduction of bacterial luciferase: occurrence, mechanism and significance. Arch Microbiol 112:73–79

Neu TR, Lawrence JR (1997) Development and structure of microbial biofilms in river water studied by confocal laser scanning microscopy. FEMS Microbiol Ecol 24:11–25

Ochsner UA, Reiser J (1995) Autoinducer-mediated regulation of rhamnolipid biosurfactant synthesis in *Pseudomonas aeruginosa*. Proc Natl Acad Sci USA 92:6424–6428

Ochsner UA, Koch AK, Reiser J (1994) Isolation and characterization of a regulatory gene affecting rhamnolipid biosurfactant synthesis in *Pseudomonas aeruginosa*. J Bacteriol 176:2044–2054

Parkinson JS (1993) Signal transduction schemes of bacteria. Cell 73:857–871

Passador L, Cook JM, Gambello MJ, Rust L, Iglewski BH (1993) Expression of *Pseudomonas aeruginosa* virulence genes requires cell-to-cell communication. Science 260:1127–1130

Passador L, Tucker KD, Guertin KR (1996) Functional analysis of the *Pseudomonas aeruginosa* autoinducer PAI. J Bacteriol 178:5995–6000

Pearson JP, Gray KM, Passador L, Tucker KD, Eberhard A, Iglewski BH, Greenberg EP (1994) Structure of the autoinducer required for expression of *Pseudomonas aeruginosa* virulence genes. Proc Natl Acad Sci USA 91:197–201

Pearson JP, Passador L, Iglewski BH, Greenberg EP (1995) A second *N*-acylhomoserine lactone signal produced by *Pseudomonas aeruginosa*. Proc Natl Acad Sci USA 92:1490–1494

Pierson LS, Keppenne VD, Wood DW (1994) Phenazine antibiotic biosynthesis in *Pseudomonas aureofaciens* 30–84 is regulated by phzR in response to cell density. J Bacteriol 176:3966–3974

Piper KR, Beck von Bodman S, Farand SK (1993) Conjugation factor of *Agrobacterium tumefaciens* regulates Ti plasmid transfer by autoinduction. Nature 362:448–450

Pirhonnen M, Flego D, Heikiheimo R, Palva ET (1993) A small diffusible signal molecule is responsible for the global control of virulence and exoenzyme production in the plant pathogen *Erwinia carotovora*. EMBO J 12:2467–2476

Powell MS, Slater NKH (1983) The deposition of bacterial cells from laminar flows onto solid surfaces. Biotechnol Bioeng 25:891–900

Ruby EG (1996) Lessons from a cooperative bacterial-animal association: the *Vibrio fischeri-Euprymna scolopes* light organ. Annu Rev Microbiol 50:591–624

Sar N, McCarter L, Simon M, Silverman M (1990) Chemotactic control of the two flagellar systems of *Vibrio parahaemolyticus*. J Bacteriol 172:334–341

Schaefer AL, Val DL, Hanzelka BL, Cronan JE, Greenberg EP (1996a) Generation of cell-to-cell signals in quorum sensing: acyl homoserine activity of a purified *Vibrio fischeri* LuxI protein. Proc Natl Acad Sci USA 93:9505–9509

Schaefer AL, Hanzelka BL, Eberhard A (1996b) Quorum sensing in *Vibrio fischeri*: probing autoinducer-LuxR interactions with autoinducer analogs. J Bacteriol 178:2897–2901

Schripsema J, Karel E, Theo B (1996) Bacteriocin small of *Rhizobium leguminosarum* belongs to the class of *N*-acyl- L-homoserine lactone molecules, known as autoinducers and as quorum sensing co-transcriptional factors. J Bacteriol 178:366–371

Seed PC, Passador L, Iglewski BH (1995) Activation of the *Pseudomonas aeruginosa* lasI gene by LasR and the *Pseudomonas* autoinducer PAI: an autoinduction regulatory hierarchy. J Bacteriol 177:654–659

Silverman M, Simon MI (1977) Bacterial flagella. Annu Rev Microbiol 31:397–419

Sonea S (1991) Bacterial evolution without speciation. In: Margulis L, Fester R (eds) Symbiosis as a source of evolutionary innovation. MIT Press, Cambridge, MA, pp 95–105

Stevens AM, Greenberg EP (1997) Quorum sensing in *Vibrio fischeri*:essential elements for activation of the luminescence genes. J Bacteriol 179:557–562

Throup JP, Camara M, Briggs GS, Winson MK, Chhabra SR (1995) Characterisation of the yenl/yenR locus from *Yersinia enterocolitica* mediating the synthesis of two *N*-acyl-homoserine lactone inducer molecules. Mol Microbiol 17:345–356

Torres-Cabassa A, Gottesman S, Frederick RD, Dolph PJ, Coplin DL (1987) Control of extracellular polysaccharide synthesis in *Erwinia stewarti* and *Escherichia coli* K-12: a common regulatory function. J Bacteriol 169:4525–4531

Wilcox M, Mitchison GJ, Smith RJ (1973) Pattern formation in the blue-green alga *Anabaena*. I. Basic mechanisms. J Cell Sci 12:707–723

Wimpenny JWT (1992) Microbial systems: patterns in time and space. Adv Microb Ecol 12: 469–522

Winson MK, Camara M, Latifi A, Fogino M, Chhabra SR, (1995) Multiple *N*-acyl-L-homoserine lactone signal molecules regulate production of virulence determinants and secondary metabolites in *Pseudomonas aeruginosa*. Proc Natl Acad Sci USA 92:9427–9431

Wolk CP (1967) Physiological basis of the pattern of vegetative growth of a blue-green alga. Proc Natl Acad Sci USA 57:1246–1251

Wolk CP (1991) Genetic analysis of cyanobacterial development. Curr Opin Gen Dev 1: 336–341

Yu C, Bassler B, Roseman S (1993) Chemotaxis of the marine bacterium *Vibrio furnissii* to sugars: a potential mechanism for initiating the chitin catabolic cascade. J Biol Chem 268:9405–9409

Zhang L, Murphy PJ, Kerr A, Tate ME (1993) *Agrobacterium* conjugation and gene regulation by *N*-acyl-homoserine lactones. Nature 362:446–448

CHAPTER 10

Function of EPS

Gideon M. Wolfaardt[1] · John R. Lawrence[2] · Darren R. Korber[3]

[1] Department of Microbiology, University of Stellenbosch, 7600 Stellenbosch, South Africa,
E-mail: gmw@land.sun.ac.za
[2] National Hydrology Research Center, 11 Innovation Blvd., Saskatoon, SK, S7N 3H5 Canada
[3] Department of Applied Microbiology and Food Science, University of Saskatchewan,
Saskatoon, SK, S7N 5A8, Canada

Keywords. EPS, Function, Coaggregation, Biofilm, Genetic transfer, Nutrient accumulation, Extracellular enzyme, Protection, Resistance, Predation, Biocide, Host defense, Microbial-mineral interaction, Bio-accumulation

1
Introduction

The production of extracellular polymeric substances (EPS) involves a significant investment of carbon and energy by microorganisms. Considering the tendency in nature to conserve rather than to waste, this expenditure of energy (in some cases more than 70% – see HARDER AND DIJKHUIZEN 1983) is likely to hold benefits to the producers of EPS, as well as those organisms associated with them. Bacteria are very efficient in converting nutrients into EPS; it has been calculated (UNDERWOOD ET AL. 1995) that a single *Azotobacter* cell can produce enough EPS to coat more than 500 particles with a 0.4 µm diameter per day. The size of a single cell is typically 1–2 µm by 0.5 µm, and often much smaller, and therefore this number is impressive. The importance of EPS has long been recognized and a variety of functions have been attributed to EPS as far as the benefits they provide to cells, either living as single organisms, in binary associations, or in heterogeneous communities. However, CHRISTENSEN AND CHARACKLIS (1990) implied that there is a lack of knowledge of the properties of EPS in biofilms, as well as their role in biofilm ecology. The aim of this chapter is to provide an overview of some of the progress that has been made in recent years to elucidate the functional role of EPS.

2
Role of EPS in Cellular Associations

2.1
Suspended and Attached Cells

Microorganisms exist in the environment as free-floating cells or attached to surfaces. The relative importance of each state differs from one environment to another, and may show a high degree of variation in a single environment when chemical and physical changes, brought about by seasonal changes, dry-wet cycles, etc., occur. Suspended (planktonic) cells dominate in environments such as the water column in oceans and deep lakes, as well as large scale industrial fermentors. Other environments are dominated by attached (sessile) bacteria. KOLENBRANDER AND ANDERSEN (1988) listed a number of environments where adherence of bacteria to surfaces enable the bacteria to become established. Traditionally, there were various obstacles to the study of attached cells, such as the difficulty in obtaining representative samples and the lack of model systems allowing direct studies of biofilms. Because of these obstacles, characterization of the activities of attached cells lagged behind that of free-living cells (GEESEY AND WHITE 1990). Recent developments, such as the application of in situ hybridization techniques to delineate aspects such as cellular rRNA content or the metabolic activity of a specific population in a biofilm community

(MØLLER ET AL. 1998) will help to overcome the difficulties traditionally associated with the study of attached cells. These methods also show potential for the study of EPS production and function under fully hydrated conditions.

Microorganisms produce EPS during both suspended and sessile growth. Using chemostat cultures to study *Staphylococcus epidermidis*, EVANS ET AL. (1994) found little difference between EPS production by suspended cells and sessile cells when growth rates were high. However, the sessile cells produced significantly more EPS than their suspended counterparts during slow growth rates. Although more work is needed to understand better the importance of EPS, previous studies have provided sufficient evidence to support the contention that EPS facilitate interactions between cells and their environment. These interactions are often essential for microbial survival; application of new technologies to better characterize EPS production and function by suspended as well as attached cells are needed to provide us with a better overall understanding of the behavior and survival of microbes in nature.

2.2
Coaggregation, Consortial Behaviour, and Floc Formation

Consortial activities are required for many microbial processes which are not possible with single species populations (GEESEY AND COSTERTON 1986). A typical example is the utilization of recalcitrant molecules as an energy source which are resistant to degradation by single species. Microorganisms maximize their metabolic capabilities through co-operative interactions between two or more populations sharing a mutual habitat. Through these interactions the individual cells contribute to the overall maintenance of population or community integrity and stability, and thereby their own survival. The number of interacting species may vary from one environment to another. For instance, the highly specific partner recognition during bacterial coaggregation (KOLENBRANDER 1989) probably involve fewer participants than the degradation of organic contaminants in the environment. EPS play an important role in these interactions by facilitating "communication" between cells through participating in cell-cell recognition, by serving as an adhesin, and the establishment of favorable micro-environments.

BOULT ET AL. (1997) referred to the importance of solid-liquid interfaces as sites of complex microbial communities in what they termed "a heterogeneous organization of cells," where individual members interact to optimize nutrient utilization and waste cycling. EPS are believed to act in various ways which promote spatial organization of cells to allow interactions between cells in these close associations. A good example where spatial organization is evident is to be found in micro-colonies. Using a capsulated bacterial strain and a non-mucoid mutant of it, ALLISON AND SUTHERLAND (1987) demonstrated the involvement of exopolymers in colony formation. They found that colony formation was only observed for the capsulated and other slime-producing strains. Interactions are important not only in biofilms, but also in aggregates such as anaerobic flocs. Similar to their role in biofilms, EPS have an important adhesive role during floc formation (CHARACKLIS AND MARSHALL 1990). MACLEOD

ET AL. (1995) demonstrated that bacteria present in granular sludge were enveloped by extensive EPS, and that the EPS completely filled the intercellular spaces within the microcolonies, thereby playing an important role in maintaining the structural integrity of granular sludge. TAGO AND AIDA (1977) discussed a mucopolysaccharide produced by *Zoogloea* sp. that was "obligatory" for floc formation. By maintaining the architecture of these structures, EPS allow members of the granules to interact in special ways. VEIGA ET AL. (1997) indicated that EPS contribute to the long term stability of granules by facilitating the adhesion between different species of methanogens and syntrophic acetogenic bacteria, while MURALIDHARAN ET AL. (1997) studied hydrogen transfer in a co-culture of two hyperthermophilic microorganisms: *Thermotoga maritima*, an anaerobic heterotroph, and *Methanococcus jannaschii*, a hydrogenotrophic methanogen. They found that the numbers of *T. maritima* increased ten-fold when co-cultured with *M. jannaschii*, and that *M. jannaschii* was able to grow in the absence of externally supplied H_2 and CO_2. However, the co-culture could not be established if the two organisms were physically separated by a dialysis membrane, suggesting that close juxtapositioning is required between the two organisms. They further found that the methanogen was only found in cell aggregates with a minimum size of 4.5 µm, and that a sharp decline existed in H_2 concentration in the proximity of a methanogen.

The spatial distribution of EPS in flocs appears to be related to the function of the EPS. For example, DE BEER ET AL. (1996) determined that approximately 50% of the total amount of EPS in sludge granules was present in a 40 µm layer at the surface, while the remainder was distributed through the rest of the granule. In contrast, the highest concentration of EPS in flocs was positioned in the center while no EPS surface layer was found. These differences are likely the result of evolutionary adaptations by these associations as the coating in the case of granules contain hydrophilic EPS which prevent the attachment of gas bubbles, thus preventing floating of the granules. In natural water bodies, these associations are usually affiliated with sediments where it is easier to maintain the strict anoxic conditions required by the obligate anaerobic organisms to grow. In flocs, anaerobic conditions in the inside are caused by oxygen consumption at the outer layers by aerobes and facultative anaerobes. NIELSEN ET AL. (1996) discussed deflocculation, the process during which prolonged exposure to anaerobic conditions results in the inhibition of EPS production, and possibly also hydrolysis and degradation of EPS. Because EPS are a major floc component, these changes lead to the collapse of floc structure. Given the important consequences of deflocculation for waste water treatment, especially in terms of the ease with which activated sludge can be de-watered, this topic has been studied extensively. However, very little has been done to describe the physiological and ecological importance of this phenomenon to microorganisms.

2.3
Biofilm Formation

Extracellular polymeric substances are viewed as important mediators in the adhesion of bacteria and other microorganisms to surfaces (SUTHERLAND

1984; MARSHALL 1985; STOTSKY 1985; QUINTERO AND WEINER 1995; BECKER 1996). Although ALLISON AND SUTHERLAND (1987) stated that little direct evidence has been presented to demonstrate the involvement of EPS in the initial adhesion process, others (GEESEY ET AL. 1977; MORRIS AND MCBRIDE 1984; MARSHALL ET AL. 1989; COWAN ET AL. 1991) suggested that capsular EPS may be involved during the initial stages of attachment. Furthermore, MCEL-DOWNEY AND FLETCHER (1988) indicated that initial attachment may occur without the requirement for the production of new adhesives when attractive interactions take place between polymers already present on the cell surface and the attachment surface. In addition to this potential role in the initial attachment of cells during biofilm formation, EPS are also involved in maintaining the structural integrity of biofilms (COSTERTON ET AL. 1981; MARSHALL 1984; VANDEVIVERE AND KIRCHMAN 1993), and therefore the overall stability of biofilm communities (UHLINGER AND WHITE 1983). In a number of instances the presence of combinations of bacteria has been reported to result in production of thicker and more stable biofilm systems. SIEBEL AND CHARACKLIS (1991) reported that biofilms formed by the combination of *Pseudomonas aeruginosa* and *Klebsiella oxytoca* were on average 40 µm thick whereas the individual organisms produced thinner biofilms. In another case the presence of *P. aeruginosa* stabilized biofilms formed by binary combinations of *Pseudomonas fluorescens* and *Klebsiella pneumoniae* (STOODLEY ET AL. 1994). It has been suggested that these results are due to stabilizing interactions between the EPS of the bacterial species forming the biofilm (SUTHERLAND 1984; MCELDOWNEY AND FLETCHER 1987; JAMES ET AL. 1995). Figure 1 illustrates the typical heterogeneity in EPS surrounding a bacterial microcolony (arrows in Fig. 1A–D) within this system. This particular microcolony type was also observed to fluoresce green following staining with a viability probe and dual channel scanning confocal laser microscopy imaging (suggesting cells possessed a functional cell membrane and were hence viable; Fig. 1E). In contrast, a number of bacteria in the same biofilm also stained red (suggesting damaged cell membranes and cell death; Fig. 1F). Such variable staining within complex biofilm systems may be explained by either predatory activity, variable states of cell metabolism, or variable probe penetration.

EPS have also been shown to play a role as an active metabolic component in biofilms (WEINER ET AL. 1995; WOLFAARDT ET AL. 1995). Like cells, EPS distribution shows much variety; using a combination of probes to differentially stain cellular nucleic acids and EPS, STEWART ET AL. (1995) showed that cell and EPS distributions did not always overlap. Thus, large areas of biofilm may actually be devoid of any cellular material, but consist mainly of EPS. For additional material on biofilm development and the role of EPS during biofilm formation, the reader is referred to LAWRENCE ET AL. (1995).

2.4
Genetic Transfer

In a review of the fate of extracellular DNA in the environment, LORENZ AND WACKERNAGEL (1994) cited a number of studies which have shown that DNA

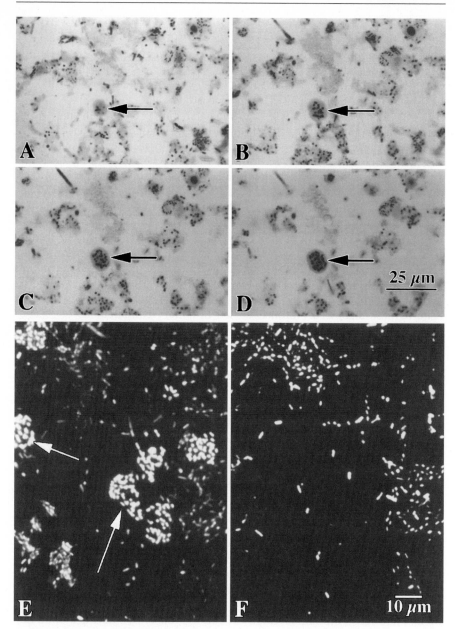

Fig. 1A–F. Scanning confocal laser micrographs illustrating the heterogeneity of EPS associated with a river-system biofilm: **A–D** dense EPS, as determined by fluorescein penetration, surrounded a bacterial microcolony (*gray zones indicated by arrows*); **E** this particular microcolony type was also observed to fluoresce green (indicating that cells were viable) following staining with a commercial viability probe (sensitive to membrane integrity) and dual-channel laser imaging; **F** different bacteria present in the same field of observation also fluoresced red (indicating cells were non-viable), although the cell metabolic state was not confirmed by alternate methods. Scale bars for the respective images are included

forms complexes with, among others, clay minerals, quartz, feldspar, humic substances, minerals in the soil environments, and suspended particulate material in water. It appears that this adsorption to surfaces affords extracellular DNA protection against DNases. Considering the importance of EPS in the attachment of cells to surfaces and the fact that attached cells produce more EPS than their planktonic counterparts (VANDEVIVERE AND KIRCHMAN 1993), it could be argued that EPS also play a role, at least indirectly, in the exchange of genetic material between cells. For instance, LORENZ ET AL. (1988) demonstrated in sand-filled flowthrough columns that competent *Bacillus subtilis* cells were able to take up extracellular chromosomal DNA. They found that the transformation efficiency was as much as 50 times higher at solid/liquid interfaces than in liquid, and up to 3200 times higher when only cells attached to mineral surfaces were considered. However, their studies on two Gram-negative bacteria did not show these increased transformation rates on solid surfaces. LEBARON ET AL. (1997) studied plasmid mobilization between *Escherichia coli* strains in biofilms as well as other environments. They concluded that bacterial adhesion is among the factors that can be related to plasmid transfer. They further suggested that the hydrodynamic conditions within biofilms could affect transfer potential. In a study of the strain specificity of plasmid sequences from different *Pseudomonas syringae* strains, MARASAS (1993) purified plasmids from "extracellular polysaccharide contaminants". TREVORS ET AL. (1987) mentioned that microbes existing in environments which allow low cell densities have a reduced possibility for cell-to-cell contact, and thus a lower frequency of conjugation. In contrast, EPS allow, together with other factors, the concentration of cells at surfaces or in aggregates, thereby increasing the opportunity for cell-to-cell contact. The research of bacterial gene transfer in the environment is still at its beginning (LORENZ AND WACKERNAGEL 1994). Very little has been done to elucidate the role of EPS in this process. Although there is some (mostly indirect) evidence that EPS play a role in genetic transfer, such as during cell-cell adhesion and altering the hydrodynamic conditions in biofilms as indicated above, the significance has not been established.

3
Role of EPS in Nutrition

Microorganisms often live under oligotrophic conditions in natural environments and to endure under these conditions they needed to evolve special adaptive strategies. It has been suggested that the production of EPS may be one way by which microorganisms ensure that they survive in environments where nutrients are available at levels below the threshold concentrations required to remain viable, or under feast-and-famine conditions characterized by fluctuations in available nutrient sources. Little work has been done to determine the extent to which nutrients in bacterial EPS can be utilized, either by the same organism, or by cross feeding between species. PATEL AND GERSON (1974) demonstrated the capacity of a *Rhizobium* strain to re-utilize its own EPS, and concluded that the enzymes involved were extracellular enzymes. However, these authors suggested that only a limited number of organisms appear to possess the enzymes required to depolymerize their own EPS, a view shared by DUDMAN (1977) who suggested that the majority of EPS-producing organisms are unable to utilize their own EPS as carbon sources through biochemical pathways. More recently, PIROG ET AL. (1997B) suggested that a whole complex of enzymes, produced by different organisms, is necessary for complete degradation of EPS. These workers (PIROG ET AL. 1997B) isolated mixed and pure cultures of microorganisms from soil samples which were able to utilize EPS from *Acinetobacter* sp. as a carbon source. Although the role of EPS in nutrition is not well understood, and despite the conclusions from earlier studies that EPS generally do not serve as reserve sources of carbon and energy, there appears to be at least two ways in which EPS may function as a food reservoir: 1) production of EPS as nutrient reserve and 2) accumulation of nutrients by EPS.

3.1
Production of EPS as a Nutrient Reserve

It has been suggested that EPS may be created for long-term storage of carbon and energy; for example, a *Rhizobium* was shown to use EPS in this fashion (PATEL AND GERSON 1974). In addition, OBAYASHI AND GAUDY (1973) demonstrated that some bacteria could utilize the polymers of other bacteria providing an opportunity for cross-feeding. In their study of the metabolic response of bacteria to wet and dry cycles, ROBERSON AND FIRESTONE (1992) noted an initial decrease in the amount of polysaccharides following wetting after a dry cycle. This decrease in the amount of polysaccharides coincided with an increase in protein concentration, leading the authors to suggest that some polysaccharide carbon was used by the bacteria for protein production – in other words, carbon was shuttled between proteins and polysaccharides as the water status changed.

3.2
Accumulation of Nutrients by EPS

Because of the economical and environmental importance of metal accumulation by EPS in waste water treatment, which involves complexation of EPS with metals (RUDD ET AL. 1983) and ion exchange (DECHO 1990), it has received more attention than the accumulation of organic nutrients by EPS. It is generally recognized that bacterial EPS can complex a variety of metals, and that different metals are bound with different binding strength. For instance, BEECH AND CHEUNG (1995) used atomic absorption spectroscopy to study the ability of EPS, released by sulfate reducing bacteria into the liquid phase of the bacterial growth media, to complex Cr, Ni, and Mo. They found that, although concentrations of these elements in EPS varied with metal species and with the sulfate reducing bacteria (SRB) isolate, levels of Mo associated with EPS were significantly higher than the measured amount of Cr or Ni, regardless of the bacterial isolate. LOAEC ET AL. (1997) studied the uptake of various heavy metals by an *Alteromonas* strain and found that Pb was preferentially complexed by EPS in a bi-component solution with Cd and Zn. DECHO (1990) suggested that the same mechanisms involved in the accumulation of metals are likely also involved in the sorption of nutrients. In contrast, CHRISTENSEN AND CHARACKLIS (1990) pointed out that, although most EPS have cation exchange properties which may enable bacteria to use EPS as a nutrient trap under oligotrophic conditions, it is unlikely that they can function effectively as a trap for soluble nutrients in environments where divalent cations are abundant. However, although the relative importance of extracellular sorption to EPS in relation to absorption into cells or adsorption onto cells has not been established (BELLIN AND RAO 1993), others (e.g., COSTERTON 1984; HANSEN ET AL. 1984) proposed that the EPS matrix of biofilms can act as an ion-exchange resin which accumulates nutrients. In addition, FREEMAN ET AL. (1995) provided experimental support for the contention that biofilms concentrate nutrients through ion exchange. These authors (FREEMAN ET AL. 1995) found that when they changed the ionic composition of waters overlying biofilms, the retention of phenolic materials by those biofilms were accordingly affected. DECHO (1990) listed a number of studies which refer to the potential of EPS in the extracellular sequestering and accumulation of nutrients.

DECHO AND LOPEZ (1993) indicated that, because of the active adsorptive microenvironment that is being created by the EPS of cells in aquatic environments, a matrix is established which can effectively concentrate dissolved organic matter (DOM) from the surrounding water. These authors also discussed the potential role of EPS in transforming DOM into a more accessible carbon source. In contrast to particulate organic carbon, which is readily accessible for ingestion, DOM is not (DECHO AND LOPEZ 1993). Because EPS contain an abundance of adsorptive sites capable of sequestering DOM, the DOM becomes incorporated in a larger nutrient reservoir. Support for this phenomenon has been based on the apparent continuing growth and activity of biofilm bacteria under starvation or nutrient limiting conditions. For example, FREEMAN AND LOCK (1995) indicated that river biofilms were remarkably resilient to depletion of organic matter from the overlying waters, suggesting that a store of energy existed within the biofilm that was able to supplement exogenous sources during periods of scarcity. FREEMAN ET AL. (1995) proposed that this nutrient sorption occurred via ion exchange processes and created a reservoir of bound available carbon in the biofilm. Similarly WOLFAARDT ET AL. (1995) found that a nine-membered biofilm community utilized EPS bound herbicide under starvation conditions.

WOLFAARDT ET AL. (1995) proposed that the accumulation of nutrients by EPS may be an adaptive mechanism which enables microbial communities to optimize utilization of potentially toxic xenobiotic compounds when provided as the sole carbon source. They applied scanning confocal laser microscopy in combination with radio-isotopes (WOLFAARDT ET AL. 1994A, 1995) to study the accumulation of chlorinated hydrocarbons in biofilms formed by a nine-membered degradative community (WOLFAARDT ET AL. 1994B). These studies demonstrated that accumulation of the herbicide diclofop methyl occurred in cell capsules as well as the EPS matrix, and that the sorption of diclofop and its breakdown products were suppressed by the presence of labile nutrients.

The structure of biofilms which accumulated chlorinated aromatic hydrocarbons (WOLFAARDT ET AL. 1994c), and the EPS composition of these biofilms
(WOLFAARDT ET AL. 1998) differed substantially from those which were cultivated on labile nutrients. These differences in EPS demonstrated the effect of
the carbon source on EPS composition formed by heterogeneous communities.
Figure 2 shows the change in binding of seven different lectins which resulted
when the same biofilm community was grown on either benzene or 2,4,6-
trichlorobenzoic acid (unpublished data). The above findings are in agreement with observations made during pure culture studies such as those by
SAMRAKANDI ET AL. (1997), which have demonstrated significant differences in
the amounts of EPS produced when biofilms were grown on sucrose or lactose,
and GANCEL AND NOVEL (1994) who showed that sugars which decreased the
growth rate of *Streptococcus salivarius* ssp. *thermophilus* increased EPS synthesis. In their studies, WOLFAARDT ET AL. (1994c, 1995) showed that the time
required for biofilms grown on chlorinated aromatic hydrocarbons to reach
maturity was between 14 and 21 days. They further showed that accumulation
of the herbicide diclofop in biofilm EPS typically reached a maximum value
after 14–21 days and this value then remained relatively constant even when the

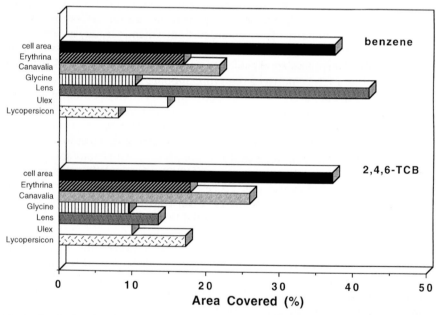

Fig. 2. Graph showing the differential binding of seven different lectins (specific for either:
D-galNAc, D-man, β-D-gal (1–4)-D-glcNAc, β-D-gal (1–3)-D-galNAc, D-glcNAc, α-D-man,
D-glc, or fucose) to the same microbial community grown on either benzene or 2,4,6-trichlo-
robenzoic acid (2,4,6-TCB). Note the large changes in the proportion of biofilm material (cells
and polymer) which bound D-man and D-glcNAc (as indicated by binding of the *Lens* and
Lycopersicon lectins, respectively) when the biofilm systems were grown on benzene and
2,4,6-TCB

community was maintained for months. This suggests that the accumulation rate of diclofop, and ring compounds formed during partial degradation of diclofop, were the same as the rate at which the ring structures of these compounds were degraded, and that there was no new net growth of polymers or that all binding sites where occupied by immobilized diclofop compounds, with no subsequent utilization. Following substitution of diclofop as the sole carbon source with a labile carbon source (trypticase soy broth) they did not observe a decrease in the amount of accumulated diclofop, indicating that the accumulated compounds were not utilized as a carbon source when another carbon source was available. In contrast, a rapid decrease in bound diclofop was observed in living biofilms after replacing the diclofop-minimal salts solution with the minimal salts solution without a carbon source, suggesting subsequent utilization of the accumulated compounds. When the biofilm bacteria were inhibited prior to switching from the diclofop-minimal salts solution to the minimal salts solution without a carbon source, there was only a small decrease in the amount of accumulated diclofop, which provided additional support for the assumption that the accumulated diclofop was utilized by biofilm bacteria.

In addition to their role in the accumulation of carbon reserves, EPS may also be important mediators for the concentration of other growth factors. DE PHILIPPIS ET AL. (1991) studied the production of EPS under different growth conditions by changing some nutritional and physical parameters known to affect the synthesis of EPS in algae and cyanobacteria. They found that among the factors that they tested (Ca^{2+}, Mg^{2+}, or PO_4^{3-} deficiencies, salinity, and pH) only Mg^{2+} shortage resulted in a significant increase of EPS production. They concluded that this enhancement of EPS production induced by Mg^{2+} deficiency demonstrated the role of anionic polysaccharides as chelating agents for cations essential to cellular metabolism.

4
Role of EPS in the Interaction of Microorganisms with their Biological, Physical, and Chemical Environment

PIROG (1997) and PIROG ET AL. (1997A) attributed a number of protective strategies to EPS. They used an *Acinetobacter* and found that EPS formed under favorable growth conditions protected the bacterium from extreme pH values, elevated temperature, drying, freezing, biocides, heavy metal ions (Cu^{2+}, Pb^{2+}, and Cr^{3+}), and detergents. In essence, these and other examples (see, e.g., DECHO 1990; LAWRENCE ET AL. 1995) of microbial EPS production to improve survival and reproductive success suggest that EPS facilitate interactions, not only among microbial populations but also between microorganisms and the environment in which they live.

4.1
EPS and the Micro-Environment

Biofilms can be viewed as ecosystems which employ a variety of strategies to regulate the exchange of molecules with their environment, thereby creating a micro-environment in which fluctuations of physical and chemical controls are less severe than in the external environment. This regulation of molecule flow is likely the result of the physical properties of the EPS matrix. A plausible description of the EPS matrix is that it has a helical conformation, forming a three-dimensional network of pores and channels (DUDMAN 1977). One advantage which is frequently attributed to biofilm-associated growth is protection against antimicrobial agents. However, the buffering capacity of biofilms will likely benefit biofilm-associated cells in numerous other ways.

The habitats of most aquatic bacteria typically have solute concentrations considerably lower than that within the cells. However, the habitat of soil-borne bacteria is characterized by dramatic fluctuations in water content, with solute concentrations often higher than within the cells. In order to prevent cells from losing their water content through plasmolysis during desiccation, or plasmoptysis (KIEFT ET AL. 1987) during rapid wetting, bacteria from these environments need special adaptations. It has been demonstrated that cells increase their internal solute concentration by synthesizing organic molecules such as amino acids (MEASURES 1975) or by the selective uptake of inorganic solutes such as K^+ (KILLHAM AND FIRESTONE 1984; KIEFT ET AL. 1987). The production of EPS is another possible survival strategy employed by bacteria to cope with fluctuations in water content. Results obtained by HARTEL AND ALEXANDER (1986) indicated that biotic factors and EPS production were not important for the survival of cowpea bradyrhizobia during drying of the soil. The authors listed other studies which have shown that EPS production did not provide *Rhizobium*, and *Bradyrhizobium* spp. better protection against soil drying. The conclusions made by HARTEL AND ALEXANDER (1986) were based on better drying survival by seven nonmucoid strains than five mucoid strains. These observations provided support for the view that bacteria, in this case the nonmucoid strains, employ a variety of mechanisms, other than the production of EPS, to survive decreases in external water content. However, the potential role of EPS in the survival of naturally occurring microbial communities during fluctuations in water content remains unclear. Indeed, as indicated by the studies of ROBERSON AND FIRESTONE (1992) using soil *Pseudomonas* sp., EPS may alter the microenvironment to enhance survival of desiccation. They found that EPS have a high affinity for water over a wide range of water potentials, and can hold up to five times their weight in water. These authors suggested that the EPS matrix may slow the rate at which bacterial colonies equilibrate with their surroundings. Such a slowing of the drying rate within the microenvironment allows the cells more time to adjust metabolically and may explain the observation by HARTEL AND ALEXANDER (1986) that the length of the drying period was important for survival. In addition, the study of OPHIR AND GUTNICK (1994) has shown that EPS do substantially increase bacterial resistance to desiccation.

The moisture holding capacity of soil organic matter and the ability of EPS to trap organic matter are other factors which should be considered. Screening for nonmucoid strains may not be the most appropriate way to evaluate the importance of EPS in the survival of mucoid strains since this approach does not allow indigenous communities to come into equilibrium with their physical and chemical environment before they are subjected to the test conditions. Typical experiments allowed strains relatively short periods to colonize soil columns (e.g., in both studies by Hartel and Alexander and Roberson and Firestone, drying followed immediately upon mixing of a cell suspension with the soil), which may not be enough time for the cells to produce the appropriate amounts and types of EPS. Other studies have demonstrated that incubation time and other environmental factors such as temperature, pH and carbon source (GANCEL AND NOVEL 1994) have an effect on the type of EPS being produced. GRINBERG ET AL. (1994) demonstrated that the ratio of acetylated and nonacetylated EPS produced by an *Acinetobacter* sp. changed with time during batch cultivation, which resulted in alteration of EPS properties. BELLEMANN ET AL. (1994) found that for the fireblight pathogen *Erwinia amylovora*, it was necessary to sustain EPS synthesis in all growth stages in order to escape plant defense reactions. When introduced into a new environment such as a sterile soil column, bacteria will likely produce EPS for functions such as adhesion and colony formation which, considering the degree of specialization that has been demonstrated for different types of EPS (QUINTERO AND WEINER 1995; HOTTER AND SCOTT 1997), may not be a factor in the survival during desiccation.

The hygroscopic properties of EPS (ROBERSON AND FIRESTONE 1992) and the resulting maintenance of a higher water content in the microenvironment around microbial communities may also be responsible for an increase in nutrient availability to the cells, and by providing a high water content, may maintain transport of nutrients to cells at diffusion rates close to that in water. CHENU AND ROBERSON (1996) demonstrated over a range of water potential values that the diffusion rate of glucose was higher in EPS and EPS-amended clays than in pure clay, and suggested that this could be attributed to the higher volumetric water content of EPS. Cell membranes need to remain hydrated to maintain selective permeability. Similarly, extracellular enzymes require a hydrated environment. EPS may play an important part in providing these conditions. When all these factors are considered it is clear that much more work is necessary to assess the role of EPS in the functioning of microbial communities in natural environments.

4.2
EPS as a Physical Barrier to Solute Translocation

Single cells, attached to a surface or in suspension, are in close contact with their environment. This means that in any non-static environment, transport of nutrients to cells and metabolites from cells will be dominated by advective flow. However, after the onset of biofilm development or aggregate formation, diffusion becomes the rate-limiting mode by which molecules are transported to and from cells. Diffusion through the cell-EPS matrix of biofilms occurs generally at a slower rate than through water and has been termed hindered diffusion (BOHRER ET AL. 1984; KORBER ET AL. 1995). LAWRENCE ET AL. (1994) used scanning confocal laser microscopy in conjunction with size-fractioned and charged dextrans to study size-dependent exclusion of molecules and hindered diffusion in pure culture as well as multi-species biofilms. They demonstrated a high degree of heterogeneity in the biofilms, where some regions of the cell-EPS matrix allowed rapid diffusion of all the test molecules used (ranging in molecular weight between 289 and 2,000,000), while only the low molecular weight molecules could migrate through adjacent regions. LAWRENCE ET AL. (1994) further concluded that the effective diffusion coefficient was a function of molecular radius, thus providing support for the concept of hindered diffusion. Similar observations regarding diffusion in biofilms were reported by DE BEER ET AL. (1997) who used a micropipette injection method to examine the diffusion of a variety of probes including fluorescein, TRITC-IgG and phycoerythrin in biofilms. BIRMINGHAM ET AL. (1995) used FRAP to monitor diffusion and binding of dextrans in oral biofilms and also reported hindered diffusion. It seems reasonable that diffusion coefficients in biofilms are approximately 80% of the values in water (WELCH AND VANDEVIVERE 1994). Despite the apparent disadvantages of existence within a diffusion-dominated EPS matrix, such as potentially reduced rates of nutrient delivery to cells and the accumulation of inhibitory waste products, the benefits appear to outweigh these limitations. BOULT ET AL. (1997) suggested that the diffusion properties and/or presence of charged reactive groups of EPS allow the maintenance of a more favorable microenvironment in an otherwise hostile environment.

The resistance of biofilms to antibiotics and biocides is of significant importance to industry and medicine, and this interest has translated into many studies to elucidate the mechanisms of biofilm resistance. Hindered diffusion and the inactivation of antimicrobial agents are the two mechanisms which have received the most attention as possible ways in which EPS are involved in increased resistance. SAMRAKANDI ET AL. (1997) noted that problems associated with diffusion through the EPS matrix that envelop biofilm bacteria are often viewed as an important reason for biofilm resistance to biocides. However, there seem to be other factors which may also lead to increased resistance, such as sorption of biocides by biofilm components. NICHOLS (1989) argued that such interactions would only have a temporary effect, because diffusion of the agent will likely continue to the base of the biofilm once all the reactive binding sites had been occupied. NICHOLS (1989) suggested that other factors, such as enzymes which deactivate the antimicrobial agent, may be more important. In a study of the penetration of antibiotics into $P.$ $aeruginosa$ aggregates, NICHOLS ET AL. (1989) concluded that β-lactamase trapped in EPS was responsible for the inactivation of β-lactam antibiotics. In their study, SAMRAKANDI ET AL. (1997) determined the susceptibility of biofilms grown on different nutrients to chlorine and monochloramine. They found that chlorine and monochloramine were equally effective in reducing viable cell counts of biofilms grown on lactose. However, when sucrose was supplied as the sole carbon source, chlorine was less effective than monochloramine. The authors related this difference in efficiency to the different amounts of EPS produced when the biofilms were grown on sucrose or lactose. Growth on sucrose resulted in significant increases of EPS production, and because chlorine has been demonstrated to have a greater tendency than monochloramine to react with organic and inorganic substances trapped in the EPS matrix, it can to a large degree be inactivated by the biofilm material. The ability of EPS in biofilms to react with chlorine is likely a key factor in determining the chlorine demand of aqueous systems. It can be expected that reaction between EPS and other antimicrobial agents will result in a similar increased demand for these compounds to obtain the minimum concentrations required to control effectively microbial growth. In addition to having practical significance for the treatment of industrial and drinking water systems, this ability to attenuate reactive compounds may be an important natural phenomenon to detoxify the environment.

GILBERT AND BROWN (1995) reviewed the potential mechanisms that may be involved in the protection of biofilms from antimicrobial agents. In addition to the direct means of protection discussed above, they also suggested that the EPS matrix may indirectly contribute to biocide resistance when other factors associated with biofilms, such as the establishment of nutrient and oxygen gradients, lead to reduced growth rates.

4.3
Role in Enzyme Reactions

The release of enzymes by microbes into their external environment forms the basis for the interaction between cells and high molecular weight exogenous substrates. CHUMAK ET AL. (1995) studied complexes formed by EPS and bacteriolytic enzymes produced by pseudomonads. They indicated that great quantities of EPS are excreted into the culture medium along with bacteriolytic enzymes. They found that, during growth in batch culture, the ratio of enzyme to EPS varied in the different growth phases, and suggested that this points to the possibility of physiological optimization of the composition of the EPS-protein complex to maximize substrate utilization. An interesting point that may also contribute to such optimization is that the secretion of bacteriolytic enzymes is initiated during the latent growth phase and continues for the duration of culture growth, while the highest rate of EPS production and accumulation occurs during the second half of the exponential growth phase. The authors pointed out that complex formation occurs when the bacteriolytic enzymes and EPS are oppositely charged, which allows electrostatic interactions to occur. The concentration of extracellular enzymes by EPS to facilitate extracellular cleavage of high molecular weight carbon sources before uptake into cells is well described in the literature (e.g., LAMED ET AL. 1983; COSTERTON 1984; DECHO 1990). The accumulated compound may serve as a carbon source during periods of nutrient limitation. Complex formation between EPS and extracellular enzymes has also been demonstrated in activated sludge. FRØLUND ET AL. (1995) utilized cation exchange to extract EPS from activated sludge and demonstrated that a very large fraction of the exoenzymes were released into the water simultaneously with the release of EPS. They proposed that a large proportion of the exoenzymes were immobilized in the sludge by adsorption in the EPS matrix, and that the exoenzymes should be considered to be an integrated part of the EPS matrix.

4.4
Role in Pathogenicity and Protection Against Host Defense Mechanisms

The ability of bacteria to establish themselves in a micro-niche where they are protected from antibiotics and host defenses is well documented. For instance, it has been demonstrated that EPS contribute to the persistence of *P. aeruginosa* in lungs of cystic fibrosis (CF) patients (PASQUIER ET AL. 1997). KOMIYAMA ET AL. (1987) presented a list of workers who indicated that the mucoid variants of *P. aeruginosa* are the most common pathogens in CF-related infections. In their studies, KOMIYAMA ET AL. (1987) found that coaggregation reactions were common between *P. aeruginosa* and bacteria indigenous to the oral cavity. They concluded that these inter-bacterial associations involve specific adhesins on the *P. aeruginosa* cell which bind to complimentary receptors on the cell surfaces of the indigenous oral bacteria. Little is known about the composition of these adhesins (MORRIS AND MCBRIDE 1984). However, it is likely that they have a high polysaccharide content, similar to the tuft fibrils of *Streptococcus sanguis* as described by HANDLEY ET AL. (1991). In the latter case, these adhesins are involved in coaggregation reactions occurring during dental plaque formation.

DEIGHTON ET AL. (1996) studied the contribution of EPS to infections by *Staphylococcus epidermidis* in the presence and absence of medical implants. They used five EPS-positive and five EPS-negative *S. epidermidis* strains in a mouse model. All ten strains produced abscesses in the absence of an implanted device, suggesting that a foreign body is not essential for the expression of virulence by *S. epidermidis*. However, they found that strains capable of producing a matrix of extracellular polysaccharide in which the cells are encased resulted in significantly more abscesses, which persisted longer than abscesses caused by EPS-negative strains. Based on these observations, they concluded that "slime" or "components of slime" appeared to delay the clearance of *S. epidermidis* from host tissues.

4.5
Protective Role Against Predation/Digestion

DECHO AND LOPEZ (1993) suggested that trophic interactions between microbial cells and consumer animals may be influenced by the different physical and chemical properties of exopolymers. For example, SNYDER (1990) demonstrated that predation by a ciliate was selective and based on the surface chemistry and EPS of the prey, and thus preferred prey were identified based on their EPS. Microbial cells in natural environments are often encased in different types of EPS. The two forms which are best described are tight capsules surrounding individual cells, and the slime matrix with a much looser tertiary configuration typically associated with the gel-like material occurring between cells in biofilms. DECHO AND LOPEZ (1993) discussed the properties of capsular and slime EPS as determined through their studies and summarized the findings from a number of other workers. They indicated that capsular EPS is significantly less digestible to consumers than slime EPS, even when extracted from the same bacterial strain. In addition, encapsulated bacteria are less efficiently digested

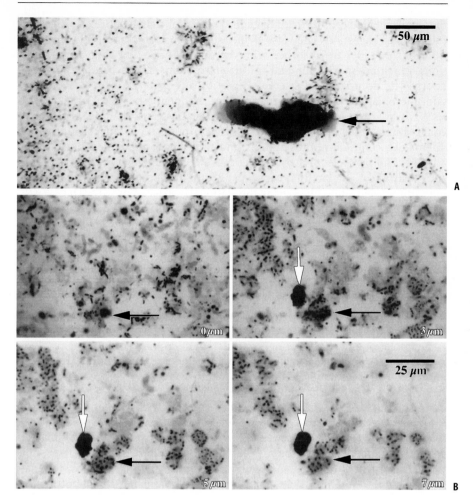

Fig. 3A. Optical thin sections showing examples of amoebae grazing in two different microbial biofilm systems. Scanning confocal laser microscopy and negative staining were used to image non-destructively the movement of amoebae on either three-day sludge biofilms, or in two-week river biofilms. A In the first case an amoeba actively grazed cells of a three-day sludge-degrading biofilm at the solid-liquid interface. Note that averaging of multiple images resulted in a blurring of the leading edges of the amoebae. There was little dense EPS evident in the sludge biofilms, as determined using fluorescein as an EPS-penetrating negative stain. B In the second example (0–7 μm), an amoeba grazed a mature river-system biofilm consisting of embedded microcolonies surrounded by dense EPS. In this case the amoebae fed above the solid-liquid interface, being active in regions >3 μm from the surface. The significance of these different strategies with respect to EPS production by biofilms are unclear. Scale bars for the respective images are included

than noncapsuled cells, providing additional evidence that capsules protect the cells against digestion. They also demonstrated that polysaccharide-rich fractions of slime EPS are absorbed with very high efficiencies while protein portions, which are more abundant in capsular polymers, are absorbed relatively poorly. The occurrence of selective digestion and increased resistance by certain microbial cells against predation have frequently been reported. DECHO AND LOPEZ (1993) cited other workers who demonstrated that microorganisms could remain intact and viable after passage through a consumer's gut, and that microalgae continue photosynthesis and nutrient accumulation during such passage. The presence of a protective capsule is often given as the reason for such resistance. The fact that capsular EPS are produced primarily during the logarithmic, and slime EPS during the stationary phase of growth (DECHO AND LOPEZ 1993), may be an indication of the functional role of EPS during each respective phase. Cells are probably more susceptible to predation during the logarithmic phase of growth, such as cells in suspension shortly after division and shedding from a biofilm, or shortly after colonization of a new surface. Thus, there is a greater need for capsular EPS under these conditions, but as an attached cell population matures, the need develops for the production of slime EPS to secure the attached cells firmly on the surface (often referred to as the irreversible step of biofilm formation) and to develop the micro-environmental conditions characteristic of biofilms.

Examples of amoebae grazing two different microbial biofilm systems obtained using scanning confocal laser microscopy and negative staining are provided in Fig. 3. In the first system (Fig. 3 A), amoebae actively grazed cells of a 3-day-old sludge-degrading biofilm at the solid-liquid interface. There was little dense EPS evident in this biofilm, as determined by the penetration of low-molecular-weight negative stain (fluorescein). In the second example, an amoeba (white arrows) can be seen grazing a mature river-system biofilm consisting of embedded microcolonies surrounded by dense EPS (Fig. 3B–E; note the reduced penetration of fluorescein in zones surrounding colonies, designated by dark arrows). In this case, the amoeba fed on cells positioned off the solid-liquid interface (at the 3–7 µm section depths). The significance of these different strategies with respect to EPS remains speculative.

4.6
Microbial-Plant Associations

Effective colonization of plant surfaces is necessary for the establishment of associations between plants and bacteria. A number of studies have demonstrated the role of EPS in plant-microbial interactions. YOU ET AL. (1995) studied the attachment of *Alcaligenes faecalis* to rice roots and found that pectinase and EPS played important roles in the association between the host plant and bacteria. They concluded that EPS were important for adsorption onto the root as well as anchoring of the attached cells. Following experiments using an EPS negative mutant of *Rhizobium loti*, HOTTER AND SCOTT (1997) concluded that EPS play a role during invasion of *Leucaena leucocephala*. The fully effective *R. loti* strain, which synthesized a flavolan binding polysaccharide, could induce nodulation, was able to invade the nodules, and could fix nitrogen, whereas the EPS-negative strain induced nodulation but not invasion or nitrogen fixation. Similar observations were made by GONZALEZ ET AL. (1996) who studied nitrogen fixing symbiosis between *Rhizobium meliloti* and *Medicago sativa* (alfalfa). In contrast, other studies, such as EGGLESTON ET AL. (1996), have shown that EPS are required not only for invasion, but also for nodulation.

EPS have also been demonstrated to protect plant pathogens against plant defense mechanisms as well as environmental factors, and are often required for successful infection. KIRALY ET AL. (1997) indicated that a number of plant pathogenic bacteria, including *Pseudomonas syringae, Erwinia amylovora, Xanthomonas campestris*, and *P. fluorescens*, have a high degree of sensitivity to reactive oxygen species. Using two chemical systems to produce reactive oxygen species (O_2^- and H_2O_2), they have demonstrated with *P. syringae* pv. *phaseolicola* that the associated disease symptoms in host plants were also suppressed by the reactive oxygen species. Bacteria became more sensitive to the reactive species when the EPS around the cells were removed by washing and centrifuging the cells. However, there was no change in the sensitivity to the killing action of the reactive oxygen species when EPS negative mutant cells were treated with this procedure. MCWILLIAMS ET AL. (1995) pointed out that many plant-pathogenic bacteria need EPS for successful infection. As an example, they discussed the bacterial wilt pathogen, *Pseudomonas solanacearum*, in which mutations that prevent EPS production reduced the ability to wilt plants. On the other hand, it is also possible that EPS producing bacteria may protect plants against pathogens such as *P. solanacearum*. FURUYA AND MATSUYAMA (1992) pretreated roots of tomato seedlings with a suspension of antibiotic producing strains of *Pseudomonas glumae*, which suppressed the severity of wilt incited by challenge inoculation with *P. solancearum*. They found that pre-treatment of tomato roots with a non-antibiotic producing strain or heat-killed cells of *P. glumae* also showed protection, suggesting that some mechanisms other than antibiotic productivity were involved in the suppression of the disease. When they treated roots of tomato seedlings in subsequent experiments with lipopolysaccharide or EPS obtained from *P. glumae* before inoculation with *P. solanacearum*, they found a remarkable disease suppression, especially by application of the EPS solution.

4.7
Effect of Environmental Conditions

It has been shown by a large number of studies that the type and amount of EPS produced by microorganisms are influenced by environmental conditions. For instance, WANG AND MCNEIL (1995) and LEE ET AL. (1997) have shown that the types and amounts of EPS were influenced by the medium pH. Other studies (e.g., MARTINEZ-CHECA ET AL. 1996; LEE ET AL. 1997; TAVERNEIER ET AL. 1997) showed the effect of growth medium composition on EPS type and quantity. Growth stage (BHOSLE ET AL. 1995; TAVERNEIER ET AL. 1997), specific growth rate (EVANS ET AL. 1994), and the type of attachment surface in the case of sessile cells (BEECH ET AL. 1991) also seem to affect EPS production. In many instances though, the challenge remains to demonstrate that altered EPS production, quantitatively or qualitatively, serves as an adaptive strategy by microbes to endure environmental fluctuations.

WOLFAARDT ET AL. (1995) discussed the role of EPS in the accumulation and subsequent mineralization of chlorinated aromatic hydrocarbons. In subsequent studies these authors (WOLFAARDT ET AL. 1998) found that the EPS composition of biofilms cultivated on aromatic hydrocarbons differed substantially from those which were cultivated on labile nutrients. A study by OTSUJI ET AL. (1994) showed that when challenged by changes in the growth medium, such as a high aromatic hydrocarbon (2,4-dichlorophenoxyacetic acid) concentration, cells of *Polianthes tuberosa* responded by increasing EPS production from 1.4 to 4.1 g l^{-1}. It thus seems possible that the production of different types and quantities of EPS are strategies employed by microbes to enable effective utilization of substrates which may not be available as nutrients in the absence of these adaptations.

5
EPS and the Macro-Environment

5.1
Stabilization of the Environment

Most of the discussion thus far has concentrated on the microenvironment. However, microbial EPS can also have a significant impact on the macro-environment. Part of soil organic matter is composed of EPS. It has been shown that soil organic matter contributes to soil aggregate stability, water retention potential, and improved water infiltration (ROBERSON ET AL. 1995). Through the production of EPS which contribute to the soil organic matter, bacteria therefore assist in the establishment of growth conditions in the soil which are more favorable to microbial and higher life forms. Similarly, microbial EPS stabilize aquatic sediments (DECHO AND LOPEZ 1993; UNDERWOOD ET AL. 1995). DADE ET AL. (1990) listed a number of workers who studied the role of microorganisms in determining the mechanical properties of sediments. In their studies, DADE ET AL. (1990) conducted a series of experiments to correlate adhesive bacterial exopolymer concentration, measured in terms of uronic acid components, with boundary shear stress required to initiate fine-sand transport in marine environments. They inferred that because permanent adhesion of marine periphytic bacteria to sedimentary surfaces involves acidic EPS, microbial exopolymers produced during bacterial adhesion to sedimentary surfaces also lead to increased grain-to-grain adhesion. Their results showed that growth of *Alteromonas atlantica* in fine sands indeed resulted in increased amounts of acidic EPS, which in turn correlated with sediment resistance to shear stress. A mathematical expression was used to calculate the efficiency of EPS in grain-to-grain adhesion (DADE ET AL. 1990), and it was shown that the particle adhesive force due to the presence of uronic acids in nanomolar amounts per gram of sediment exceeded the submerged weight of fine quartz particles by an order of magnitude. Even more impressive was their estimation that the particle-specific adhesive force was seven orders of magnitude greater than particle-specific uronic acid weight. As a second potentially important mechanism of biological control of sediment erosion they suggested that the presence of long-chain sugar exopolymers released into the area immediately adjacent to the sediment-water interface may result in thickened viscous-dominated regions and subsequently, drag reduction. YALLOP ET AL. (1994) also demonstrated that the stability of sediments was greatly increased by the presence of a microbial mat. They found that cohesive sediment stabilization was primarily facilitated by the secretion of EPS while both EPS binding and network formation by cyanobacteria were responsible for increased sediment stability within non-cohesive sediments. Areas with visible biofilms were more resistant to sediment erosion than adjacent areas without obvious biofilm development.

5.2
Microbial-Mineral Interactions

Mineral dissolution at water-rock interfaces is an important process which controls the mineral and solution composition of natural waters and soils. WELCH AND VANDEVIVERE (1994) described the possible roles of EPS in mineral dissolution. They showed that polymers composed primarily of neutral sugars, starch, cellulose, gum xanthan, and poly-sucrose had no detectable effect on mineral dissolution. Even though the hydroxyl groups of the sugar monomers may, through the formation of hydrogen bonds with oxygen, weakly adsorb to mineral surfaces, they do not remove elements such as Al and Si when they desorb. Alginate, another simple polysaccharide which has acid functional groups, inhibited dissolution. When fresh microbial EPS were tested, WELCH AND VANDEVIVERE (1994) found increased dissolution rates and attributed these increases to two mechanisms: 1) an increase in the solubility of the mineral by soluble organic ligands; and 2) the formation of metal-organic complexes at the mineral surface, thereby weakening metal-oxygen bonds to the bulk material.

It is also possible that surface chemistry and charge can be altered by adsorbed EPS. Ligands with a charge similar to the EPS complex will be repelled, while low molecular weight organic ligands may preferentially react with the EPS complex and not with the mineral surface. Dissolution may be inhibited when EPS adsorb to the mineral surface to form a diffusion barrier. CHEN ET AL. (1995) suggested that acidic polysaccharides may sorb to soil surfaces to produce an organic coating that may bind dissolved trace metals, and BOULT ET AL. (1997) proposed that iron immobilization from acid mine drainage-contaminated streams by streambed biofilms is likely due to the attraction and physical trapping of colloid iron hydroxide by the EPS matrix of the biofilms. EPS have also been shown to play a role in the desorption of metals from sand and may play a role in the mobilization of trace metals in groundwater. CHEN ET AL. (1995) studied the EPS from 13 bacterial isolates for their effects on Pb adsorption on aquifer sand. They demonstrated that all the EPS used in their study showed an ability to release absorbed Cd and Pb; in some cases, reduction of metal adsorption of more than 90% was achieved. Using column experiments to determine the distribution for Cd through sand in the presence and absence of EPS, they further found that EPS can be used to enhance metal mobility in porous media. To be effective in the mobilization of organic or inorganic molecules through a porous medium, a carrier must be able to bind the molecules while having a relatively high mobility in the medium. The EPS produced by at least one of the isolates studied by CHEN ET AL. (1995) displayed the ability to complex Cd, had a relatively low sorption to sand surfaces, and released Cd and Pb from the sand. JENKINS ET AL. (1994) showed that suspensions containing methanotrophic cells, or soluble EPS produced by the methanotrophic cells, enhanced the transport of aromatic hydrocarbons and metals in aquifer material. These authors used the term "facilitated transport," which they defined as rates of contaminant transport greater than those calculated on the basis of hydrodynamic flow of groundwater and sorption of the contaminant to the porous medium.

Lithothrophic bacteria are able to obtain energy through the oxidation of minerals. It has been suggested that this process is dependent on close interaction between bacteria and the mineral. LAWRENCE ET AL. (1997) discussed colonization and weathering of sulfide minerals by *Thiobacillus ferrooxidans*. The oxidation of sulfide minerals by *T. ferrooxidans* can occur via either direct or indirect mechanisms. For indirect oxidation, which occurs through the chemical action of ferric iron produced during bacterial growth, surface colonization is not required. However, the direct mechanisms require that the cells be in close association with the mineral surface because the enzymes involved are not released outside the cell. The role of EPS in establishment of such close interactions has not been established. CHAKRABARTI AND BANERJEE (1991) and SOUTHAM AND BEVERIDGE (1993) suggested that ionic and hydrophobic interactions are involved. It is therefore likely that charged and hydrophobic regions of EPS play a role in bacterial-mineral interactions, and that EPS-mediated attachment of cells to minerals is of similar importance as EPS-mediated attachment of cells to metals during corrosion. Studies on the role of SRB in metal corrosion have shown that EPS produced by SRB facilitate irreversible cell attachment and the binding of metal ions, which in turn contribute to the formation of metal concentration cells and galvanic coupling (LITTLE ET AL. 1996; ZINKEVICH ET AL. 1996). In the case of colonization and corrosion of copper, EPS may also be involved in the detoxification of copper ions. It was shown that the marine bacterium *Oceanospirillum* produces large amounts of EPS when grown on copper which allowed non-copper tolerant bacteria to colonize the copper surface (LITTLE ET AL. 1996).

5.3
Controls on Flow

WU ET AL. (1997) showed that the accumulation of microbial biomass and EPS in porous media greatly influenced water flow. They discussed the importance of this phenomenon to the transport of chemicals in subsurface environments, as well as on groundwater recharge, water treatment, land disposal of waste water and septic tank effluent, enhanced oil recovery, and in situ bioremediation. They further stressed the need for a quantitative understanding of the relationship between the amount of microbial biomass and reduction in hydraulic conductivity (k) to enable predictions of the fate and transport of biologically reactive contaminants in soil and groundwater systems. WU ET AL. (1997) measured a 1.5-order of magnitude reduction of k-values over a three week period after the application of dextrose-nutrient solutions to stimulate microbial growth. Because of a lack of data on EPS, these authors analyzed only the correlation between attached microbial cells and the reduction in k. They obtained a regression equation relating attached microbial biomass and the logarithmic ratio of k, but found no correlation between suspended and attached biomass. The high correlation coefficient between attached cells and increased k-values can be interpreted as indicative of a high correlation between the presence of EPS and cell attachment. To illustrate the significance of EPS, the authors referred to work by others describing the role of "sticky exopolysaccharides" which often act as cementing agents between microbial cells and particle surfaces. In another study, DENNIS AND TURNER (1998) evaluated the potential use of EPS-producing bacteria to create low-permeability waste containment barriers by measuring the hydraulic conductivity of a compacted silty sand. In the absence of bacteria, values of k ranged from 10^{-5} to 10^{-6} cm s^{-1}, but when bacteria and a nutrient solution were added to the soil columns, permeability was significantly reduced (final k of 10^{-8} cm s^{-1}).

Modification of transport parameters through porous materials is also of relevance to industries such as oil exploration, and the use of bacteria to plug selectively water injection wells has received much attention (e.g., SHAW ET AL. 1985; CUSACK ET AL. 1987, 1992; CUNNINGHAM ET AL. 1997). In essence, this technique uses bacteria to plug selectively oil depleted zones within a reservoir to divert displacing fluids into oil-rich zones. LAPPAN AND FOGLER (1994) described the underlying mechanism for this technique as the extracellular breakdown of nutrients and the subsequent production of EPS. These authors further discussed the selective supply of the nutrients to bacteria as a strategy to manipulate EPS production. For instance, by varying the ratio of yeast extract to sucrose in the growth medium in which they cultivated *Leuconostoc mesenteroides*, they could control the amount of EPS produced per cell.

Another way in which the production of EPS alters the macro-environment is by their rheological properties. DE PHILIPPIS ET AL. (1991) studied *Cyanospira capsulata*, a heterocystous cyanobacterium possessing a thick polysaccharide capsule, in batch cultures and found an increased viscosity in the medium due to the continuous release of a soluble polysaccharide into the culture medium. Cyanobacteria are known to modulate the production of EPS as a response to growth conditions. Eubacteria have also been shown to produce a highly viscous, anionic EPS (see, e.g., DASINGER ET AL. 1994; MEADE ET AL. 1994).

5.4
Role of EPS in the Bio-Accumulation of Contaminants

Naturally-occurring microbial communities affect the fate of organic contaminants which enter the environment in various ways. By utilizing these contaminants as sources of carbon and energy (LAPPIN ET AL. 1985; ASCON-CABRERA

AND LEBEAULT 1993), microbes contribute to the mineralization of contaminants to CO_2 and H_2O. However, refractory contaminants are often not completely mineralized which may result in accumulation of these molecules in microbial biomass and, when predation of bacteria occurs, subsequent transfer through food chains in natural systems. Although the relative importance of contaminant-accumulation by bacteria, and the mechanisms responsible are not well understood, it is possible that organic contaminants are accumulated in EPS by the same mechanisms proposed for the accumulation of naturally-occurring nutrients. Our knowledge of these mechanisms is based primarily on observations made in a relatively small number of laboratory studies which addressed the role of microbial cultures in the bioaccumulation of organic molecules. Earlier work on this topic was done by LESHNIOWSKY ET AL. (1970) who reported the ability of two isolates from a floc-forming bacterial community to concentrate and accumulate the pesticide aldrin from solution, GELLER (1979) who studied sorption of atrazine by three bacterial isolates, and GRIMES AND MORRISON (1975) who studied accumulation of chlorinated hydrocarbons by isolated cell lines. The latter study suggested that cell surface area played an important role in microbial sorption of pesticides. More recently, WOLFAARDT ET AL. (1994A) described the accumulation of the herbicide diclofop methyl by bacterial EPS and subsequent transfer during predation by protista.

Microbial EPS may also affect the fate of inorganic contaminants in the environment. MITTELMAN AND GEESEY (1985) demonstrated that EPS which accumulated copper could serve as a route of entry for this metal into environmental food chains. The incorporation of the accumulated metals in food chains, and transfer to higher organisms by biomagnification, have also been discussed by PATRICK AND LOUTIT (1976).

6
Conclusions

In many instances one can predict with a significant level of confidence what benefits bacteria derive from the modification of their micro- or macro-environments through the production of EPS. For example, the protection cells enjoy in biofilms, the advantages of close associations in flocs and microcolonies, and the establishment of a coupled oxidation-reduction process which provides cells with energy. However, other things may be more difficult to explain, such as the role of bacterial EPS in facilitated transport. What is the basis for a relatively low sorption of certain types of EPS to surfaces, while other types of EPS show a strong tendency to adhere, or what benefits, if any, do naturally-occurring bacteria derive in slowing down the dissolution of minerals? Also, why do cells alter their EPS composition during different growth conditions? Many of these factors are of significance to microbial competitiveness and, more generally, environmental processes. Development of a fundamental understanding of the physiological and ecological reasons as to why microorganisms produce EPS with such a great variety of chemical, physical and functional properties is key to this problem.

References

Allison DG, Sutherland IW (1987) The role of exopolysaccharides in adhesion of freshwater bacteria. J Gen Microbiol 133:1319–1327

Ascon-Cabrera M, Lebeault JM (1993) Selection of xenobiotic-degrading microorganisms in a biphasic aqueous-organic system. Appl Environ Microbiol 59:1717–1724

Becker K (1996) Exopolysaccharide production and attachment strength of bacteria and diatoms on substrates with different surface tensions. Microb Ecol 32:23–33

Beech IB, Cheung CWS (1995) Interactions of exopolymers produced by sulphate-reducing bacteria with metal ions. Int Biodet Biodegr 35:59–72

Beech IB, Gaylarde CC, Smith JJ, Geesey GG (1991) Extracellular polysaccharides from *Desulfovibrio desulfuricans* and *Pseudomonas fluorescens* in the presence of mild and stainless steel. Appl Microbiol Biotechnol 35:65–71

Bellemann P, Bereswill S, Berger S, Geider KV (1994) Visualization of capsule formation by *Erwinia amylovora* and assays to determine amylovoran synthesis. Int J Biol Macromol 16:290–296

Bellin CA, Rao PSC (1993) Impact of bacterial biomass on contaminant sorption and transport in a subsurface soil. Appl Environ Microbiol 59:1813–1820

Bhosle NB, Sawant SS, Garg A, Wagh AB (1995) Isolation and partial chemical analysis of exopolysaccharides from the marine fouling diatom *Navicula subinflata*. Bot Mar 38:103–110

Birmingham JJ, Hughes NP, Treloar R (1995) Diffusion and binding measurements within oral biofilms using fluorescence photobleaching recovery methods. Phil Trans Soc Lond B Biol Sci 350:325–343

Bohrer MP, Patterson GD, Carrol PJ (1984) Hindered diffusion of dextran and ficoll in microporous membranes. Macromolecules 11:70–73

Boult S, Johnson N, Curtis C (1997) Recognition of a biofilm at the sediment-water interface of an acid mine drainage-contaminated stream, and its role in controlling iron flux. Hydrol Processes 11:391–399

Chakrabarti BK, Banerjee PC (1991) Surface hydrophobicity of acidophilic heterotrophic bacterial cells in relation to their adhesion on minerals. Can J Microbiol 37:692–696

Characklis WG, Marshall KC (1990) Biofilms, a basis for an interdisciplinary approach. In: Characklis WG, Marshall KC (eds) Biofilms. Wiley, New York, pp 3–15

Chen J-H, Lion LW, Ghiorse WC, Shuler ML (1995) Mobilization of adsorbed cadmium and lead in aquifer material by bacterial extracellular polymers. Wat Res 29:421–430

Chenu C, Roberson EB (1996) Diffusion of glucose in microbial extracellular polysaccharide as affected by water potential. Soil Biol Biochem 28:877–884

Christensen BE, Characklis WG (1990) Physical and chemical properties of biofilms. In: Characklis WG, Marshall KC (eds) Biofilms. Wiley, New York, pp 93–130

Chumak NE, Stepnaya OA, Chermenskaya TS, Kulaev IS, Nesmeyanova MA (1995) Specific features of secretion of bacteriolytic enzymes and polysaccharide by bacteria of the family Pseudomonadaceae. Microbiology 64:44–50

Costerton JW (1984) The formation of biocide-resistant biofilms in industrial, natural and medical systems. Dev Ind Microbiol 25:363–372

Costerton JW, Marrie TJ, Cheng K-J (1981) The bacterial glycogalyx in nature and disease. Annu Rev Microbiol 35:299–324

Cowan MM, Warren TM, Fletcher M (1991) Mixed species colonization of solid surfaces in laboratory biofilms. Biofouling 3:23–34

Cunningham A, Warwood B, Sturman P, Horrigan K, James G, Costerton JW, Hiebert R (1997) Biofilm processes in porous media-practical applications. In: Amy PS, Haldeman DL (eds) The microbiology of the terrestrial deep subsurface. CRC Press, Boca Raton, Florida, USA, pp 325–344

Cusack F, Brown DR, Costerton JW, Clementz DM (1987) Field and laboratory studies of microbial/ fines plugging of water injection wells: mechanism, diagnosis and removal. J Petroleum Sci Engin 1:39–50

Cusack F, Singh S, McCarthy C, Grieco J, De-Rocco M, Nguyen D, Lappin-Scott H, Costerton JW (1992) Enhanced oil recovery: three-dimensional sandpack simulation of ultramicrobacteria resuscitation in reservoir formation. J Gen Microbiol 138:647–655

Dade WB, Davis JD, Nichols PD, Nowell ARM, Thistle D, Trexler MR, White DC (1990) Effects of bacterial exopolymer adhesion on the entrainment of sand. Geomicrobiol J 8:1–16

Dasinger BL, McArthur HAI, Lengen JP, Smogowicz AA, Miller JW, O'Neill JJ, Horton D, Costa JB (1994) Composition and rheological properties of extracellular polysaccharide 105-4 produced by *Pseudomonas* sp. strain ATCC 53923. Appl Environ Microbiol 60:1364–1366

De Beer D, O'Flaharty V, Thaveesri J, Lensand P, Verstraete W (1996) Distribution of extracellular polysaccharides and flotation of anaerobic sludge. Appl Microbiol Biotechnol 46:197–201

De Beer D, Stoodley P, Lewandowski Z (1997) Measurement of local diffusion coefficients in biofilms by microinjection and confocal microscopy. Biotechnol Bioeng 53:151–158

Decho AW (1990) Microbial exopolymer secretions in ocean environments: their role(s) in food webs and marine processes. Oceanogr Mar Biol Ann Rev 28:73–153

Decho AW, Lopez GR (1993) Exopolymer microenvironments of microbial flora: multiple and interactive effects on trophic relationships. Limnol Oceanogr 38:1633–1645

Deighton MA, Borland R, Capstick JA (1996) Virulence of *Staphylococcus epidermidis* in a mouse model: significance of extracellular slime. Epidemiol Infect 117:267–280

Dennis ML, Turner JP (1998) Hydraulic conductivity of compacted soil treated with biofilm. J Geotech Geoenviron Engin 124:120–127

De Philippis R, Sili C, Tassinato G, Vincenzini M, Materassi R (1991) Effects of growth conditions on exopolysaccharide production by *Cyanospira capsulata*. Bioresource Technol 38:101–104

Dudman WF (1977) The role of surface polysaccharides in natural environments. In: Sutherland IW (ed) Surface carbohydrates of the prokaryotic cell. Academic Press, London, pp 357–414

Eggleston G, Huber MC, Liang RT, Karr AL, Emerich DW (1996) *Bradyrhizobium japonicum* mutants deficient in exo- and capsular polysaccharides cause delayed infection and nodule initiation. Mol Plant-Microbe Interactions 9:419–423

Evans E, Brown MRW, Gilbert P (1994) Iron chelator, exopolysaccharide and protease production in *Staphylococcus epidermidis*: a comparative study of the effects of specific growth rate in biofilm and planktonic culture. Microbiology 140:153–157

Freeman C, Lock MA (1995) The biofilm polysaccharide matrix: a buffer against changing organic substrate supply. Limnol Oceanog 40:273–278

Freeman C, Chapman PJ, Gilman K, Lock MA, Reynolds B, Wheater HS (1995) Ion exchange mechanisms and the entrapment of nutrients by river biofilms. Hydrobiology 297:61–65

Frølund B, Griebe T, Nielsen PH (1995) Enzymatic activity in the activated-sludge floc matrix. Appl Microbiol Biotechnol 43:755–761

Furuya N, Matsuyama N (1992) Biological control of the bacterial wilt of tomato with antibiotic-producing strains of *Pseudomonas glumae*. J Faculty Agric Kyushu Univ 37: 159–171

Gancel F, Novel G (1994) Exopolysaccharide production by *Streptococcus salivarius* ssp. *thermophilus* cultures: 1 conditions of production. J Dairy Sci 77:685–688

Geesey GG, Costerton JW (1986) The microphysiology of consortia within adherent bacterial populations. In: Megusar F, Gantar M (eds) Perspectives in microbial ecology. Mladinski Knjiga, Ljubljana, Yugoslavia, pp 238–242

Geesey GG, White DC (1990) Determination of bacterial growth and activity at solid-liquid interfaces. Annu Rev Microbiol 44:579–602

Geesey GG, Richardson WT, Yoemans HG, Irvin RT, Costerton JW (1977) Microscopic examination of natural bacterial populations from an alpine stream. Can J Microbiol 23:1733–1736

Geller A (1979) Sorption and desorption of atrazine by three bacterial species isolated from aquatic systems. Arch Environ Contam Toxicol 8:713–720

Gilbert P, Brown MRW (1995) Mechanisms of the protection of bacterial biofilms from anti-microbial agents. In: Lappin-Scott HM, Costerton JW (eds) Microbial biofilms. Cambridge University Press, Cambridge, pp 118–130

Gonzalez JE, York GM, Walker GC (1996) *Rhizobium meliloti* exopolysaccharides: synthesis and symbiotic function. Gene 179:141–146

Grimes DJ, Morrison SM (1975) Bacterial bioconcentration of chlorinated hydrocarbon insecticides from aqueous systems. Microb Ecol 2:43–59

Grinberg TA, Pirog TP, Pinchuk GE, Buklova VN, Malashenko R (1994) Change of composition and properties of exopolysaccharides synthesized by *Acinetobacter* sp. in the course of batch cultivation. Mikrobiologiya 63:1015–1019

Handley PS, Hesketh LM, Moumena RA (1991) Charged and hydrophobic groups are localized in the short and long tuft fibrils on *Streptococcus sanguis* strains. Biofouling 4:105–111

Hansen JB, Doubet RS, Ram J (1984) Alginase enzyme production by *Bacillus circulans*. Appl Environ Microbiol 47:704–709

Harder W, Dijkhuizen L (1983) Physiological responses to nutrient limitation. Annu Rev Microbiol 37:1–23

Hartel PG, Alexander M (1986) Role of extracellular polysaccharide production and clays in the desiccation tolerance of cowpea bradyrhizobia. Soil Sci Soc Am J 50:1193–1198

Hotter GS, Scott DB (1997) The requirement for exopolysaccharide precedes the requirements for flavolan binding polysaccharide in nodulation of *Leucaena leucocephala* by *Rhizobium loti*. Arch Microbiol 167:182–186

James GA, Beaudette L, Costerton JW (1995) Interspecies bacterial interactions in biofilms. J Ind Microbiol 15:257–262

Jenkins MB, Chen J-H, Kadner DJ, Lion LW (1994) Methanotrophic bacteria and facilitated transport of pollutants in aquifer material. Appl Environ Microbiol 60:3491–3498

Kieft TL, Soroker E, Firestone MK (1987) Microbial biomass response to a rapid increase in water potential when dry soil is wetted. Soil Biol Biochem 19:119–126

Killham H, Firestone MK (1984) Salt stress control of intracellular solutes in streptomycetes indigenous to saline soils. Appl Environ Microbiol 47:301–306

Kiraly Z, Elzahaby HM, Klement Z (1997) Role of extracellular polysaccharide (EPS) slime of plant pathogenic bacteria in protecting cells to reactive oxygen species. J Phytopathol 145:59–68

Kolenbrander PE (1989) Surface recognition among oral bacteria: multigeneric coaggregations and their mediators. Crit Rev Microbiol 17:137–159

Kolenbrander PE, Andersen RN (1988) Intergeneric rosettes: sequestered surface recognition among human periodontal bacteria. Appl Environ Microbiol 54:1046–1050

Komiyama K, Habbick BF, Gibbons RJ (1987) Interbacterial adhesion between *Pseudomonas aeruginosa* and indigenous oral bacteria isolated from patients with cystic fibrosis. Can J Microbiol 33:27–32

Korber DR, Lawrence JR, Lappin-Scott HM, Costerton JW (1995) The formation of microcolonies and functional consortia within biofilms. In: Lappin-Scott HM, Costerton JW (eds) Microbial biofilms. Cambridge University Press, Cambridge, pp 15–45

Lamed R, Setter E, Bayer EA (1983) Characterization of a cellulose-binding, cellulase-containing complex in *Clostridium thermocellum*. J Bacteriol 156:828–836

Lappan RE, Fogler HS (1994) *Leuconostoc mesenteroides* growth kinetics with application to bacterial profile modification. Biotechnol Bioengin 43:865–873

Lappin HM, Greaves MP, Slater JH (1985) Degradation of the herbicide Mecoprop [2-(2-methyl-4-chlorophenoxy) propionic acid] by a synergistic microbial community. Appl Environ Microbiol 49:429–433

Lawrence JR, Wolfaardt GM, Korber DR (1994) Monitoring diffusion in biofilm matrices using scanning confocal laser microscopy. Appl Environ Microbiol 60:1166–1173

Lawrence JR, Korber DR, Wolfaardt GM, Caldwell DE (1995) Behavioural strategies of surface colonizing bacteria. Adv Microb Ecol 14:1–75

Lawrence JR, Kwong YTJ, Swerhone GDW (1997) Colonization and weathering of natural sulfide mineral assemblages by *Thiobacillus ferrooxidans*. Can J Microbiol 43:178–188

Lebaron P, Bauda P, Lett MC, Duval-Iflah Y, Simonet P, Jacq E, Frank N, Roux B, Baleux B, Faurie G, Hubert JC, Normand P, Prieur D, Schmitt S, Block JC (1997) Recombinant plasmid mobilization between *E. coli* strains in seven sterile microcosms. Can J Microbiol 43:534–540

Lee IY, Seo WT, Kim GJ, Kim MK, Ahn SG, Kwon GS, Park YH (1997) Optimization of fermentation conditions for production of exopolysaccharide by *Bacillus polymyxa*. Bioproc Engin 16:71–75

Leshniowsky WO, Dugan PR, Pfister RM, Frea JI, Randles CI (1970) Aldrin: removal from lake water by flocculent bacteria. Science 169:993–994

Little B, Wagner P, Angell P, White D (1996) Correlation between localized anodic areas and *Oceanospirillum* biofilms on copper. Int Biodet Biodegr 37:159–162

Loaec M, Olier R, Guezennec J (1997) Uptake of lead, cadmium and zinc by a novel bacterial exopolysaccharide. Wat Res 31:1171–1179

Lorenz MG, Wackernagel W (1994) Bacterial gene transfer by natural genetic transformation in the environment. Microbiol Rev 58:563–602

Lorenz MG, Aardema BW, Wackernagel W (1988) Highly efficient genetic transformation of *Bacillus subtilis* attached to sand grains. J Gen Microbiol 134:107–112

MacLeod FA, Guiot SR, Costerton JW (1995) Electron microscopic examination of the extracellular polymeric substances in anaerobic granular biofilms. World J Microbiol Biotechnol 11:481–485

Marasas CN (1993) Plasmids in *Pseudomonas syringae* pv. *morsprunorum* race 1. Phytophylactica 25:53–58

Marshall KC (ed) 1984. Microbial adhesion and aggregation. Springer, Berlin Heidelberg New York

Marshall KC (1985) Mechanism of bacterial adhesion at solid-water interfaces. In: Savage DC, Fletcher M (eds) Bacterial adhesion. Mechanisms and physiological significance. Plenum, New York, pp 133–161

Marshall PA, Loeb GI, Cowan MM, Fletcher M (1989) Response of microbial adhesives and biofilm matrix polymers to chemical treatments as determined by interference reflection microscopy and light section microscopy. Appl Environ Microbiol 55:2827–2831

Martinez-Checa F, Calvo C, Caba MA, Ferrer MR, Bejar V, Quesada E (1996). Effect of growth conditions on viscosity and emulsifying activity of V2-7 biopolymer from *Volcaniella eurihalina*. Microbiologia 12:55–60

McEldowney S, Fletcher M (1987) Adhesion of bacteria from mixed cell suspension to solid surfaces. Arch Microbiol 148:57–62

McEldowney S, Fletcher M (1988) Effect of pH, temperature, and growth conditions in the adhesion of a gliding bacterium and three nongliding bacteria to polystyrene. Microb Ecol 16:183–195

McWilliams R, Chapman M, Kowalczuk KM, Hersberger D, Sun JH, Kao CC (1995) Complementation analyses of *Pseudomonas solanacearum* extracellular polysaccharide mutants and identification of genes responsive to EpsR. Mol Plant-Microbe Interactions 8:837–844

Meade MJ, Tanenbaum SW, Nakas JP (1994) Optimization of novel extracellular polysaccharide production by an *Enterobacter* sp. on wood hydrolysates. Appl Environ Microbiol 60:1367–1369

Measures JC (1975) Role of amino acids in osmoregulation of nonhalophilic bacteria. Nature 257:398–400

Mittelman MW, Geesey GG (1985) Copper-binding characteristics of exopolymers from a freshwater-sediment bacterium. Appl Environ Microbiol 49:846–851

Møller S, Sternberg C, Anderson JB, Christensen BB, Ramos JL, Givskov M, Molin S (1998) In situ gene expression in mixed-culture biofilms: evidence of metabolic interactions between community members. Appl Environ Microbiol 64:721–732

Morris EJ, McBride BC (1984) Adherence of *Streptococcus sanguis* to saliva-coated hydroxyapatite: evidence for two binding sites. Infect Immun 43:656–663

Muralidharan V, Rinker KD, Hirsh IS, Bouwer EJ, Kelly RM (1997) Hydrogen transfer between methanogens and fermentative heterotrophs in hyperthermophilic cocultures. Biotechnol Bioeng 56:268–278

Nichols WW (1989) Susceptibility of biofilms to toxic compounds. In: Characklis WG, Wilderer PA (eds) Structure and function of biofilms. Wiley, New York, pp 321–331

Nichols WW, Evans MJ, Slack MPE, Walmsley HL (1989) The penetration of antibiotics into aggregates of mucoid and non-mucoid *Pseudomonas aeruginosa*. J Gen Microbiol 135: 1291–1303

Nielsen PH, Frølund B, Keiding K (1996) Changes in the composition of extracellular polymeric substances in activated sludge during anaerobic storage. Appl Microb Biotechnol 44:823–830

Obayashi AW, Gaudy AF (1973) Aerobic digestion of extracellular microbial polysaccharides. J Water Pollut Control Fed 45:1584–1594

Ophir T, Gutnick DL (1994) A role for exopolysaccharides in the protection of microorganisms from desiccation. Appl Environ Microbiol 60:740–745

Otsuji K, Honda Y, Sugimura Y, Takei A (1994) Production of polysaccharides in liquid cultures of *Polianthes tuberosa* cells. Biotechnol Lett 16:943–948

Pasquier C, Marty N, Dournes JL, Chabanon G, Pipy B (1997) Implication of neutral polysaccharides associated to alginate in inhibition of murine macrophage response to *Pseudomonas aeruginosa*. FEMS Microbiol Lett 147:195–202

Patel JJ, Gerson T (1974) Formation and utilization of carbon reserves by *Rhizobium*. Arch Microbiol 101:211–220

Patrick FM, Loutit M (1976) Passage of metals in influents, through bacteria to higher organisms. Wat Res 10:333–335

Pirog TP (1997) Role of *Acinetobacter* sp. exopolysaccharides in protection against heavy metal ions. Microbiol 66:284–288

Pirog TP, Grinberg TA, Malashenko Y (1997a) Protective functions of exopolysaccharides produced by an *Acinetobacter* sp. Microbiol 66:279–283

Pirog TP, Grinberg TA, Malashenko Y, Yu R (1997b) Isolation of microorganism-producers of enzymes degrading exopolysaccharides from *Acinetobacter* sp. Mikrobiologiya 33:550–555

Quintero EJ, Weiner RM (1995) Evidence for the adhesive function of the exopolysaccharide of *Hyphomonas* strain MHS-3 in its attachment to surfaces. Appl Environ Microbiol 61:1897–1903

Roberson EB, Firestone MK (1992) Relationship between desiccation and exopolysaccharide production in a soil *Pseudomonas* sp. Appl Environ Microbiol 58:1284–1291

Roberson EB, Sarig S, Shennan C, Firestone MK (1995) Nutritional management of microbial polysaccharide production and aggregation in an agricultural soil. Soil Sci Soc Am J 58: 1587–1594

Rudd T, Sterritt RM, Lester JN (1983) Mass balance of heavy metal uptake by encapsulated cultures of *Klebsiella aerogenes*. Microb Ecol 9:261–272

Samrakandi MM, Roques C, Michel G (1997) Influence of trophic conditions on exopolysaccharide production: bacterial biofilm susceptibility to chlorine and monochloramine. Can J Microbiol 43:751–758

Shaw JC, Bramhill B, Wardlaw NC, Costerton JW (1985) Bacterial fouling in a model core system. Appl Environ Microbiol 49:693–701

Siebel MA, Characklis WG (1991) Observations of binary population biofilms. Biotech Bioeng 37:778–789

Snyder RA (1990) Chemoattraction of a bacterivorous ciliate to bacteria surface compounds. Hydrobiology 215:205–213

Southam G, Beveridge TJ (1993) Examination of lipopolysaccharide (O-antigen) populations of *Thiobacillus ferrooxidans* from two mine tailings. Appl Environ Microbiol 59:1283–1288

Stewart PS, Murga R, Srinivasan R, de Beer D (1995) Biofilm structural heterogeneity visualized by three microscopic methods. Wat Res 29:2006–2009

Stoodley P, de Beer D, Lewandowski Z (1994) Liquid flow in biofilm systems. Appl Environ Microbiol 60:2711–2716

Stotsky G (1985) Mechanisms of adhesion to clays, with reference to soil systems. In: Savage DC, Fletcher M (eds) Bacterial adhesion. Mechanisms and physiological significance. Plenum, New York, pp 195–253

Sutherland IW (1984) Microbial exopolymers – their role in microbial adhesion in aqueous systems. CRC Crit Rev Microbiol 10:173–201

Tago Y, Aida K (1977) Exocellular mucopolysaccharide closely related to bacterial floc formation. Appl Environ Microbiol 34:308–314

Taverneier P, Portais JC, Saucedo JEN, Courtois J, Courtois B, Barbotin JN (1997) Exopolysaccharide and poly-beta-hydroxybutyrate coproduction in two *Rhizobium meliloti* strains. Appl Environ Microbiol 63:21–26

Trevors JT, Barkay T, Bourquin W (1987) Gene transfer among bacteria in soils and aquatic environments: a review. Can J Microbiol 33:191–198

Uhlinger DJ, White DC (1983) Relationship between physiological status and formation of extracellular polysaccharide glycocalyx in *Pseudomonas atlantica*. Appl Environ Microbiol 45:64–70

Underwood GJC, Paterson DM, Parkes RJ (1995) The measurement of microbial carbohydrate exopolymers from intertidal sediments. Limnol Oceanogr 40:1243–1253

Vandevivere P, Kirchman DL (1993) Attachment stimulates exopolysaccharide synthesis by a bacterium. Appl Environ Microbiol 59:3280–3286

Veiga MC, Jain MK, Wu W-M, Hollingsworth RI, Zeikus JG (1997) Composition and role of extracellular polymers in methanogenic granules. Appl Environ Microbiol 63:403–407

Wang Y, McNeil B (1995) pH effects on exopolysaccharide and oxalic acid production in cultures of *Sclerotium glucanicum*. Enzyme Microb Technol 17:124–130

Weiner R, Langille S, Quintero EJ (1995) Structure, function and immunochemistry of bacterial exopolysaccharides. J Ind Microbiol 15:339–346

Welch SA, Vandevivere P (1994) Effect of microbial and other naturally occurring polymers on mineral dissolution. Geomicrobiol J 12:227–238

Wolfaardt GM, Lawrence JR, Headley JV, Robarts RD, Caldwell DE (1994a) Microbial exopolymers provide a mechanism for bioaccumulation of contaminants. Microb Ecol 27:279–291

Wolfaardt GM, Lawrence JR, Robarts RD, Caldwell DE (1994b) The role of interactions, sessile growth and nutrient amendment on the degradative efficiency of a bacterial consortium. Can J Microbiol 40:331–340

Wolfaardt GM, Lawrence JR, Robarts RD, Caldwell SE, Caldwell DE (1994c) Multicellular organization in a degradative biofilm community. Appl Environ Microbiol 60:434–446

Wolfaardt GM, Lawrence JR, Robarts RD, Caldwell DE (1995) Bioaccumulation of the herbicide diclofop in extracellular polymers and its utilization by a biofilm community during starvation. Appl Environ Microbiol 61:152–158

Wolfaardt GM, Lawrence JR, Robarts RD, Caldwell DE (1998) In situ characterization of biofilm exopolymers involved in the accumulation of chlorinated organics. Microb Ecol 35:213–223

Wu JQ, Gui SX, Stahl P, Zhang RD (1997) Experimental study on the reduction of soil hydraulic conductivity by enhanced biomass growth. Soil Sci 162:741–748

Yallop ML, de Winder B, Paterson DM, Stal LJ (1994) Comparative structure, primary production and biogenic stabilization of cohesive and non-cohesive marine sediments inhabited by microphytobenthos. Estuarine Coastal and Shelf Science 39:565–582

You CB, Lin M, Fang XJ, Song W (1995) Attachments of *Alcaligenes* to rice roots. Soil Biol Biochem 27:463–466

Zinkevich V, Bogdarina I, Kang H, Hill MAV, Tapper R, Beech IB (1996) Characterization of exopolymers produced by different isolates of marine sulfate-reducing bacteria. Int Biodet Biodegr 37:163–172

CHAPTER 11

Polysaccharases in Biofilms – Sources – Action – Consequences!

Ian W. Sutherland

Institute of Cell and Molecular Biology, Edinburgh University, Mayfield Road, Edinburgh EH9 3JH, UK, *E-mail: I.W.Sutherland@ed.ac.uk*

Keywords. Phage, Polysaccharase, Polysaccharide lyase

1
Polysaccharases in Biofilms

As such a large proportion of the structure of biofilms is composed of polysaccharides secreted by the constituent micro-organisms, the presence of enzymes (polysaccharases) acting on these polymers will inevitably have a very marked effect on the structure and on the integrity of the biofilm. It is also possible that glycosidases capable of cleaving exposed terminal monosaccharide residues may modify both polysaccharides and glycoproteins present in biofilms. Enzymes will derive from a variety of sources and may well differ considerably in their effects. It has also to be remembered that in multi-species biofilms, the collective action of several different enzymes may result in the degradation or alteration of polysaccharides which are resistant to discrete enzymes. Thus, the growth of different enzyme-secreting species in close proximity with intimate cell:cell contact may permit synergistic action of the enzyme mixture within the confines of the biofilm. The effects of polysaccharases may well be moderated if a mixture of polysaccharides is present and removal of one polymer leaves others with similar physical properties intact. The presence of other chemical compounds absorbed to the polysaccharides may also have a moderating influence on enzyme action. Thus simultaneous release of biosurfactants could well affect enzyme activity either positively (enhancing degradation) or negatively (inhibiting destruction of the substrate).

2
Nature of Polysaccharases

Microbial exopolysaccharides may be degraded either by polysaccharide hydrolases or by polysaccharide lyases. The latter cleave the linkage between a neutral monosaccharide and the C_4 of a uronic acid with simultaneous introduction of a double bond at the C_4 and C_5 of the uronic acid. Both types of enzyme are commonly found to degrade exopolysaccharides as well as eukaryotic polymers (SUTHERLAND 1995). Either type of enzyme may be endo- or exo-acting leading to rapid or slow breakdown of the polymer chain respectively. As far as homopolysaccharides are concerned, a wider range of enzymes is usually available. As has been pointed out by MISHRA AND ROBBINS (1995), β-D-glucanases with a variety of specificities can be used to elucidate the structures of this group of polysaccharides. Thus, even for the linear 1,3-β-D-glucan curdlan, four different enzymes will degrade the polysaccharide. Two exo-glucanases release D-glucose and laminarabiose respectively. Two endoglucanases differ in their mode of action. One attacks randomly, releasing glucose, laminarabiose, laminaratriose and other oligosaccharides in the β-1,3-linked series. The other yields the pentasaccharide laminarapentaose. Similarly, degradation of cellulose or xylans requires the concerted action of various endo- and exo-β-D-glucanases (GILBERT AND HAZLEWOOD 1993).

3
Sources of Polysaccharases

Polysaccharases may derive from three major sources – endogenously from polysaccharide-synthesising micro-organisms, exogenously from a wide range of other micro-organisms and, finally, from bacteriophage particles or phage-induced bacterial lysates.

3.1
The Endogenous Production of Polysaccharases

Many of the recent molecular studies on exopolysaccharide synthesis have revealed that glycanases or polysaccharide lyases are gene products associated with the biosynthesis of the exopolysaccharide itself. Such enzymes have now been found in a wide range of exopolysaccharide-synthesising bacterial species as can be seen in Table 1. In most examples which have been adequately studied, the genes for the enzymes formed part of the operonic systems or gene cassettes regulating synthesis, polymerisation and excretion of the exopolysaccharide (e.g. GLUCKSMAN ET AL. 1993; MATTHYSSE ET AL. 1995; SUTHERLAND AND

KENNEDY 1996). It is not yet clear whether these enzymes are always expressed. If the enzymes are present in the exopolysaccharide-synthesising bacteria, they may only be released slowly as cells lyse. This may nevertheless cause a rapid reduction in both the polymer mass and the solution viscosity. It is also possible that some release occurs during cell division when it is known that for some types of bacteria, considerable turnover of wall or matrix material may occur. This was reported by XUN ET AL. (1990) for *Methanosarcina mazei*. This bacterium released a "disaggregatase" at certain stages of growth, resulting in cell separation from the normal cellular aggregates. The enzyme responsible was characterised as a polysaccharase acting on the trisaccharide repeat unit of the matrix polymer 'methanochondroitin'. The structure of this polysaccharide is known to be … -(1 → 4)-D-GlcpA-(1 → 3)-D-GalpNAc-(1 → 3)-D-GalpNAc …, but the exact site of enzyme cleavage was not identified.

In the case of alginates from *Pseudomonas fluorescens* or *Pseudomonas putida*, the mass fell approximately 50% per 24 h at 30 °C in shaken planktonic cultures due to the action of endogenous poly-D-mannuronate-specific lyases after cell growth had ceased (CONTI ET AL. 1994). Similar action would probably be expected to occur when these bacteria were present in biofilms. Within the confines of the biofilms, enzyme action could also be greater in its effect. The result of the alginase (alginate lyase) action is to reduce greatly the aqueous solution viscosity and the binding properties of the exopolysaccharide. In alginate-synthesising strains of *Pseudomonas aeruginosa*, the *algL* gene similarly controls an alginate lyase capable of limited action on alginate from this bacterium (SCHILLER ET AL. 1993).

It appears to be extremely rare for a bacterial species to be both capable of extensive breakdown of an extracellular polysaccharide synthesised by the same species and able to use the degradation products as a carbon and energy source, i.e. to have the ability to use the exopolysaccharide as an effective extracellular carbon and energy reserve. This mechanism has only been clearly demonstrated in *Cellulomonas flavigena* (VOEPEL AND BULLER 1990), and is probably not of major significance in the destruction of exopolysaccharide present in biofilms. However, in recent years a significant number of different bacterial species have been found to possess genes for polysaccharide-degrading enzymes closely associated with the genes responsible or synthesis and export of their exopolysaccharides (Table 1). It has been assumed that these enzymes must play some role in either molecular mass determination or export of the exopolysaccharide. Such enzymes would not normally come into direct contact with their substrates as the enzymes, at least in Gram-negative bacteria, of have normally been found within the periplasm. However, if there is extensive cell lysis either through autolysis on cell ageing, through lytic phage action or as a result of protozoal grazing, release of such enzymes could then lead to action on the exopolysaccharides. This was demonstrated for the two *Pseudomonas* spp. by CONTI ET AL. (1994) under the physiological conditions tested.

Table 1. Glycan depolymerases associated with exopolysaccharide synthesis

Bacterial species	Enzyme	Mol Mass	Reference
Acetobacter xylinum	Endoglucanase (CM-Cellulase)	35.6 kDa	STANDAL ET AL. (1994)
Agrobacterium tumefaciens	Endoglucanase (CM-Cellulase)	–	MATTHYSSE ET AL. (1995)
Azotobacter chroococcum	Alginate lyase (polymannuronate lyase)	43 kDa	PECINA AND PANEQUE (1994)
Azotobacter vinelandii	Alginate lyase (polymannuronate lyase)	–	KENNEDY ET AL. (1992)
Cellulomonas flavigena	1,3-β-D-Glucanase		VOEPEL AND BULLER(1990)
Escherichia coli K5	N-acetyl-heparosan lyase	70–89kDa	LEGOUX ET AL. (1996)
Pseudomonas aeruginosa	Alginate lyase (polymannuronate lyase)	39kDa	SCHILLER ET AL. (1993)
Pseudomonas fluorescens	Alginate lyase (polymannuronate lyase)	–	Hughes K (unpublished)
Pseudomonas fluorescens (marginalis)	Galactoglucanase	–	OSMAN ET AL. (1993)
Pseudomonas mendocina	Alginate lyase (polymannuronate lyase)	–	SENGHA ET AL. (1989)
Pseudomonas putida	Alginate lyase (polymannuronate lyase)	–	CONTI ET AL. (1994)
Rhizobium meliloti	Succinoglycan depolymerase		GLUCKSMAN ET AL. (1993)
Rhizobium meliloti	Polyglucuronic acid hydrolase		COURTOIS ET AL. (unpublished)
Sphingomonas spp.	Gellan lyase		SUTHERLAND AND KENNEDY (1996)
Streptococcus equi	Hyaluronidase		

3.2
The Exogenous Production of Polysaccharases

Polysaccharases may be produced to enable micro-organisms to degrade polysaccharide substrates and utilise their component monomers as carbon and energy sources. The enzymes are extracellular and thus have ready access to the polysaccharides in biofilms that cement cells to each other and to the substratum. The enzymes themselves usually represent a complex mixture of activities, capable of degrading the polysaccharide substrate to oligosaccharides and then acting on these smaller fragments to reduce them to smaller oligosaccharides or to monosaccharides which can then enter the uptake and utilisation systems of the enzyme-producing bacteria. The enzyme systems needed for degradation of cellulose provide a good example of the complexity. Typically, such mixtures comprise several examples of each of three types of enzyme:– β-1,4-endoglucanases cleaving internal β-1,4-glucosidic bonds; cellobiohydrolases releasing cellobiose from the non-reducing terminus of the cellulose molecule; and β-D-glucosidase degrading the cellulose so formed (GILBERT AND HAZLEWOOD 1993). The source of exopolysaccharide-degrading enzyme mixtures may either be single microbial species or more commonly mixed cultures. It has proved relatively rare to find a pure bacterial culture capable of degrading an exopolysaccharide following normal enrichment procedures. One example was the succinoglycan-degrading bacterial species *Cytophaga arvensicola* (OYAIZU ET AL. 1982). A pure bacterial culture secreting one or more xanthan-degrading enzyme has also been found by CADMUS ET AL. (1982). More commonly, as was demonstrated in several laboratories, a complex mixture of polysaccharide-degrading micro-organisms has been obtained. Mixed microbial cultures were found to synthesise a complex enzyme mixture which achieved extensive degradation (e.g. HOU ET AL. 1986). Generally speaking, such mixed cultures produce a range of both polysaccharases and glycosidases. Because of the close proximity of diverse groups of microbial species, these enzyme activities and those acting directly on the ordered state of the polysaccharide can be expected to have the greatest effect on biofilms. Many of the other enzymes produced in mixed culture only act on the initial degradation products, causing further breakdown of the oligosaccharides to monosaccharides, disaccharides or trisaccharides, fragments utilisable by the microbial cells.

3.3
Bacteriophage

Many of the bacteria which are surrounded by polysaccharide slime or capsules are also hosts for virulent bacteriophage. As demonstrated by STIRM AND FREUND-MOELBERT (1971) and EICHHOLTZ ET AL. (1975), the phages themselves vary greatly in their structures. Preliminary results for phages acting on several Gram-negative bacterial strains initially isolated from natural and industrial biofilms show a similar picture (Skillman LC, unpublished results). Some had small icosahedral heads and small, barely visible tails. Others conformed to Bradley type "C" with long slightly curved tails resembling the coliphages T1 and T5. If the phage is to have access to cell surface receptors, the virus must remove the polymer that occludes the bacterial cell envelope. To do so, many of the phages possess associated polysaccharide-degrading enzymes. As with polysaccharide-degrading enzymes in general, these enzymes may act either hydrolytically, cleaving specific linkages in the polysaccharides, or they may be polysaccharide lyases acting by eliminative cleavage at a monosaccharide-uronic acid linkage and introducing an unsaturated bond at the C_4 and C_5 of the non-reducing uronic acid terminal. In these bacteriophage particles which have been carefully examined by electron microscopy, the polysaccharide depolymerase activity has often been shown to be associated with small spikes attached to the viral base-plate. Several studies have separated the spikes and demonstrated enzyme activity. In addition to the phage-associated enzyme, further activity is found in the soluble proteins in the cell lysates following viral maturation.

Other bacteriophage possess or induce lysozyme-like or other muralytic activity. The enzymes may also be closely associated with the viral particles but in many phages it is probably predominantly found as a free form following cell lysis. In the *Escherichia coli* phage 29 system, the lysozyme is found in addition to an exopolysaccharide depolymerase and was not associated with the virus particles (EICHHOLTZ ET AL. 1975). If lysozymes or other muralytic enzymes normally associated with cell division are released during that process, they too may affect bacteria within a biofilm through destruction of the peptidoglycan in the cell wall. Subsequent release of intracellular polysaccharases or polysaccharide lyases would then affect biofilm exopolysaccharides. However, few bacteria have peptidoglycan sufficiently exposed and accessible to the enzyme for it to be destroyed in neighbouring heterologous bacteria. Similarly, phages destroying their lipopolysaccharide (LPS) receptors are well known and in one example the tail spike protein has been fully characterised and functions in both adhesion to the host cell surface and receptor destruction (BAXA ET AL. 1996; STEINBACHER ET AL. 1997). The enzyme in this phage is an endorhamnosidase that cleaves the 1,3-α-O-glycosidic bond between L-rhamnose and D-galactose, yielding octasaccharide fragments (two repeat units) from the LPS side-chains as the major product. The phage-induced enzymes acting on exopolysaccharide substrates have revealed a very wide range of specificities as indicated by the examples in Table 2. Although several "endoglucosidases", "endogalactosidases" or "endorhamnosidases" are listed, each is distinct in its specificity. It is perhaps of interest that many of the phage-induced enzymes appear to preferentially attack either (1 → 3) α- or (1 → 3) β-bonds. Further, the residue targeted is very often adjacent to an anionic residue such as glucuronic acid, which may be part of the main chain or attached as a side-chain. The enzymes are usually highly specific. It is rare for one such enzyme to act on more than one polysaccharide substrate. Recent work in our laboratory using a phage originally isolated on a Pseudomonad host isolated from a freshwater biofilm, has shown that this phage is unusual. It can also form plaques on *Enterobacter cloacae* NCTC 5920 (Lopez MJ and Sutherland IW, unpublished results). On each host, the plaques are surrounded by very large haloes. Assays using viscometry and measurement of the reducing sugar released clearly demonstrated enzyme action on the poly-

Table 2. Examples of phage-associated polysaccharide depolymerases

Bacterial host/Polysaccharide	Enzyme action	Reference
	Endoglycanases	
Klebsiella type K43	*Endogalactosidase*	AEREBOE ET AL. (1993)
Klebsiella type K51	*Endogalaclosidase*	CHAKRABORTY (1985)
Klebsiella sp.	*Endogalaclosidase*	YUREWICZ ET AL. (1971)
Klebsiella serotype K25	*Endoglucosidase*	NIEMANN ET AL. (1977)
Klebsiella serotype N60	*Endoglucosidase*	DiFABIO ET AL. (1984)
Klebsiella serotype K63	*Endoglucosidase*	DUTTON AND MERRIFIELD (1982)
Escherichia coli serotype 29	*Endoglucosidase*	FEHMEL ET AL. (1975)
Klebsiella penumoniae SKI	*Endoglucanase (Endoglucosidase)*	CESCUTTI AND PAOLETTI (1994)
Klebsiella serotype K6	*Endoglucanase*	ELSÄSSER-BEILE AND STIRM (1981)
Escherichia coli Type 8	*Endomannosidase*	PREHM AND JANN (1976)
Klebsiella type serotype K30	*Endomannosidase*	RAVENSCROFT ET AL. (1988)
Escherichia coli Type 44	*Endo-N-acetyl-β-D-galactosaminidase*	DUTTON EL AL. (1988)
Acetobacter methanolicus	*Endorhamnosidase*	GRIMMEKE ET AL. (1994A,B)
Pseudonionas syringae pv. morsprunorum	*Endorhamnosidase*	SMITH ET AL. (1994)
	Neuraminidases/Sialidases	
Escherichia coli	*neuraminidase*	HALLENBECK ET AL. (1987)
Escherichia coli	*Endo-N-acetyltieuraninidase*	KWIATKOWSKI ET AL. (1983)
Escherichia coli K1 Phage E	*endosialidase*	LONG ET AL. (1995)
Escherichia coli	*KDO[a]-KDO glycanase*	NIMMICH (1997)
Escherichia coli	*KDO[a]-KDO glycanase*	ALTMANN ET AL. (1986, 1987)
	Lyases	
Azotobacter vinelandii	*Alginate lyase*	DAVIDSON ET AL. (1977)
Streptococcus	*Hyaluronidase*	NIEMANN ET AL. (1976)
Klebsiella serotype K5	*Polysaccharide lyase*	VAN DAM ET AL. (1985); HÄNFLING ET AL. (1996)
Rhizobium spp.	*Polysaccharide lyase*	MCNEIL ET AL. (1986)

[a] KDO = ketodeoxyoctonic acid.

saccharides from both bacteria. Reducing sugar release was greatest from the polysaccharide of the original host. This result appears to be exceptional and it is more likely that phage particles or associated enzymes released from one bacterial species within a biofilm will not normally have any direct affect on other closely associated bacteria. The effect will be indirect through destruction of the bacterial cells or their associated exopolysaccharides.

Any bacterial cells on or near the surface of a biofilm are also potentially subject to high velocity attack by *Bdellovibrio*. It has recently been shown that the presence of capsules and slime provide the bacterial cell with no protection against penetration by the *Bdellovibrio* (KOVAL AND BAYER 1997). The *Bdellovibrio* passed directly through the capsule which then reformed behind it as the endoparasite penetrated the cell envelope. As can be seen from Fig. 1, the action of these endoparasites did not in itself destroy any exopolysaccharide present. The exopolysaccharide capsule reformed after removal of the shear force exerted by the *Bdellovibrio* in its passage into the periplasm of the host bacterium. The action of the parasite will certainly have the potential to cause considerable perturbation of the biofilm surface and there might be some limited release of periplasmic enzymes during entry. The action of the *Bdellovibrio* could, however, *indirectly* lead to enzymic degradation of the extracellular polysaccharide. During the process of invading the cell, or following release of mature *Bdellovibrio* from the bdelloplast, localised release of enzymes could occur. If

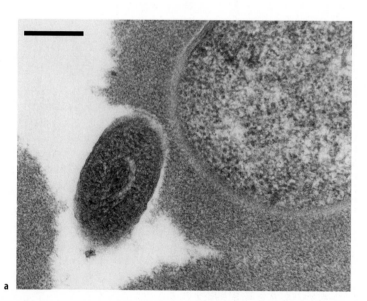

Fig. 1a–c. *Bdellovibrio bacteriovorus* attack on capsulate *Escherichia coli* cells. The capsules are apparently not degraded by the high velocity attack of the smaller bacteria. c Bdelloplast seen after *Bdellovibrio* bacteriovirus attack on capsulate *Escherichia coli* cells. The capsules have regained their original structure after entry of the *Bdellovibrio* into the host and are apparently not degraded (KOVAL AND BAYER 1977)

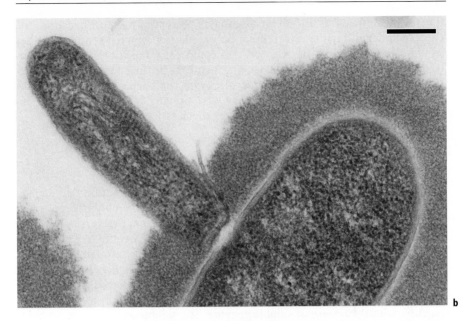

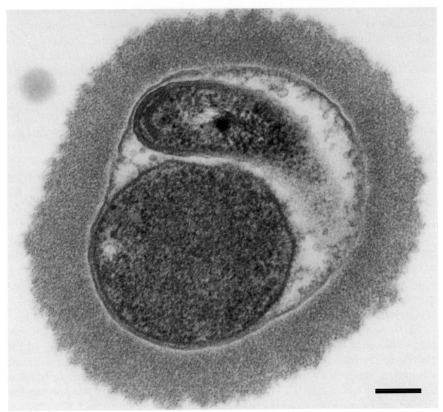

Fig. 1b, c (continued)

there were sufficient intracellular glycanases present in the host bacterial cell, capable of degrading the extracellular polysaccharide which it synthesises, they would almost certainly be released and could act on exopolysaccharide surrounding the infected bacterial host. The disturbance caused by *Bdellovibrio* action may also cause local dissolution of exopolysaccharide or may expose the polymers to enzymic attack when they were previously 'protected' by the intimate association with other polymers or surfaces.

4
Effects of Enzymes on Biofilms

The action of any polysaccharase or similar enzyme on a biofilm will depend very much on the nature of the biofilm. In particular, if the polymer molecules present act synergistically to provide increased adhesion, destruction of a key component may have a much more significant effect. The effect of the phage enzyme attached to the viral base plate is to carve a path through polysaccharide capsules as demonstrated by BAYER ET AL. (1979) for *E. coli* K29. Thus if isolated cells within the outer reaches of a biofilm are so infected, they will provide a focus for the generation of further phage particles on lysis from within together with the release of soluble enzyme. The soluble enzymes are probably likely to have a much more drastic effect on the biofilm than the bacteriophage. Studies by HUGHES (1997) have demonstrated that, for relatively young monoculture biofilms with *Enterobacter agglomerans* that synthesise an exopolysaccharide with very low water solubility, addition of phage enzyme to the biofilm caused very rapid release of cells from the substratum. The addition of even a small number of phage under conditions where they can attach and infect their hosts could thus be expected to have a very drastic effect. This would be seen both at the site of infection on the biofilm itself and on any biofilms of similar composition downstream, when both phage particles and soluble enzyme are released in significant quantities.

In a study of attached growth of alginate-producing *P. aeruginosa*, BOYD AND CHAKRABARTY (1994) observed that increased expression of alginate lyase caused alginate degradation and increased cell detachment. As the enzyme probably has limited activity against its substrate, it probably caused cell detachment through reduced mass and viscosity. This would correspond to the observations of CONTI ET AL. (1994) in other alginate-synthesising *Pseudomonas* spp. The effect of the enzyme would however depend very greatly on the nature of the biofilm. JASS ET AL. (1995) found that while a strain of *P. fluorescens* yielded confluent biofilms over the surface tested, *P. putida* formed distinct microcolonies. Thus, enzyme action on the former would have much greater effect in terms of increasing the available surface for new biofilm formation.

Much depends on the nature of the enzyme action. Endo-acting polysaccharases will cause a rapid reduction in the degree of polymerisation (DP) of the polymer substrate. The glycan chains may however remain associated with one another or with other macromolecules to some extent after a single bond is cleaved. Only when extensive bond breakage has occurred will the polysaccharide be totally dispersed and the integrity of the biofilm destroyed. The effect of enzyme action is not always detrimental to the biofilm. The properties of the polysaccharide may actually be enhanced, as is the case with the levans and dextrans associated with dental plaque. Although fructans may be found associated with dental plaque and fructanases are also produced, the action of the enzymes probably does not greatly affect the structure of the biofilms. The fructans are relatively labile and rapidly degraded (BURNE ET AL. 1987). *Streptococcus mutans* secretes an extracellular endo-dextranase that may play a more significant role in modifying the physicochemical properties of mutans (LAWMAN AND BLEIWEIS 1991). The enzymes attack α-1,6 linkages. The net effect is thus to increase the proportion of α-1,3 linkages and enhance the hydrophobicity of the polysaccharide, rendering it increasingly insoluble in water.

Exo-acting enzymes only cause a slow reduction in DP and a slow gradual release of oligosaccharide products. Unlike the well documented degradation of cellulose in which there is combined attack of endo- and exo-1,4-β-D-glucanases (GILKES ET AL. 1991), most exopolysaccharides subjected to enzymic degradation are subjected to the action of a single endo-acting enzyme. Perhaps in the biofilm community the potential exists for combined action of endo- and exo-acting glycanases emanating from different micro-organisms within the biofilm. There could thus be synergy between several enzymes leading to total degradation of the polysaccharide substrate. This might also explain the rarity of discovery of extensive polysaccharide-degrading activity associated with single bacterial species as opposed to the finding of mixed cultures on enrichment with exopolysaccharides as sole carbon and energy source (e.g. HOU ET AL. 1986). In synergistic action of this type, endo-acting enzymes will produce more free chain ends for the subsequent attack by exo-enzymes. The net result will be a complex mixture of monosaccharide and oligosaccharide products. It should also be remembered that glycosidases present in a biofilm might alter the epitopes of carbohydrate-containing molecules. Although action may be slow and incomplete, their presence may lead to alteration of the exposed surfaces over a prolonged period. Such enzymes could cleave exposed monosaccharide termini from either exopolysaccharides or glycoproteins. Removal of the terminal sugars would also lead to possible changes in physical properties. Thus, if exposure to a β-D-glucuronidase removed terminal uronic acid residues, ion binding and interchain associations could be affected. This has been studied for a *Xanthomonas campestris* mutant partially defective in the terminal β-linked D-mannosyl residues of the xanthan-type polysaccharide (TAIT AND SUTHERLAND 1989). Further removal of the exposed D-glucuronosyl residues enhanced the solution viscosity.

Other enzymes derive from bacteriophage infection and provide a means of enabling the phage to reach receptors on the cell surface. The enzymes are either integral components of the viral particle (usually in the form of small spikes attached to the base-plate) or are soluble proteins released during cell lysis on phage maturation. Most of the enzymes are endo-acting glycanases causing very rapid, random cleavage of the exopolysaccharide chains with resultant loss of viscosity and integrity.

Many microbial exopolysaccharides are acylated, the commonest substituents being ketal-linked pyruvate or ester-linked acetyl groups. Removal of the acyl groups, especially acetate (SUTHERLAND 1997) may greatly affect the properties of the polysaccharides. There may be enhancement of gelation if acetyl groups are removed! Esterase activity capable of removing these substituents has not yet been demonstrated although deacetylation of pectin has recently been reported in a strain of *Erwinia chrysanthemi* (SHEVCHIK AND HUGOU-VIEUX-COTTE-PATTAT 1997). Phage-induced esterases removing acetyl groups from LPS are also known (IWASHITA AND KANEGASAKI 1976), while SHABTAI AND GUTNICK (1985) demonstrated esterase activity against emulsan, the surface active lipopolysaccharide synthesised by *Acinetobacter calcoaceticus* strain RAG-1. Clearly the effect of enzymes of this type on biofilm composition and architecture will depend on their specificity and on the proximity and amount of available substrate. The emulsan esterase appeared to be showing specificity for this unusual polymer but also cleaved nitrophenyl esters. Now that several prokaryotic exopolysaccharides are known to carry sulphate groups, sulphatases similar in action to those acting on glycosaminoglycans (SHAKLEE ET AL. 1985) may also play a role in biofilms if appropriate substrates are available. In persistent biofilms, even enzymes with relatively low activity may influence the nature of the biofilm and act slowly on substrates.

Studies on degradative enzyme activities within biofilms have not been extensively reported. However, FRØLUND ET AL. (1995) investigated enzyme activity within activated sludge and discovered that the activities of different enzymes fluctuated within the bulk sludge. Peptidase was the dominant activity noted. Among the enzymes assayed were aminopeptidases, dehydrogenases, lipases and phosphatases. The carbohydrate-degrading enzymes examined were galactosidases and α-D- and β-D-glucosidases, β-D-glucuronidase and chitinase. Clearly, in such complex mixtures of micro-organisms, a wide and very variable range of enzyme activities can be expected. Similar results can be expected when mixed biofilms are examined.

5
Application of Enzymes to Biofilms

The removal of many types of biofilm encountered in industrial, domestic or medical environments could be aided by enzyme treatment. Equally, combined application of enzymes and antimicrobial agents might be much more effective than use of either type of compound per se. Because of the large number of different types of polysaccharide which might be present and the very high specificity of polysaccharases and polysaccharide lyases, a wide range of different enzymes could be required if they were to be effective. A recent study to determine the efficacy of enzymes in the removal of attached micro-organisms used mixed biofilms containing *Staphylococcus epidermidis*, *P. aeruginosa* and *P. fluorescens* on steel and polypropylene substrata (JOHANSEN ET AL. 1997). Oral bacteria on saliva-coated hydroxyapatite were also examined. Application of a complex mixture of commercial enzymes released the biofilm from the steel and polypropylene surfaces without significant bactericidal activity. The enzyme activities known to be present were those of pectinases, arabinase, cellulase, hemicellulase, β-glucanases and xylanase. With such a complex mixture it remains unclear how many of these activities were needed for biofilm removal. It was also possible that other, unrecognised enzyme activities might have been present. Oxidoreductases in combination with the enzymes yielded bactericidal action as well as biofilm removal. A combination of dextranase and mutanase (α-1,3-D-glucanase) was most effective in releasing bacteria from the simulated plaque coating of hydroxyapatite disks pre-treated with saliva. Each of these two enzymes when used on its own did have some effect, the dextranase being superior to the mutanase in effecting bacterial cell release from the biofilm.

In any environment in which complex mixed biofilms are found, the structure and integrity of the biofilm will undoubtedly depend on different types of macromolecules with polysaccharides and proteins likely to play a dominant role. The use of complex enzyme mixtures in which both polysaccharase and protease activities are present is thus more likely to be successful in practice than the use of single enzymes.

References

Aereboe M, Parolis H, Parolis, LAS (1993) *Klebsiella* K43 capsular polysaccharide:primary structure and depolymerisation by a viral borne endoglycanase. Carbohydr Res 248: 213–223

Altmann F, Kwiatkowski B, Stirm S (1986) A bacteriophage associated glycanase cleaving β-pyranosidic linkages of 3-deoxy-D-amnno-2-octulosonic acid (KDO). Biochem Biophys Res Commun 136:329–335

Altmann F, Maerz L, Stirm S, Unger FM (1987) Two additional phage-associated glycan hydrolases cleaving ketosidic bonds of 3-deoxy-D-manno-octulosonic acid in capsular polysaccharides of *Escherichia coli*. FEBS Lett 221:145–149

Baxa U, Steinbacher S, Miller S, Weintraub A, Huber R, Seckler R (1996) Interactions of phage P22 tails with their cellular receptor, *Salmonella* O-antigen polysaccharide. Biophys J 71:2040–2048

Bayer ME, Thurow H, Bayer MH (1979) Penetration of the polysaccharide capsule of *Escherichia coli* (Bi161/42) by bacteriophage K29. Virol 94:95–118

Boyd A, Chakrabarty AM (1994) Role of alginate lyase in cell detachment of *Pseudomonas aeruginosa*. Appl Environ Microbiol 60:2355–2359

Burne RA, Schilling K, Bowen WH, Yasbin RE (1987) Expression, purification and characterization of an exo-β-D-fructosidase of *Streptococcus mutans*. J Bacteriol 169:4507–4517

Cadmus MC, Jackson LK, Burton KA, Plattner RD, Slodki ME (1982) Biodegradation of xanthan gum by *Bacillus* sp. Appl Environ Microbiol 44:5–11

Cescutti P, Paoletti S (1994) On the specificity of a bacteriophage borne endoglycanase for the native capsular polysaccharide produced by *Klebsiella pneumoniae* SK1 and its derived polymers. Biochem Biophys Res Commun 198:1128–1134

Chakraborty AK (1985) Depolymerization of capsular polysaccharide by glycanase activity of *Klebsiella* bacteriophage 51. Ind J Biochem 22:22–26

Conti E, Flaibani A, O'Regan M, Sutherland IW (1994) Alginate from *Pseudomonas fluorescens* and *Pseudomonas putida*: production and properties. Microbiology 140:1128–1132

Davidson IW, Lawson CJ, Sutherland IW (1977) An alginate lyase from *Azotobacter vinelandii* phage. J Gen Microbiol 98:223–229

DiFabio JL, Dutton GGS, Parolis H (1984) Preparation of a branched heptasaccharide by bacteriophage depolymerization of *Klebsiella* K60 capsular polysaccharide. Carbohydr Res 126:261–269

Dutton GGS, Merrifield EH (1982) Acylated oligosaccharides from *Klebsiella* K63 capsular polysaccharide: depolymerization by partial hydrolysis by bacteriophage-borne enzymes. Carbohydr Res 103:107–128

Dutton GGS, Lam Z, Lim AVS (1988) N-acetyl-β-D-galactosaminidase activity of E. coli phage 44 and the sequencing of E. coli K44 capsular polysaccharide by mass spectrometry. Carbohydr Res 183:123–125

Eichholtz H, Freund-Mölbert E, Stirm S (1975) Escherichia coli capsule bacteriophages. J Virol 15:985–993

Elsässer-Beile U, Stirm S (1981) Substrate specificity of the glycanase activity associated with particles of Klebsiella bacteriophage no 6. Carbohydr Res 88:315–322

Fehmel F, Feige U, Niemann H, Stirm S (1975) Escherichia coli capsule bacteriophages VII. Bacteriophage 29-host capsular polysaccharide interactions. J Virol 16:591–601

Frølund B, Griebe T, Nielsen PH (1995) Enzymatic activity in the activated sludge floc matrix. Appl Microbiol Biotechnol 43:755–761

Gilbert HJ, Hazlewood GP (1993) Bacterial cellulases and xylanases. J Gen Microbiol 139:187–194

Gilkes NR, Claeyssens M, Aebersold R, Henrissat B, Meinke A (1991) Structural and functional relationships in two families of β-1,4-glycanases. Eur J Biochem 202:367–377

Glucksman MA, Reuber TL, Walker GC (1993) Genes needed for the modification, polymerization, export and processing of succinoglycan by Rhizobium meliloti: a model for succinoglycan biosynthesis. J Bacteriol 175:7045–7055

Grimmeke H-D, Knirel YA, Kiesel B, Voges M, Rietschel ET (1994a) Structure of the Acetobacter methanolicus MB 129 capsular polysaccharide and of oligosaccharides resulting from degradation by bacteriophage Acm. Carbohydr Res 259:45–58

Grimmeke H-D, Knirel YA, Shashkov AS, Kiesel B, Lauk W, Voges M (1994b) Structure of the capsular polysaccharide and the O-side chain of the lipolysaccharide from Acetobacter methanolicus. Carbohydr Res 253:277–282

Hallenbeck PC, Vimr ER, Yu F, Bassler B, Troy FA (1987) Purification and properties of a bacteriophage-induced endo-N-acetylneuraminidase. J Biol Chem 262:3553–3561

Hänfling P, Shashkov AS, Jann B, Jann K (1996) Analysis of the enzymatic cleavage (β-elimination) of the capsular K5 polysaccharide of E. coli by the K5-specific coliphage: a re-examination. J Bacteriol 178:4747–4750

Hou CT, Barnabe N, Greaney K (1986) Biodegradation of xanthan by salt-tolerant aerobic micro-organisms. J Ind Microbiol 1:31–37

Hughes KA (1997) Bacterial biofilms and their exopolysaccharides. PhD thesis, Edinburgh University

Iwashita S, Kanegasaki S (1976) Deacetylation reaction catalyzed by Salmonella phage. J Biol Chem 251:5361–5365

Jass J, Costerton JW, Lappin-Scott HM (1995) Assessment of a chemostat-coupled modified Robbins device to study biofilms. J Ind Microbiol 15:283–289

Johansen C, Falholt P, Gram L (1997) Enzymatic removal and disinfection of bacterial biofilms. Appl Environ Microbiol 63:3724–3728

Kennedy L, McDowell K, Sutherland IW (1992) Alginases from Azotobacter species. J Gen Microbiol 138:2465–2471

Koval SF, Bayer ME (1997) Bacterial capsules – no barrier against Bdellovibrio. Microbiol 143:749–753

Kwiatkowski B, Boschek B, Thiele H, Stirm S (1983) Substrate specificity of two bacteriophage associated endo-N-acetylneuraminidases. J Virol 45:367–374

Lawman P, Bleiweis AW (1991) Molecular cloning of the extracellular endodextranase of Streptococcus salivarius. J Bacteriol 173:495–504

Legoux R, Lelong P, Jourde C, Feuillerat C, Capdeville J, Sure V, Ferran E, Kaghad M, Delpech B, Shire D, Ferrara P, Loisin G, Salomé M (1996) N-acetyl-heparosan lyase of E. coli K5: gene:gene cloning and expression. J Bacteriol 178:7260–7264

Long GS, Bryant JM, Taylor PW, Luzio JP (1995) Complete nucleotide sequence of the gene encoding bacteriophage E endosialidase: implications for K1E endosialidase structure and function. Biochem J 309:43–55

Matthysse AG, White S, Lightfoot R (1995) Genes required for cellulose synthesis in Agrobacterium tumefaciens. J Bacteriol 177:1069–1075

McNeil M, Darvill J, Darvill AG, Albersheim P, van Veen R, Hooykas P, Schilepoort R, Dell A (1986) The discernible, structural features of the acidic polysaccharides secreted by different *Rhizobium* species are the same. Carbohydr Res 146:307–326

Mishra C, Robbins PW (1995) Specific beta-glucanases as tools for polysaccharide structure determination. Glycobiology 5:645–654

Niemann H, Birch-Andersen A, Kjems E, Mansa B, Stirm S (1976) Streptococcal bacteriophage 12/12-borne hyaluronidase and its characterization as a lyase. Acta Pathol Microbiol Scand 84:145–153

Niemann H, Kwiatkowski B, Westphal, U, Stirm S (1977) *Klebsiella* serotype-13 capsular polysaccharide: primary structure and depolymerization by a bacteriophage-borne glycanase. J Bacteriol 130:366–374

Nimmich W (1997) Degradation studies on *Escherichia coli* capsular polysaccharides by bacteriophages. FEMS Microbiol Lett 153:105–110

Osman SF, Fett WF, Irwin PL, Bailey DG, Parris N, O'Connor JV (1993) Isolation and characterization of an exopolysaccharide depolymerase from *Pseudomonas marginalis* HTO41B. Curr Microbiol 26:299–304

Oyaizu H, Komagata K, Amemura A, Harada T (1982) A succinoglycan-decomposing bacterium, *Cytophaga arvensicola* sp. nov. J Gen Appl Microbiol 28:369–388

Pecina A, Paneque A (1994) Detection of alginate lyase by activity staining after SDS PAGE and subsequent renaturation. Anal Biochem 217:124–127

Prehm P, Jann K (1976) Enzymatic action of coliphage 8 and its possible role in infection. J Virol 19:940–949

Ravenscroft N, Jackson GE, Joao H, Stephen AM (1988) Spectroscopic analysis of oligosaccharides produced by bacteriophage-borne enzyme action on *Klebsiella* K36 polysaccharide. S Afr J Chem 41:42

Schiller NL, Monday SR, Boyd C, Keen NT, Ohman DE (1993) Characterization of the *Pseudomonas* alginate lyase gene (algL): cloning, sequencing and expression in *E. coli*. J Bacteriol 175:780–789

Sengha SS, Anderson, AJ, Hacking AJ, Dawes E. (1989) The production of alginate by *Pseudomonas mendocina* in batch and continuous culture. J Gen Microbiol 135:795–804

Shabtai Y, Gutnick DL (1985) Exocellular esterase and emulsan release from the cell surface of *Acinetobacter calcoaceticus*. J Bacteriol 161:1176–1181

Shaklee PN, Glaser, JH, Conrad HE (1985) A sulfatase specific for glucuronic acid 2-sulfate residues in glycosaminoglycans. J Biol Chem 260:9146–9149

Shevchik VE, Hugouvieux-Cotte-Pattat N (1997) Identification of a bacterial pectin acetyl esterase in *Erwinia chrysanthemi* 3937. Mol Microbiol 24:1285–1301

Smith ARW, Zamze SE, Hignett RC (1994) Morphology and hydrolytic activity of A7, a typing phage of *Pseudomonas syringae* pv. *morsprunorum*. Microbiol 140:905–913

Standal R, Iversen T, Coucheron DH, Fjaervik E, Blatny JM, Valla S (1994) A new gene required for cellulose production and a gene encoding cellulolytic activity in *Acetobacter xylinum* are colocalised with the bcs operon. J Bacteriol 176:665–672

Steinbacher S, Mille S, Baxa U, Budisa N, Weintraub A, Seckler R, Huber R (1997) Phage P22 tailspike protein – crystal structure of the head-binding domain at 2.3 angstrom, fully refined structure of the endorhamnosidase at 1.56 angstrom resolution, and the molecular basis of O-antigen recognition and cleavage. J Mol Biol 267:865–880

Stirm S, Freund-Moelbert E (1971) *Escherichia coli* capsule bacteriophages II. Morphology. J Virol 8:330–342

Sutherland IW (1995) Polysaccharide lyases. FEMS Microbiol Rev 16:323–347

Sutherland IW (1997) Microbial exopolysaccharides – structural subtleties and their consequences. Pure Appl Chem 69:1911–1917

Sutherland IW, Kennedy L (1996) Polysaccharide lyases from gellan-producing *Sphingomonas* spp. Microbiol 142:867–872

Tait MI, Sutherland IW (1989) Synthesis and properties of a mutant type of xanthan. J Appl Bacteriol 66:457–460 van Dam JEG, Halbeek H, Kamerling JP, Vliegenhart JFG, Snippe H, Jansze M, Willers JMN (1985) A bacteriophage-associated lyase acting on *Klebsiella* serotype K5 capsular polysaccharide. Carbohydr Res 142:338–343

Voepel KC, Buller CS (1990) Formation of an extracellular energy reserve by *Cellulomonas flavigena* strain KU. J Ind Microbiol 5:131–138

Xun L, Mah RA, Boone DR (1990) Appl Environ Microbiol 56:3693–3698

Yamazaki M, Thorne L, Mikolajczak MJ, Armentrout RW, Pollock TJ (1996) Linkage of genes essential for synthesis of a polysaccharide capsule in *Sphingomonas* Strain S88. J Bacteriol 178:2676–X2687

Yurewicz EC, Ghalambor MA, Duckworth DH, Heath EC (1971) Catalytic and molecular properties of a phage induced capsular polysaccharide depolymerase. J Biol Chem 246: 5607–5616

Extracellular Enzymes Within Microbial Biofilms and the Role of the Extracellular Polymer Matrix

Monica Hoffman · Alan W. Decho

Department of Environmental Health Sciences, School of Public Health, University of South Carolina, Columbia, SC. 29208, USA, *E-mail: adecho@sph.sc.edu*

Keywords. Extracellular enzymes, Extracellular polymers, Biofilms

1
Introduction: Importance of Extracellular Enzymes to Bacterial Cells and Organic Matter Processing

Bacteria lie at the base of the food web, and occupy an important position in the recycling of organic matter in a range of natural and artificial systems. For example, organic matter turnover in ocean systems is largely mediated by bacterial processes (POMEROY 1984). An important rate-limiting step in understanding bacterial activities and the more general transformation of organic matter is the extracellular hydrolysis of large molecular-weight organic matter by bacteria (HOPPE 1991).

Passive transport through bacterial membranes is restricted to very small and chemically simple compounds. It was estimated that bacteria can directly utilize just 5–10% of the dissolved organic matter, primarily amino acids and small peptides (ALBERTSON ET AL. 1990). For larger molecules, cells require the involvement of specialized enzymes, called permeases. Under these conditions, one might expect a minor contribution of bacterial growth to the total microplankton biomass, which is not the case. This indicates that bacteria are more likely to make use of the high molecular weight DOM fraction by extracellular hydrolyses to smaller fragments. The general paradigm is that these microbial enzymes hydrolyze large organic macromolecules to smaller oligomers or monomers, which can then be directly taken up and utilized by cells (AZAM AND CHO 1987). In recent years several studies have demonstrated that heterotrophic bacteria produce hydrolytic enzymes that interact with substrates beyond the boundary of the cell wall, and the activity of these enzymes is a rate-limiting factor in the aquatic turn-over of particulate organic matter. Also, there is a direct relation between the presence of hydrolytic enzymes and uptake of their low molecular weight digestion products (HOPPE ET AL. 1988).

Hydrolysing enzymes that degrade substrates outside the cell fall into one of two general categories: 1) ectoenzymes (i.e., any enzyme that is secreted and actively crosses the cytoplasmic membrane, but remains associated with the cell; 2) extracellular enzymes (i.e., occur as free forms dissolved in water and/or are adsorbed to surfaces other than those of the producer cells) (CHRÓST 1991). Extracellular enzymes may be actively excreted to the environment, may start as ectoenzymes that are later washed-out, or may be intracellular enzymes that are released after lysis of cells. In the case of some enzymes, it was noted that Gram-negative bacteria tend to release less enzymes in soluble form, compared to Gram-positive bacteria, when all other factors are similar. This difference is currently explained by the more complex structure of Gram-negative cell wall (CEMBELLA ET AL. 1984).

For ectoenzymes, hydrolysis and uptake are often closely coupled owing to the direct association of enzymes with cells (HOLLIBAUGH AND AZAM 1983). A logistical difficulty facing microbial cells in the use of extracellular enzymes, however, is that both enzymes and hydrolysis products may be easily lost to the surrounding medium through diffusion. In order for extracellular hydrolysis processes to be energy-efficient for cells, the hydrolysis products generated by enzyme activities must be quickly taken up by cells before they are lost to the surrounding environment (KARP-BOSS ET AL. 1996). Since most heterotrophic bacteria are osmotrophic (i.e., nutrient uptake occurs from solution) and of relatively small size (0.5–1.5 μm), nutrient accessibility is governed largely by molecular diffusion (AZAM AND AMMERMAN 1984). This implies that both enzymes and their hydrolysis products must remain relatively close to cells, in order that diffusive transfer to cells of hydrolysis products can operate with a net gain. Studies of water-column aggregate systems (KARNER AND HERNDL 1992; SMITH ET AL. 1992) and data from other aggregate systems have shown that much of the organic matter that is hydrolyzed by extracellular enzymes is lost by diffusion (and turbulence) to the surrounding water, and not taken up by aggregate cells (AZAM AND CHO 1987; KARL ET AL. 1988; KARNER AND HERNDL 1992). Therefore, efficient localization of enzymes and hydrolysis products close to cells is needed to reduce the metabolic costs of maintaining high levels of extracellular enzymes in the bulk phase, and to reduce diffusive losses of products away from cells (DUCKLOW AND CARLSON 1992).

2
The Microbial Biofilm and Extracellular Polymers

When a surface is placed in seawater, colonization by microbial cells immediately follows (FLETCHER 1997). Attachment triggers a number of phenotypic changes in cells (DAVIES ET AL. 1993; VANDEVIVERE AND KIRCHMAN 1993; GOODMAN AND MARSHALL 1995). Upon attachment, cells surround themselves with a matrix of large polymeric molecules called "extracellular polymers," to form a "microbial biofilm" (COSTERTON ET AL. 1995). Biofilms and their associated extracellular polymer matrix are a common feature of aquatic systems. In ocean systems they form the sticky amorphous matrix of phytoplankton aggregates, marine snow, and transparent extracellular polymer particles (TEP) (ALLDREDGE ET AL. 1993), which are abundant in many water-column environments. Biofilm extracellular polymers also form the amorphous coatings on sediments and other particulate surfaces, where they enhance the trophic availability of dissolved organic carbon (DECHO AND LOPEZ 1993). Microbial biofilms also occur at hydrothermal vents, microbial mats, and in association with epiphytic and epizoic microbial communities (for review see DECHO 1990). Biofilms are also common in artificial systems in the fouling of heat exchangers, the corrosion of pipelines, and the infection of medical implant devices (LAPIN-SCOTT AND COSTERTON 1989). While the extracellular polymers have been recognized to secure the attachment of microbial cells to surface, they appear to serve many important roles for the regulation of microbial activities at surfaces (GEESEY 1982).

Comparative studies of enzyme activities in attached and "free" bacteria have shown that, following attachment, enzymatic activity is increased (JONES AND LOCK 1991). In a biofilm structure cells, enzymes, and substrates are in an advantageous proximity. Calculations suggest that as long as the substrate or the hydrolysis products stay within 500 μm of the cell, the energy of synthesizing and exporting enzymes appears justified (WETZEL 1991).

The biofilm may act as an important microbial adaptation to enhance the overall processing of organic matter by surface-associated cells. This will occur using some of the following mechanisms which may act in concert: 1) sorption and concentration of larger dissolved organic matter (DOM); 2) physical trapping and utilization of colloidal organic matter; 3) localization of extracellular enzymes, and prolonging their stability; 4) localization of hydrolysis products in proximity to cells; 5) protection of extracellular enzymes against potentially toxic contaminants; and 6) buffering of cells against sudden environmental fluctuations or stresses.

3
Extracellular Polymers as a "Sorptive Sponge"and "Colloidal Trap" for High-Molecular-Weight Organic Matter

Though the extracellular processing of large molecules is a necessary step in microbial utilization of organic matter, it appears at first glance to be a very inefficient process. Available substrate concentrations are low, some of them being insoluble, complexed with other chemicals present in the water column, or adsorbed to surfaces. Also, given the fragile chemical nature of many enzymes, denaturation may occur from the same influences, and/or degradation by chemical (other proteases, ionic strength, salinity) and physical factors (e.g., pH, temperature, UV irradiation).

The physical/chemical properties of biofilm extracellular polymers may act as a "sorptive sponge" to bind and concentrate dissolved organic molecules, facilitating their later utilization by cells (Fig. 1). Most bacterial extracellular polymers examined thus far are composed largely of anionic polysaccharides,

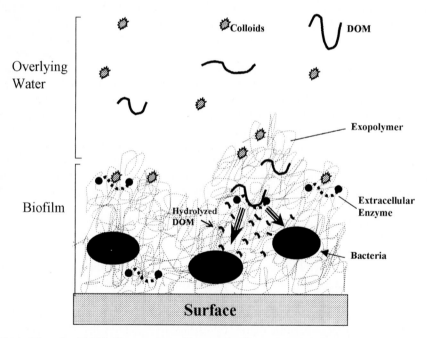

Fig. 1. The microbial biofilm consists of bacterial cells enveloped by a matrix of large polymeric molecules (i.e., extracellular polymers). The biofilm secures the attachment of cells to a surface, but may also enhance the abilities of cells to access dissolved and colloidal organic matter. The high sorptive nature of anionic extracellualr polymers facilitates the binding and concentration of organic molecules in proximity to cells. Extracellular enzymes, which are also localized close to cells, then efficiently hydrolyze the sorbed organic matter. The proximity of extracellular hydrolyses to cells facilitates efficient uptake of hydrolysis products, and reduces diffusional losses of products to the surrounding water

with lesser amounts of proteins (KENNEDY AND SUTHERLAND 1987) and hydrophobic residues (NEU 1996). They possess many potentially reactive groups (e.g., carboxyl, hydroxyl, amine, phosphate) which can provide a range of different binding affinities for dissolved organic matter (DOM) and ions (GEESEY AND JANG 1989). Biologically important types of dissolved organic molecules which can act as substrates for bacteria include proteins, peptides, and amino acids (MAYER ET AL. 1995), and organic colloids (KEPKAY 1994). These organics can be sorbed to a biofilm and serve as a nutrient source to sustain bacterial activities. Recent studies by SAMUELSSON AND KIRCHMAN (1990) and TAYLOR (1995) showed that when protein molecules are bound to particle surfaces they became significantly more accessible to heterotrophic bacteria, with significantly higher remineralization rate constants than for dissolved forms. The sorptive concentration of large DOM to biofilm extracellular polymers represents a potentially effective means for bacteria to sequester and concentrate organic molecules close to the cell.

Biofilms may also effectively trap colloidal-sized particles, and in doing so increase their accessibility to bacteria. A large pool of colloidal organic particles (10 nm – 0.45 μm) has recently been demonstrated to exist in marine systems (KOIKE ET AL. 1990; WELLS AND GOLDBERG 1991, 1993, 1994) and other aquatic systems (LEPPARD ET AL. 1996). Colloids are now considered to outweigh any particulate carbon source in the ocean (KEPKAY 1994). Empirical studies, however, have shown that colloids in solution are not readily available to bacteria because of similar size and electrostatic properties (JOHNSON AND KEPKAY 1992). Association of colloids with surfaces (e.g., surface of a bubble) increases their availability to bacteria (KEPKAY AND JOHNSON 1989). The fibrillar and polyanionic nature of extracellular polymers at the biofilm surface may act as an effective physical trap for colloidal particles (STOLZENBACH 1989), making this abundant organic source potentially more available to biofilm enzymes and cells when compared to free-living microbial cells. Physical trapping of colloids by biofilms and their subsequent utilization by microbial cells, therefore, represents an important pathway for investigations of microbial transformations of organic matter. The chemical nature and turnover rates of the colloidal reservoir remains unknown (DUCKLOW AND CARLSON 1992). However, given their high concentrations and biomass in marine systems, there are important implications for colloids in the cycling of organic matter, trace metals, and the transfer of carbon and nitrogen to food webs (HOLLIBAUGH ET AL. 1991; SHIMETA AND JUMARS 1991).

4
Localization of Enzymes by Extracellular Polymers: the "Lectin-Localization Model"

Extracellular enzymes hydrolyze organic molecules ranging from very large biopolymers to small peptides. A wide range of extracellular enzymes, from hydrolytic enzymes to lyases (i.e., eliminases), have been isolated from various bacteria (HASE AND FINKELSTEIN 1993; SUTHERLAND 1995 for reviews; WONG-MADDEN AND LANDRY 1995).

The chemical properties of extracellular polymers make them a potentially efficient matrix for localizing both extracellular enzymes and hydrolysis products. Under ideal conditions, a biofilm should facilitate an efficient coupling of both extracellular enzyme hydrolysis and the rapid cellular uptake of hydrolysis products. The biofilm matrix, owing to its inherent chemical properties, appears to facilitate the tight binding of large (unhydrolyzed) molecules, while smaller molecules (i.e., hydrolysis products) appear to diffuse through the matrix with varying facility (LAWRENCE ET AL. 1994). Diffusive movement of hydrolysis products permits these molecules to be taken up by cells through osmotrophy, while the stronger binding of large molecules to extracellular polymers permits these potential energy sources to be concentrated by the biofilm. Experimental evidence indicates that the highly-hydrated polysaccharide gels of biofilms facilitate diffusion of smaller molecules, with biofilm diffusion coefficients close to those of water. Diffusion of larger molecules appears to be greatly reduced, however, probably due to greater charge densities and generally stronger sorption, all else being constant (CHRISTENSEN AND CHARACKLIS 1990). Different retention properties of biofilms with respect to molecular size would allow macromolecules to be tightly bound by extracellular polymers where they can be hydrolyzed by extracellular enzymes, followed by a diffusive transfer of smaller hydrolysis products to cells.

Enzymes need to be able to move within the extracellular polymer matrix in order to access substrate molecules. The localization mechanism of extracellular enzymes within biofilms, however, is not well understood as yet. Preliminary evidence indicates that extracellular enzymes often show heterogeneous spatial distributions within biofilms, and suggests that they may be localized within specific regions or "microdomains" having specific chemical properties (Decho, in press). Enzymes are generally large molecules (10–100 kd, dia. > 25 Å). However, the active sites of enzymes are very small, and therefore most of the amino acid residues are not in direct contact with the hydrolytic substrate during enzymatic attack (STRYER 1995). An exciting new idea is emerging in carbohydrate chemistry which involves "lectins." The lectin hypothesis (SHARON AND LIS 1993) suggests that the higher-order structure of carbohydrate polymers may be very important in mediating the specific binding of certain proteins and carbohydrates. Lectins are glycoproteins which recognize and bind to specific saccharide sequences or isolated sugars. They generally have two binding sites, one for a terminal sugar, the other for an internal sequence sugar. Lectin binding is reversible and often provides a highly specific recognition mechanism by which cells may recognize other cells and/or molecules.

The lectin-binding model also offers a fundamental mechanism by which extracellular enzymes (i.e., proteins) can associate reversibly with specific regions of the extracellular polymer (i.e., polysaccharide) matrix of a biofilm. This exciting possibility would also permit the extracellular polymer matrix to partition extracellular enzymes from toxic metals within the biofilm matrix. This partitioning of enzymes and metals could occur because the binding of toxic metals to polysaccharides would alter their higher order structure (and specific lectin recognition sites), and thus interfere with the lectin localization mechanism, and therefore enzymes would not localize in regions where toxic metals have been localized.

A second mechanism of extracellular enzyme localization may occur through cation bridges via carboxyl groups on extracellular polymer acid polysaccharides and enzymes. Many enzymes possess carboxyl groups away from their active sites (STRYER 1995). In order to reach substrate molecules, enzymes could potentially attach to or detach from polysaccharides by small changes in pH. However, the binding to polysaccharides through more-toxic transition metal complexes would also occur. Since these often form more stable multidentate complexes(GEESEY AND JANG 1989), such associations could limit enzyme mobility within the biofilm and potentially denature the enzyme.

The mechanisms of these processes can be approached experimentally using biofilm polymers having different binding properties as a localizing agent for various extracellular enzymes and organic matter. If a biofilm operates efficiently (i.e., successfully localizing enzymes and hydrolysis products), then bacterial conversion efficiencies of organic matter could be expected to be quite high (>50%). In contrast, when biofilms operate inefficiently or are absent, much of the extracellular enzyme activity and DOC products may be lost to the surrounding water. This would result in relatively low conversion efficiencies of DOC. Recent studies of natural riverine biofilms have shown that a wide array of potentially refractory organic molecules (i.e., humic acids) associate with extracellular polymers under natural conditions (FREEMAN ET AL. 1990; FREEMAN AND LOCK 1992; NEU AND LAWRENCE 1997). The binding of these organic molecules to biofilms has negative effects on overall microbial activities (FREEMAN AND LOCK 1992). It is not yet known how the binding of these molecules affects enzyme activities.

5
"Extracellular Polymer Microdomains" and the Stability of Extracellular Enzymes

The extracellular environment in proximity to most free-living bacteria is characterized by frequent fluctuations in a variety of environmental parameters: osmotic conditions (i.e., salinity changes), micrometer-scale pH changes (due to high respiratory activities of other microorganisms), and nutrient fluxes. These fluctuations have the potential to stress both microbial cells and inhibit the activities of extracellular enzymes.

The stability of an enzyme can either be an inherent property of the protein, or be conferred by the extracellular microenvironment (LUDLOW AND CLARK 1991). Most enzymes, by their inherent chemical properties, are designed to remain active only within a relatively narrow pH range, which is generally ±1 pH units of their pK_a (STRYER 1995). Outside of this range their activity is markedly decreased or inhibited. In many model enzyme systems which have been examined thus far, the degree of hydration appears to be a major parameter which can influence enzyme stability (LUDLOW AND CLARK 1991). A highly hydrated environment (e.g., distilled water) will often result in denaturation of an enzyme, while the presence of certain ions or macromolecules are required to maintain stability.

The protective microenvironment of a cell allows intracellular enzymes to exist in a relatively constant ionic environment and hydration state. Outside of the cell, hydration states and ionic concentrations are not constant. The extra-cellular polymer gels of biofilms may contain varying degrees of hydration, de-pending on the binding and concentration of ions within the matrix (TAM AND VERDUGO 1981). Therefore, certain regions of a biofilm may afford a more sta-bilized (i.e., constant) hydration and ionic environment for enzymes. Also, within the protected microenvironment of a cell, specialized chaparonin pro-teins assist in renaturation of enzymes, once denaturation has begun (ELLIS 1996). It is not known how extracellular enzymes survive the ionic fluctuations and denaturing agents which exist outside of the cell. For example, the binding of transition metals may render substrate molecules refractory to enzymatic hydrolyses (GEESEY AND JANG 1989; GEESEY ET AL. 1992), especially since such binding reduces the hydration of the substrate molecule and many enzymatic cleavages are the result of hydrolysis reactions. Also, most enzymes require the presence of cations (e.g., Ca^{++}) within a narrow concentration range in order to maintain their activity. Cation concentrations which are too high or too low may inactivate or denature the enzymes. In contrast, other enzymes are quite resilient to stressors such as heat or ionic or pH fluctuations. These enzymes are thought to owe their resilience to an increased numbers of H-bonds within the enzyme molecule (LUDLOW AND CLARK 1991). This has been especially noted in enzymes which are stable at high temperatures ($> 90\,^{\circ}C$). However, even when enzymes remain stable (i.e., are not denatured) in the presence of such stres-sors, their flexibility and efficiency of hydrolysis is often reduced when outside the range of their optimal environmental conditions.

The biofilm matrix may provide bacteria with a protective microenviron-ment for conducting extracellular processes. The biofilm matrix may act to sta-bilize bacteria against the fluctuating and less than ideal conditions typical of natural systems (SUTHERLAND 1980). The physical/chemical properties of ex-tracellular polymers may act as a protective matrix from which cells may con-duct physiological activities under more stabilized conditions than their free-living counterparts (COSTERTON ET AL. 1987). Recent studies in riverine sys-tems by Lock and colleagues have indicated that bulk cellular activities within biofilms remain relatively stable even when nutrient fluctuations occur in wa-ters overlying the biofilm (FREEMAN AND LOCK 1995). This allows more con-stant heterotrophic activities to occur in the biofilm.

Extracellular enzymes contained within a biofilm, may remain active (i.e., pro-tected from denaturation) for longer periods than enzymes released into the wa-ters overlying the biofilm. The impact of biofilm formation on enzyme stability is suggested by the protective effects of enzyme complexation to dissolved organic acids (WETZEL 1991). Upon binding to polyphenolic acids, the enzymes are more resistant to degradation by physical factors or other proteases present in the wa-ter column. Although the enzymes are most often inactivated, binding is rever-ible. Evidence from mammalian systems indicates that oligosaccharides can be effective in stabilizing enzymes and reducing their denaturation rates over time (ALPIN AND HUGHES 1982). The abundant exopolysaccharides of a biofilm po-tentially represent an ideal substrate for stabilization of extracellular enzymes.

In a study following the effect of nutrient availability on exoprotease activity, Albertson and co-workers found that within 4 h of nutrient and energy deprivation, cultures of *Pseudomonas* sp. showed an increased enzymatic activity, coincidental with increased production of exopolysaccharide material. The investigators suggest that the polysaccharide may support the enzymes, either by scavenging for substrate molecules at low concentration, or by anchoring the enzyme in the vicinity of the cells (ALBERTSON ET AL. 1990). Nitrogen limitation induces increased production of polysaccharides. Also, under nitrogen-limiting conditions, protease activity is stimulated. By supplementing natural waters with inorganic nitrogen, CHRÓST (1991) reported an inhibition of peptidase activity, but since the study did not consider exopolysaccharides the increased activity might reflect the switch to more complex sources of nitrogen, once the easily accessible ones were depleted (CHRÓST 1991).

6
Environmental Influences on Enzyme Activity

The nutritional conditions in the aquatic environment are extremely diverse and unstable. Bacteria may respond to these changes by modulating hydrolytic enzymes production (synthesis or activity). Both environmental and laboratory studies have shown that the presence of abundant and large (polymeric) substrates induces the synthesis of some of the proteolytic ectoenzymes, while small peptides act to inhibit their synthesis (CHRÓST 1991).

Other chemical characteristics of the environment, besides the nutrient status, play an important role in determining enzymatic activities in the environment. Proteins have a charge distribution pattern that allows them to interact with both positively and negatively charged substances in the water column. Part of the highly refractory dissolved organic matter (i.e., DOM), and often a major component of DOM, consists of polyphenolic acids which are generally of plant origin (WETZEL 1991). At the pH of most natural waters (pH 5-8) the carboxyl groups of most polyphenolic acids are ionized and can interact electrostatically with the protonated amino groups of proteins. Upon interaction with proteins, polyphenolic acids confer on the protein a more hydrophobic character, which can lead to precipitation of the protein (Haslam 1988). Even when precipitation does not occur, the phenol-enzyme complexation decreases the hydrolytic activity. These studies suggest that the binding of refractory organic molecules to labile DOM may decrease the accessibility of the DOM to hydrolytic lysis by enzymes.

7
Biofilm Induction and Regulation of Extracellular Enzymes

Bacterial cells within a biofilm are often in close proximity and, as such, must foster pathways of communication and cooperation in order for successful propagation (WIMPENNY 1992; CALDWELL ET AL. 1997). A density-dependent chemical communication between cells of the same or different bacterial species has been demonstrated in certain types of bacteria. This process has been called "quorum sensing" (FUQUA ET AL. 1996 for review). In quorum sensing, bacteria produce specific chemical signals called autoinducers. An autoinducer is a diffusible compound which accumulates in the surrounding environment during growth of cells (FUQUA ET AL. 1994). These "signal" molecules are key elements for intercellular communication in bacteria, and freely diffuse to other cells where they will enter into the cells (via diffusion) in response to local concentration gradients. At low cell densities an autoinducer passively diffuses out of cells down a concentration gradient, while at high cell densities it accumulates at intracellular concentrations equivalent to the extracellular concentration. Therefore, high densities of similar (signal-producing) cells will result in a locally high concentration of the autoinducer signal. When the concentration of the signal exceeds a threshold, activation of the response-regulator proteins occurs and is followed by alterations in gene expression. This represents a unique but highly efficient system of communication (and regulation) for bacteria in proximity to each other. The autoinduction process potentially allows cells having the same or similar physiological capabilities to regulate expression of these activities. Quorum sensing signals have been demonstrated to be involved in a range of extracellular enzyme biosyntheses in bacteria (JONES ET AL. 1993; PIRHONEN ET AL. 1993; CHATTERJEE ET AL. 1995; CUI ET AL. 1995).

7.1
Induction of Extracellular Enzymes at the Physiological Level

It is suggested that a basal, constitutive level of enzyme is continually excreted. When a substrate is present in the environment, it is degraded by the enzyme and low-molecular weight hydrolysis products are generated. These products induce further enzyme synthesis. It is wasteful to the individual organism to produce excess amounts of assimilable nutrients in its local environment. Therefore, feedback mechanism can slow or repress synthesis of new enzymes under conditions of excess.

The rate of both extracellular and exocellular enzyme production has been shown to be associated with physiological functions coupled to growth or physiological state. In laboratory cultures, enzyme production in cells has been observed during lag phase, exponential phase, and early stationary phase (VOTRUBA ET AL. 1987). Patterns of extracellular enzyme production often differ between closely related species of bacteria. Two distinct patterns of extracellular enzyme production occur in the Gram-positive bacteria *Bacillus subtilus* and *Bacillus amyloliquefaciens*. *B. subtilus* produces the exoenzyme alpha amylase at a low level throughout growth. *B. amyloliquefaciens* produces high levels of the enzyme at the end of exponential growth (COLEMAN ET AL. 1987).

7.2
Regulation of Enzyme Activity by Extracellular Factors

Feedback mechanisms, such as catabolite repression, may be important in regulating enzyme production by bacteria based on cues originating in their extracellular environment. This process has been recognized and utilized for a long time by industry as a means of regulating exoenzyme and extracellular enzyme production in bacteria (COLEMAN ET AL. 1987). In catabolite repression, enzyme production is reduced or halted when certain hydrolysis products or "catabolites" become abundant in proximity to cells. Catabolite repression appears to inhibit enzyme production in *B. subtilus*. Addition of glucose to the culture medium causes a dramatic reduction in exoenzyme formation (CHRÓST 1990). Detailed information about the exact pathway of repression is still needed. Some studies suggest the involvement of cAMP and its receptor in the transfer of information between the environment and the cells. Extracellular cAMP (released by lysed algae or resulting from "sloppy feeding") would indicate the simultaneous presence of potential substrates, leached from the surrounding cells. The synthesis of many ectoenzymes is also repressed by the presence of degradation end-products in excess of a certain concentration. In some bacteria amino acids and small peptides were shown to induce enzyme synthesis. It is hypothesized that in this case the enzymes are produced constitutively (at a low level) (CHRÓST 1990). The low-molecular weight products of hydrolyses accumulate to a certain level, enter the cell, and trigger enzyme synthesis.

References

Albertson NH, Nyström T, Kjelleberg S (1990) Exoprotease activity of two marine bacteria during starvation. Appl Environ Microbiol 56:218–223

Alldredge AL, Passow U, Logan B (1993) The existence, abundance and significance of large transparent extracellular polymer particles in the ocean. Deep Sea Res I 40:1131–1140

Alpin JD, Hughes RC (1982) Complex carbohydrates of the extracellular matrix structures, interactions and biological roles. Biochim Biophys Acta 694:375–418

Azam F, Ammerman JW (1984) Cycling of organic matter by bacterioplankton in pelagic marine ecosystems: microenvironmental considerations. In: Fasham MJR (ed) Flows of energy and materials in marine ecosystems. Plenum Press, New York, pp 345–360

Azam F, Cho BC (1987) Bacterial utilization of organic matter in the sea. In: Fletcher M, Gray TRG, Jones JG (eds) Ecology of microbial communities. Cambridge University Press, pp 261–281

Caldwell DE, Wolfaardt GE, Korber DR, Lawrence JR (1997) Do bacterial communities transcend contemporary theories of ecology and evolution? Adv Microbial Ecol 15:1–86

Cembella AD, Antia NJ, Harrison PJ (1984) The utilization of inorganic and organic phosphorus compounds as nutrients by eukaryotic microalgae: a multidisciplinary perspective: pt I. CRC Crit Rev in Microbiol 10:317–391

Chatterjee A, Cui Y, Lui Y, Dumenyo CK, Chatterjee AK (1995) Inactivation of *rsmA* leads to overproduction of extracellular pectinases, cellulases, and proteases in *Erwinia carotovora* subsp. *carotovora* in the absence of the starvation/cell density-sensing signal, N-(3-oxohexanoyl)-L-homoserine lactone. Appl Environ Microbiol 61:1959–1967

Christensen BE, Characklis WG (1990) Physical and chemical properties of biofilms. In: Characklis WG, Marshall KC (eds) Biofilms. Wiley, New York, pp 93–130

Chróst (1990) Microbial ectoenzymes in aquatic environments. In: Overbeck J, Chróst RJ (eds) Aquatic microbial ecology. Springer, Berlin Heidelberg New York, pp 47–78

Chróst RJ (1991) Environmental control of the synthesis and activity of aquatic microbial ectoenzymes. In: Microbial enzymes in aquatic environments. Springer, Berlin Heidelberg New York, pp 29–59

Coleman G, Abbas Ali B, Sutherland J, Fyfe F, Finley A (1987) A comparison of the characteristics of extracellular protein secretion by a Gram-positive and Gram-negative bacterium.

In: Chaloupka J, Krumphanzl V (eds) Extracellular enzymes of microorganisms. Plenum Press, New York, pp 13–22

Costerton JW, Cheng KJ, Geesey GG, Ladd T, Nickel JC, Dasgupta M, Marie TJ (1987) Bacterial biofilms in nature and disease. Annu Rev Microbiol 41:435–464

Costerton JW, Lewandowski Z, Caldwell DE, Korber DR, Lappin-Scott HM (1995) Microbial biofilms. Annu Rev Microbiol 49:711–745

Cui Y, Chatterjee A, Liu Y, Dumenyo CK, Chatterjee AK (1995) Identification of a global repressor gene rsmA of Erwinia carotovora subsp. carotovora that controls extracellular enzymes, N-(3-oxohexanoyl)-L-homoserine lactone, and pathogenicity in soft-rotting Erwinia spp. J Bacteriol 177:5108–5115

Davies DG, Chakrabarty AM, GG Geesey (1993) Exopolysaccharide production in biofilms: substratum activation of alginate gene expression by Pseudomonas aeruginosa. Appl Environ Microbiol 59:1181–1186

Decho AW (1990) Microbial extracellular polymer secretions in ocean environments: their role(s) in food webs and marine processes. Oceanogr Mar Biol Annu Rev 28:73–153

Decho AW (in press) Extracellular polymer microdomains as a structuring agent for heterogeneity within microbial biofilms. In: Riding R (ed) Microbial sediments. Springer, Berlin Heidelberg New York

Decho AW, Lopez GR (1993) Exopolymer microenvironments of microbial flora: multiple and interactive effects on trophic relationships. Limnol Oceanogr 38:1633–1645

Ducklow HW, Carlson CA (1992) Oceanic bacterial production. Adv Microbial Ecol 12:113–181

Ellis RJ (1996) The chaperonins. Academic Press, New York

Fletcher M (1997) Bacterial attachment in aquatic environments: a diversity of surfaces and adhesion strategies. In: Fletcher M (ed) Bacterial adhesion: molecular and ecological diversity. Wiley-Liss, New York, pp 1–24

Freeman C, Lock MA (1992) Recalcitrant high molecular weight material, an inhibitor of microbial metabolism in river biofilms. Appl Environ Microbiol. 58:2030–2033

Freeman C, Lock MA (1995) The biofilm polysaccharide matrix: a buffer against changing organic substrate supply? Limnol Oceanogr 40:273–278

Freeman C, Lock MA, Marxsen J, Jones SE (1990) Inhibitory effects of high molecular weight dissolved organic matter upon metabolic processes in biofilms from contrasting rivers and streams. Freshwater Biol 24:159–166

Fuqua WC, Winans SC, Greenberg EP (1994) Quorum sensing in bacteria: the LuxR-LuxI family of cell density-responsive transcriptional regulators. J Bacteriol 176:269–275

Fuqua C, Winans SC, Greenberg EP (1996) Census and consensus in bacterial ecosystems: the Lux-R-LuxI family of quorum sensing transcriptional regulators. Annu Rev Microbiol 50:727–751

Geesey GG (1982) Microbial exopolymers: ecological and economic considerations. ASM News 48:9–14

Geesey GG, Jang L (1989) Interactions between metal ions and capsular polymers. In: Beveridge TJ, Doyle RJ (eds) Metal ions and bacteria. Wiley, NewYork, pp 325–358

Geesey GG, Bremer PJ, Smith JJ, Muegge M, Jang LK (1992) Two-phase model for describing the interactions between copper ions and exopolymers from Alteromonas atlantica. Can J Microbiol 38:785–793

Goodman AE, Marshall KC (1995) Genetic responses of bacteria at surfaces. In: Lappin-Scott HM, Costerton JW (eds) Microbial biofilms. Cambridge University Press, pp 80–98

Hase CC, Finkelstein RA (1993) Bacterial extracellular zinc-containing metalloproteases. Microbiol Rev 57:823–837

Haslam E (1988) Plant polyphenols (syn vegetable tannins) and chemical defense – a reappraisal. J Chem Ecol 14:1789–1805

Hollibaugh JT, Azam F (1983) Microbial degradation of dissolved protein in seawater. Limnol Oceanogr 28:1104–1116

Hollibaugh JT, Buddemeier RW, Smith SV (1991) Contributions of colloidal and high molecular weight dissolved material to alkalinity and nutrient concentrations in shallow marine and estuarine systems. Mar Chem 34:1–27

Hoppe HG (1991) Microbial extracellular enzyme activity: a new key parameter in aquatic ecology. In: Chróst RJ (ed) Microbial enzymes in aquatic environments. Springer, Berlin Heidelberg NewYork, pp 60–95

Hoppe HG, Kim S-J, Gocke K (1988) Microbial decomposition in aquatic environments: combined process of extracellular enzyme activity and substrate uptake. Appl Environ Microbiol 54:784-790

Johnson BD, Kepkay PE (1992) Colloidal transport and bacterial utilization of oceanic DOC. Deep Sea Res 39:855–869

Jones SE, Lock MA (1991) Peptidase activity in river biofilms by product analysis. In: Chróst RJ (ed) Microbial enzymes in aquatic environments. Springer, Berlin Heidelberg New York, pp 144-154

Jones S, Yu B, Bainton NJ, Birdsall M, Bycroft BW, Chhabra SR, Cox AJR, Golby P, Reeves PJ, Stephens S, Winson MK, Salmond GSAB, Williams P (1993) The lux autoinducer regulates the production of exoenzyme virulence determinants in *Erwinia carotovora* and *Pseudomonas aeruginosa*. EMBO J 12:2477-2482

Karl DM, Knauer GA, Martin JH (1988) Downward flux of particulate organic matter in the ocean: a particle decomposition paradox. Nature (London) 332:438-441

Karner M, Herndl GJ (1992) Extracellular enzyme activity and secondary production in free-living and marine snow associated bacteria. Mar Biol 113:341-347

Karp-Boss L, Boss E, Jumars PA (1996) Nutrient fluxes to planktonic osmotrophs in the presence of fluid motion. Oceanogr Mar Biol Annu Rev 34:71-107

Kennedy AFD, Sutherland IW (1987) Analysis of bacteria exopolysaccharides. Biotechnol Biochem 9:12-19

Kepkay AFD (1994) Particle aggregation and biological reactivity of colloids. Mar Ecol Prog Ser 109:293-304

Kepkay PE, Johnson BD (1989) Coagulation on bubbles allows the microbial respiration of oceanic dissolved organic carbon. Nature (London) 385:63-65

Koike I, Hara S, Terauchi K, Kogure K (1990) Role of sub-micrometer particles in the ocean. Nature (London) 345:242-244

Lappin-Scott HM, Costerton JW (1989) Bacterial biofilms and surface fouling. Biofouling 1:323-342

Lawrence JR, Wolfaardt GM, Korber DR (1994) Determination of diffusion coefficients in biofilms using confocal laser microscopy. Appl Environ Microbiol 60:1166-1173

Leppard GS, Heissenberger A, Herndl GJ (1996) Ultrastructure of marine snow. I. Transmission electron microscopy methodology. Marine Ecol Prog Ser 135:289-298

Ludlow JM, Clark DS (1991) Engineering considerations for the application of extremophiles in biotechnology. Crit Revs Biotechnol 10:321-345

Mayer LM, Schick LL, Sawyer T, Plante CJ, Jumars PA, Self RL (1995) Bioavailable amino acids in sediments: a biomimetic, kinetics-based approach. Limnol Oceanogr 40:511-520

Neu TR (1996) Significance of bacterial surface-active compounds in the interaction of bacteria with surfaces. Microbiol Rev 60:151-166

Neu TR, Lawrence JR (1997) Development and structure of microbial biofilms in river water studied by confocal laser scanning microscopy. FEMS Microbial Ecol 24:11-25

Pirhonen M, Flego D, Heikiheimo R, Palva ET (1993) A small diffusible signal molecule is responsible for the global control of virulence and exoenzyme production in the plant pathogen *Erwinia carotovora*. EMBO J 12:2467-2476

Pomeroy LR (1984) Significance of microorganisms in carbon and energy flow in marine ecosystems. In: Klug MJ, Reddy CA (eds) Current perspectives in microbial ecology. Amer Soc Microbiol, Washington DC, pp 405-411

Priest FG (1987) Regulation of extracellular enzyme synthesis in Bacilli. In: Chaloupka J, Krumphanzl V (eds) Extracellular enzymes of microorganisms. Plenum Press, New York, pp 3X-12

Samuelsson MO, Kirchman DL (1990) Degradation of adsorbed protein by attached bacteria in relationship to surface hydrophobicity. Appl Environ Microbiol 56:3643-3648

Sharon N, Lis H (1993) Carbohydrates in cell recognition. Sci Am January:82-89

Shimeta J, Jumars PA (1991) Physical mechanisms and rates of particle capture by suspension feeders. Oceanogr Mar Biol Annu Rev 29:191–257

Smith DC, Simon M, Alldredge AL, Azam F (1992) Intense hydrolytic enzyme activity on marine aggregates and implications for rapid particle dissolution. Nature (London) 359: 139–142

Stolzenbach KD (1989) Particle transport and attachment. In: Characklis WG, Wilderer PA (eds) Structure and function of biofilms. Wiley, Chichester, pp 33–47

Stryer L (1995) Biochemistry, 4th edn. WH Freeman, New York

Sutherland IW (1980) Polysaccharides in the adhesion of marine and freshwater bacteria. In: Derkeley RCW, Lynch JM, Melling J, Rutter PR, Vincent B (eds), Microbial adhesion to surfaces. Ellis Horwood, Chichester, pp 228–329

Sutherland IW (1995) Polysaccharide lyases. FEMS Microbiol Rev 16:323–347

Tam PY, Verdugo P (1981) Control of mucus hydration as a Donnan equilibrium process. Nature 292:340–342

Taylor G (1995) Microbial degradation of sorbed and dissolved protein in seawater. Limnol Oceanogr 40:875–885

Vandevivere P, Kirchman DL (1993) Attachment stimulates exopolysaccharide synthesis by a bacterium. Appl Environ Microbiol 59:3280–3286

Votruba J, Pazlarova J, Chaloupka J (1987) Modelling of physiological control of production of hydrolytic enzymes. In: Chaloupka J, Krumphanzl V (eds) Extracellular enzymes of microorganisms. Plenum Press, New York, pp 23–28

Wells ML, Goldberg ED (1991) Occurrence of small colloids in seawater. Nature (London) 353:342–344

Wells ML, Goldberg ED (1993) Colloid aggregation in seawater. Mar Chem 41:353–358

Wells ML, Goldberg ED (1994) The distribution of colloids in the North Atlantic and southern oceans. Limnol Oceanogr 39:286–302

Wetzel RG (1991) Extracellular enzymatic interactions: storage, redistribution and interspecific communication. In: Chróst RJ (ed) Microbial enzymes in aquatic environments. Springer, Berlin Heidelberg New York, pp 6–28

Wimpenny JWT (1992) Microbial systems: patterns in time and space. Adv Microb Ecol 12:469–522

Wong-Madden ST, Landry D (1995) Purification and characterization of novel glycosidases from the bacterial genus Xanthomonas. Glycobiol 5:19–28

CHAPTER 13

Interaction Between Extracellular Polysaccharides and Enzymes

Jost Wingender[1] · Karl-Erich Jaeger[2] · Hans-Curt Flemming[1]

[1] Department of Aquatic Microbiology, University of Duisburg and IWW Center for Water Research, Geibelstrasse 41, D-47057 Duisburg, *E-mail: hh239wi@uni-duisburg.de* (Wingender), *E-mail: 100606.3337@compuserve.com* (Flemming)
[2] Chair Biology of Microorganisms, Ruhr-University Bochum, D-44780 Bochum, Germany

Keywords. Extracellular enzymes, EPS, Polysaccharide, Alginate, Lipase, Secretion, Biofilm, Interaction

1
Extracellular Enzymes in Biofilms

Different forms of microbial aggregates such as biofilms and flocs in natural and engineered environments have in common that the constituent cells of these structures are embedded in a hydrated matrix of extracellular polymeric substances (EPS). These EPS are supposed to consist predominantly of polysaccharides and proteins, but other macromolecules such as nucleic acids, (phospho)lipids, and humic substances have also been found in varying amounts as components of the EPS matrix in microbial aggregates such as in wastewater biofilms and activated sludges (URBAIN ET AL. 1993; JAHN AND NIELSEN 1996; GEHRKE AND SAND, this volume).

Polysaccharides and proteins have been assigned mainly structural functions in mediating the formation and mechanical stability of the three-dimensional, gel-like matrix of biofilms. However, as extracellular enzyme activities are commonly observed in biofilms, it can be assumed that part of the proteins consist of enzyme molecules. In general, the secretion of enzymes represents a widespread property among bacteria and fungi (PRIEST 1992). Sometimes, released enzymes are categorized according to their location after transport across the cytoplasmic membrane of the cell. The term "ectoenzyme" refers to any enzyme that remains associated with its producer cell either in the periplasmic space or bound at the cell surface, whereas the term "extracellular enzyme" is used for any enzyme that has lost contact with its producer cell and may occur in a cell-free form in the surrounding environment or adsorbed to other living or inert surfaces (CHRÓST 1991). In the following, the term "extracellular enzyme" is used in a more comprehensive sense for any enzyme which acts on substrates either in a cell surface-associated manner or in a free form with respect to its function as a secreted enzyme.

There may be transitions between cell association and the free state of an extracellular enzyme with no differences in the properties of the enzyme from these different locations; examples are alkaline phosphatase of *Bacillus licheniformis* (WETZEL 1991) or lipase of *Pseudomonas aeruginosa* (JAEGER ET AL. 1996). Mostly from laboratory studies using pure cultures grown in liquid media it is known that in planktonic populations the ratios between cell-bound and free enzymes depend on many factors such as growth phase, nutrient concentration, nutrient composition, association with lipopolysaccharide (LPS) as an integral part of the outer membrane of Gram-negative bacteria (JAEGER ET AL. 1996), or accumulation within the thick cell walls of Gram-positive bacteria and fungi. However, a different situation exists between single planktonic cells in the bulk liquid phase and aggregated cells immobilized in biofilms and flocs. After secretion from single free-living cells, extracellular enzymes can dissolve, readily diffuse away from the producing cell, and be diluted into the aqueous environment. In cell aggregates the mobility of enzymes leaving the cell surface may be restricted by diffusion limitation within the viscous gel-like EPS matrix, by immobilization through interactions with structural polymers, e.g., by forming enzyme-polysaccharide complexes (see below), or by binding to inorganic and/or organic particulate material incorporated in the biofilm (e.g., BURNS 1989; VETTER ET AL. 1998). Thus, release of extracellular enzymes from microbial aggregates into the surrounding environment may be delayed or even prevented, resulting in a local accumulation of enzymes within biofilms and flocs.

Extracellular enzyme activities in biofilms or flocs may have different origins. Enzymes can be actively secreted by transport across membranes onto the bacterial surface or into the medium from viable and intact bacterial cells within the biofilm. Discrete mechanisms of enzyme release have been described for Gram-negative and Gram-positive bacteria (see below). Liberation of enzymes also results from damage and disintegration of biofilm bacteria due to noxious physicochemical influences of the environment, but may also be caused by viral lysis or protozoan grazing. In addition, extracellular enzymes from the surrounding medium can be trapped in biofilms. Whatever the source of extracellular enzymes, their site of biosynthesis, secretion, and final location may be spatially separated.

Enzyme activities have been determined experimentally in biofilms and flocs, occurring in natural and engineered environments or grown under laboratory conditions. Enzyme activities were measured either directly on whole flocs and biofilms adherent to surfaces or alternatively, on homogenized biofilm suspensions after detaching the biofilm from solid supports by scraping and sonication techniques. Sometimes a fractionation of biofilms and flocs was performed by such methods as centrifugation, filtration, or extraction of EPS using ion-exchange resins (e.g., SINSABAUGH ET AL. 1991; FRØLUND ET AL. 1995; CONFER AND LOGAN 1998). These techniques were applied to separate inorganic particulate material, cellular aggregates, and single cells from the cell-free aqueous phase in order to distinguish between particle-sorbed/cell-bound enzymes and cell-free extracellular enzymes, which are obtained by 0.2-µm membrane filtration as the final step in the fractionation procedures. On the basis of enzymatic analyses of the fractions obtained by these techniques, it has been reported that many enzyme activities in biofilms were often associated with the cells and/or other particulate material. In wastewater biofilms, most of the leucine aminopeptidase and α-glucosidase activities occurred in contact with cells, whereas only a minor fraction of these hydrolytic activities (no more than 3%) was found to be located in the cell-free (< 0.2 µm) fraction (CONFER AND LOGAN 1998). In epilithic biofilms, phenol oxidase, peroxidase, and phosphatase activities were largely particle bound; in contrast, the largest pool of carbohydrase activities were in the aqueous phase (SINSABAUGH ET AL. 1991). In activated sludge, 75% to 98.5% of phosphatase, glycosidase, and aminopeptidase activities were found to be located in the flocs (TEUBER AND BRODISCH 1977). However, the separation/extraction procedures employed do not necessarily allow discrimination between enzymes directly bound to integral components of the cell surface and those cell-free enzymes immobilized within that fraction of the EPS matrix, which may not be removable, e.g., by centrifugation/filtration, but may still remain associated with cells.

The main focus has been on the analysis of biofilm enzymes degrading biopolymers such as proteins and polysaccharides or enzymes cleaving oligomeric fragments of macromolecules such as peptides, oligosaccharides, or phosphomonoester compounds. Hydrolytic activities in biofilms were frequently determined using substrate analogs in the form of synthetic substrates; enzyme activities were monitored by an increase in absorbance or fluorescence due to the enzymatic cleavage of ester, glucoside, or peptide (amide) bonds with the concomitant release of chromogenic groups or fluorophores such as 4-nitrophenol, 4-nitroaniline, β-naphthylamine, methylumbelliferone, or fluorescein from the artificial substrate molecules (e.g., FRØLUND ET AL. 1995; JONES AND LOCK 1989; LEMMER ET AL. 1994; SINSABAUGH ET AL. 1991). The activity of oxidoreductase enzymes was determined photometrically by following an increase in absorbance on reduction of tetrazolium salts (BLENKINSOPP AND LOCK 1990; WUERTZ ET AL. 1998) or L-3,4-dihydroxyphenylalanine (SINSABAUGH ET AL. 1991). Sometimes enzyme activities in biofilms were studied by analyzing the enzymatic degradation products from natural occurring or model substrates. Thus, the proteolytic activity of river biofilms was determined by monitoring the release of amino acids from proteins using high pressure liquid chromatography (JONES AND LOCK 1991). Size-fractionated hydrolysis products resulting from the enzymatic degradation of bovine serum albumin by wastewater biofilms were quantified with a colorimetric protein assay (CONFER AND LOGAN 1997). The proteolysis of various cytokines by laboratory biofilms of *Porphyromonas gingivalis* was demonstrated by using SDS-PAGE, Western blotting and antibodies recognizing both the intact cytokines and their breakdown products (FLETCHER ET AL. 1998).

Enzyme assays are often performed at relatively high substrate concentrations under laboratory conditions, so that potential rather than actual enzyme activities are measured. Most extracellular enzymes identified in biofilms belong to the main class of hydrolases, whereas enzymes from other main classes of enzymes (e.g., oxidoreductases, lyases, transferases, or isomerases) were rarely considered with respect to their presence in biofilms. Examples of hydrolytic enzymes in flocs and biofilms include proteases, polysaccharidases (e.g., amylases, cellulases), lipases, esterases, aminopeptidases, glycosidases, and phosphatases.

The primary function of these enzymes is supposed to be the extracellular degradation of macromolecules to low-molecular-weight-products, which are small enough to be transported into the cells to be available as carbon and energy sources for microbial metabolism. Thus, they are essential for providing nutrients for immobilized bacteria in biofilms by extracellular digestion of macromolecules, which are too large to be taken up directly by the bacterial cell. Since in many ecosystems organic material either dissolved or particulate is predominantly macromolecular, heterotrophic bacteria within biofilms are largely dependent on the presence of degradative extracellular enzymes in their microenvironment providing utilizable compounds for growth and maintenance of metabolic activity.

Extracellular enzymes may sometimes also be involved in biosynthetic processes. Extracellular polysaccharides such as dextrans or levans of *Leuconostoc mesenteroides* and *Streptococcus mutans* are polymerized from glucose and fructose, respectively, by extracellular glycosyltransferases (SUTHERLAND 1990). In *Azotobacter vinelandii* extracellular calcium-dependent mannuronan C-5-epimerases catalyze the conversion of mannuronate residues to guluronate residues within homopolymeric mannuronan molecules after their polymerization and secretion to yield the final extracellular polysaccharide alginate (ERTESVÅG AND VALLA 1998).

Another function of extracellular enzymes produced by biofilm bacteria may be the partial depolymerization of structural polymers of the biofilm matrix with subsequent release of bacterial cells, allowing the spreading of the biofilm bacteria and the colonization of new environments. Thus, enhanced expression of alginate lyase in *P. aeruginosa* led to the degradation of the slime polysaccharide alginate and increased detachment of cells from agar-grown biofilms, suggesting a role of this enzyme in the sloughing of immobilized bacteria from solid surfaces (BOYD AND CHAKRABARTY 1994). Similarly, the presence of EPS-degrading enzymes was demonstrated in dense biofilms of *Pseudomonas fluorescens* grown on glass surfaces (ALLISON ET AL. 1998); the enzymic degradation of EPS was supposed to be responsible for the observed detachment of biofilm cells under starvation conditions during prolonged incubation. *Methanosarcina mazei* was shown to produce an extracellular disaggregatase that caused their cell aggregates to disperse into single cells (XUN ET AL. 1990).

The enzyme appeared to be an endo-polysaccharide hydrolase that degraded the matrix polymer methanochondroitin, suggesting a function of this enzyme in the liberation of cells from cellular aggregates during certain stages in its growth cycle (XUN ET AL. 1990).

The relationship between extracellular enzyme activity and microbial abundance in biofilms has been shown to vary depending on the enzyme studied. In river biofilms, phenol oxidase, peroxidase, and phosphatase activities that were largely particle-bound often correlated with microbial biomass, whereas carbohydrase activities with their main activities in the aqueous phase of the biofilms were not correlated with biomass (SINSABAUGH ET AL. 1991). Good correlations have been shown between fixed bacterial biomass and proteolytic activity (LAURENT AND SERVAIS 1995) or esterase activity (JONES AND LOCK 1989; DE ROSA ET AL. 1998) in different aquatic environments (river and drinking water biofilms). Since these enzyme activities were considered near-constant properties of biofilm bacteria, the determination of protease and esterase activities measured by the hydrolysis of L-leucyl-β-naphthylamide and fluorescein diacetate, respectively, were proposed as rapid and simple methods for measuring bacterial biomass on surfaces. (LAURENT AND SERVAIS 1995; DE ROSA ET AL. 1998). From a practical point of view this may be important in technical water systems, where the monitoring, control, and reduction of biofilm growth is necessary. Thus, the measurement of potential extracellular proteolytic activity has been used to quantify fixed bacterial biomass in various French drinking-water distribution systems, where a clear correlation between biodegradable organic matter in the water and fixed bacterial abundance was demonstrated, when free chlorine as disinfectant was absent (SERVAIS ET AL. 1995).

2
Mechanisms of Enzyme and Polysaccharide Secretion

Secretion involves the transport of enzyme proteins and polysaccharides across the cytoplasmic membrane of Gram-positive bacteria or across the inner and outer membranes of Gram-negative bacteria. After secretion, proteins and polysaccharides are located extracellularly either permanently or transitorily bound to cell surface structures or cell-free in the aqueous phase of the environment. Since secretion and interaction between extracellular enzymes and polysaccharides may be linked processes, a short overview on the bacterial secretions pathways of both polymer classes is given. A number of recent publications have reviewed the biosynthesis and secretion of bacterial proteins (e.g., BINET ET AL. 1997; FILLOUX ET AL. 1998; HUECK 1998) as well as capsular and slime polysaccharides (e.g., SUTHERLAND 1990; ROBERTS 1995, 1996; GACESA 1998; BECKER 1998).

2.1
Secretion of Enzymes

Extracellular enzymes are synthesized in the cytoplasm and must therefore be translocated through bacterial membranes to reach their final destination. At present, three main secretion pathways have been identified (BINET ET AL. 1997; PUGSLEY 1993; SALMOND AND REEVES 1993; FILLOUX ET AL. 1998), which were designated ABC (ATP-binding cassette) pathway (type I pathway), general secretory pathway (type II pathway), and contact site-dependent secretion pathway (type III pathway). Several secretion pathways can be used by a single species, when different extracellular enzymes are produced. *P. aeruginosa* is an example of a biofilm-forming bacterium, which employs all three pathways to secrete various enzymes and toxins.

2.1.1
The ABC Pathway

This secretion pathway is a one-step process, bypassing the periplasm. An ABC exporter machinery consists of three different proteins, which form a pore-like complex, directly connecting the cytoplasm with the extracellular medium (BINET ET AL. 1997). Lipases from *P. fluorescens* (DUONG ET AL. 1994) and *Serratia marcescens* (AKATSUKA ET AL. 1995; LI ET AL. 1995) are secreted via this pathway. In *S. marcescens* it consists of the inner membrane protein LipB containing an ATP-binding cassette (ABC) protein, which confers the substrate specificity on the system. Additional components include LipC as a membrane fusion protein, which can be associated with both the inner and the outer membrane and LipD as an outer membrane protein. Both lipase and metalloprotease are secreted through this ABC-transporter (AKATSUKA ET AL. 1997); however, the S-layer protein SlaA, which can presumably form an essential component of the EPS matrix, seems to be its natural substrate (KAWAI ET AL. 1998).

2.1.2
The General Secretory Pathway

In Gram-negative bacteria, secretion of proteins via this pathway occurs as a two-step process – the initial signal peptide-mediated transport across the inner membrane and the subsequent translocation of the transient periplasmic intermediate across the outer membrane. Figure 1 summarizes schematically the present knowledge on the mechanism of protein secretion via the type II pathway, exemplified by lipase secretion in *P. aeruginosa*.

Many proteins secreted by both Gram-positive and Gram-negative bacteria possess an N-terminal signal sequence. It consists of one or more positively charged amino acids followed by a segment of ten or more hydrophobic amino acids, which can adopt an α-helical conformation, thereby mediating insertion into the lipid bilayer of the cytoplasmic membrane. After translocation of the protein through the inner membrane, the signal sequence is cleaved off by a specific signal peptidase, which recognizes a conserved cleavage site consisting of the amino acids alanine-X-alanine (VAN HEIJNE 1986). Secretion proceeds through the so-called Sec-translocase. In *Escherichia coli*, this is a multisubunit protein complex which consists of the soluble dimeric SecA and a membrane-embedded complex formed by SecY, E, D, G, and F. The protein to be secreted is kept in a translocation-competent unfolded state by means of the chaperone

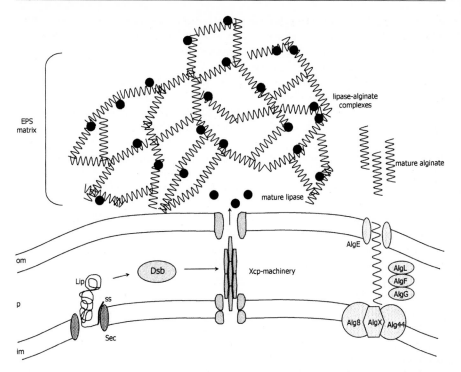

Fig. 1. Schematic representation of synthesis and secretion of enzymes and polysaccharides and their extracellular interactions within the EPS matrix of a biofilm. As an example, secretion of lipase and alginate by *P. aeruginosa* are shown. Lipase is secreted through the inner membrane (im) via the Sec-translocase. The signal sequence (ss) is cleaved off, and the enzyme folds into its enzymatically active conformation in the periplasm (p). This process is assisted by specific foldases and enzymes catalyzing the formation of disulfide bonds (Dsb). Secretion through the outer membrane (om) is mediated by the Xcp secretion machinery consisting of 12 different proteins. The first polymeric product of alginate biosynthesis is polymannuronate assembled at the inner membrane. In the periplasm, O-acetylation mediated by the three gene products (algF, algI, algJ) as well as epimerization of mannuronate to guluronate catalyzed by a single gene product (AlgG) yield the final polymer. The function of the periplasmic alginate lyase (AlgL) is still unknown; it has been supposed to control chain length of alginate molecules or provide oligosaccharide primers for the polymerization reaction (MAY AND CHAKRABARTY 1994). The pore protein AlgE seems to be involved in the transport of nascent alginate molecules across the outer membrane (REHM ET AL. 1994). Finally, a gel-like network of alginate molecules is formed extracellularly, to which lipase molecules and perhaps other enzymes are bound (WICKER-BÖCKELMANN ET AL. 1987)

SecB (DUONG ET AL. 1997). A similar Sec-translocase exists in *Bacillus* species (SIMONEN AND PALVA 1993). Specific foldase proteins have been identified in *Pseudomonas* species, which assist correct folding of lipase (JAEGER ET AL. 1994) and elastase (BRAUN ET AL. 1996) in the periplasm. The main function of these chaperone proteins is to help their cognate enzymes in overcoming a kinetic barrier along their folding pathway. In addition, they may also function as temporary competitive inhibitors preventing enzyme activity inside the bacterial cell. Proteins to be secreted often contain disulfide bonds which are formed in the periplasm. In *E. coli*, a complex system consisting of Dsb proteins mediates disulfide bond formation (RAINA AND MISSIAKIS 1997). DsbA is a thiol: disulfide oxidoreductase which oxidizes the Cys-SH residues to form a disulfide bond; DsbC is a thiol:disulfide isomerase which isomerizes wrongly oxidized disulfide bonds (MISSIAKIS AND RAINA 1997). DsbA and DsbC are also involved in the formation of disulfide bonds of *P. aeruginosa* lipase (URBAN A, LEIPELT M, EGGERT T, AND JAEGER K-E, submitted for publication).

After folding into an enzymatically active conformation in the periplasm, proteins to be secreted are transported through the outer membrane by means of a complex machinery called "secreton" consisting of up to 14 different proteins forming the so-called main terminal branch of the general secretion pathway (PUGSLEY 1993). In *P. aeruginosa*, lipase is secreted through such a secreton (JAEGER ET AL. 1996; FILLOUX ET AL. 1998), here encoded by 12 *xcp* genes organized in two divergently transcribed operons (FILLOUX ET AL. 1998; TOMMASSEN ET AL. 1992). The Xcp proteins are located both in the inner and the outer membrane with XcpQ forming a multimeric pore with a diameter of 95 Å (BITTER ET AL. 1998). Similar multicomponent secretons have been identified in *P. putida*, *Klebsiella oxytoca*, *Aeromonas hydrophila*, *Vibrio cholerae*, *E. coli*, and a number of phytopathogenic bacteria (PUGSLEY 1993; FILLOUX ET AL. 1998).

2.1.3
The Contact Site-Dependent Pathway

Animal pathogens including *Yersinia* sp., *Shigella flexneri*, *Salmonella typhimurium*, and *P. aeruginosa* have developed a complex secretion system consisting of at least 20 different proteins (CORNELIS AND WOLF-WATZ 1997; HUECK 1998). Most of them are located in the inner membrane, although an outer membrane pore protein analogous to the secretins of the type II secretion machinery is also present. Interestingly, this pathway functions independently from the Sec-translocase, the secreted proteins do not contain a signal sequence, and no periplasmic intermediates can be isolated. The type III secretion system is designated to deliver bacterial proteins directly into the cytosol of a eukaryotic cell which means that the assembly of the secretion apparatus is regulated by the contact with a eukaryotic target cell. It will be interesting to study whether cell-to-cell contacts mediated by EPS in a biofilm may also be able to exert a signal for assembling a functional type III secretion complex.

A novel mechanism of enzyme secretion, which differs from the specific pathways mentioned above, was first discovered in *P. aeruginosa* and seems to be common among Gram-negative bacteria (KADURUGAMUWA AND BEVERIDGE 1995; BEVERIDGE ET AL. 1997; LI ET AL. 1998). This nonspecific enzyme release occurs through formation of outer membrane-derived vesicles (blebs). Surface blebbing can be observed during normal growth and represents a process by which cellular macromolecules, including periplasmic compounds and membrane components (enzymes, nucleic acids, LPS, phospholipids), are shed into the extracellular space in the form of membrane vesicles. Release of cellular material by this mechanism may be the result of metabolic turnover processes. Alternatively, membrane vesicles, into which hydrolytic enzymes are packaged (e. g., peptidoglycan hydrolases), may serve to degrade surrounding cells in the biofilm ("predatory vesicles" – see KADURUGAMUWA AND BEVERIDGE 1995; BEVERIDGE ET AL. 1997), providing nutrients for vesicle-forming biofilm bacteria.

2.2
Secretion of Polysaccarides

The most common mechanism of polysaccharide biosynthesis is intracellular; only the neutral polysaccharides dextran or levan are known to be synthesized extracellularly (SUTHERLAND 1990). Intracellular biosynthesis starts from activated monosaccharides (nucleoside diphosphate sugars or less often nucleoside monophosphate sugars). In this form the sugars are sequentially transferred to a lipid acceptor (polyisoprenyl phosphate) on which the assembly of sugar residues to complete repeating units and their polymerization occurs. Noncarbohydrate substituents may be added to the oligosaccharide repeating units attached to the lipid carrier. Thus, the biosynthesis of xanthan, produced by the plant-pathogenic bacterium X*anthomonas campestris* pv. campestris, occurs via the stepwise assembly of repeating pentasaccharide units and their subsequent polymerization to yield the complete polysaccharide; acetyl and pyruvyl residues from acetyl-CoA and phosphoenolpyruvate as donors, respectively, are attached at the lipid-linked pentasaccharide level (reviews: SUTHERLAND 1990; BECKER ET AL. 1998).

In Gram-negative bacteria, the polymeric chains are further secreted across the outer membrane to the cell surface or extruded into the surrounding medium. The process of polysaccharide secretion may occur through Bayer adhesion zones between inner and outer membranes. Alternatively, pore proteins in the outer membranes of Gram-negative bacteria have been implicated in the secretion of polysaccharides. Thus, *P. aeruginosa* produces a porin-like outer membrane protein, AlgE, that was only detected in mucoid (alginate-producing) variants, where it has been supposed to be involved in the secretion of alginate to the cell surface (GRABERT ET AL. 1990; REHM ET AL. 1994 – see Fig. 1). During or after polymerization polysaccharide chains may be modified in the periplasm or after secretion in the extracellular environment. In mucoid strains of *P. aeruginosa*, alginate is O-acetylated at the C2- and/or C3-position on its mannuronate residues in the periplasmic space; extracellular epimerization of polymannuronate to the final alginate by an extracellular epimerase has been shown in *A. vinelandii* (recent reviews of alginate biosynthesis: REHM AND VALLA 1997; GACESA 1998; DAVIES, this volume).

3
Interaction of Enzymes with Extracellular Polysaccharides

Interactions between enzymes and polysaccharides within the extracellular matrix may be enzymatic or nonenzymatic. As mentioned above, extracellular enzyme-substrate interactions occur primarily in the process of polysaccharide degradation for nutrient aquisition, but may also be involved in the release of biofilm matrix-enclosed organisms or in the modification of polysaccharides as exemplified by the epimerization of polymannuronate to alginate. In the following, only nonenzymatic interactions between enzymes and polysaccharides in the slime matrix of biofilm bacteria are considered.

In general, it is assumed that synthesis and release of enzyme proteins and polysaccharides via specific secretory pathways normally occur independently from each other, so that interaction between them is expected to occur only after transport across the bacterial membranes. However, interaction between protein and polysaccharide may already occur during secretion. In *Micrococcus sodonensis*, the appearance of extracellular enzyme activities of alkaline phosphatase, nuclease, and protease has been shown to be dependent on the cosecretion of at least one of several polysaccharides produced by this bacterium (BRAATZ AND HEATH 1974). From a more detailed study of alkaline phosphatase, it was hypothesized that the polysaccharide protected the polypeptide chains of the enzyme from proteolytic degradation during a vulnerable stage of enzyme secretion (BRAATZ AND HEATH 1974).

On the basis of the observation of enzyme abundance in biofilms, it has been assumed that extracellular enzymes may accumulate and become stabilized in biofilms by binding at the cell surface of biofilm bacteria or by interacting with nonenzymatic EPS such as polysaccharides of the biofilm matrix (LOCK ET AL. 1984; BURNS 1989; JONES AND LOCK 1989; SINSABAUGH ET AL. 1991). There have been occasional reports in the literature about the interactions between enzymes and polysaccharides, but little work has been done to study such interactions in biofilm bacteria.

The Gram-negative bacterium *P. aeruginosa* produces a number of extracellular products, among them various polysaccharides and at least ten different enzymes, four of which are lipolytic: a hemolytic and a nonhemolytic phospholipase C, a lipase (JAEGER ET AL. 1996), and an esterase. The best-studied extracellular polysaccharide of *P. aeruginosa* is alginate, which is overproduced in mucoid strains on short-term incubation on agar media (GOVAN 1990). Alginate is supposed to be a key factor in the formation and persistence of *P. aeruginosa* biofilms on human mucosal tissues (e.g., in chronic lung infections of cystic fibrosis patients; see GOVAN AND DERETIC 1996) and in aquatic environments (GROBE ET AL. 1995). Mucoid strains of *P. aeruginosa* provide a suitable model to study interactions between EPS, since they are characterized by the simultaneous production of extracellular polysaccharides and different enzymes. In particular the interaction of alginate with the extracellular lipase of *P. aeruginosa* has been studied in greater detail (WINGENDER 1990).

Evidence for an interaction between alginate and the extracellular lipase first came from in vitro experiments, demonstrating an enhancement of extracellular lipase activity of typically nonmucoid wild-type strains of *P. aeruginosa*, when the bacteria were incubated in the presence of exogenous alginates (WINGENDER AND WINKLER 1984). Although a number of different polysaccharides other than alginate were also effective in in vitro stimulation of extracellular lipase activity in *P. aeruginosa* (SCHULTE ET AL. 1982), only the observed in vitro effect of alginate seemed to be relevant under natural conditions and was considered to mimic the in vivo situation of mucoid strains, since among the lipase-stimulating polysaccharides only alginate was known to be produced by mucoid bacteria as a major slime component of biofilms. For this reason the effect of stimulation of extracellular lipase activity by alginate was investigated in more detail in both nonmucoid and mucoid strains of *P. aeruginosa* (WINGENDER AND WINKLER 1984; WINGENDER ET AL. 1987; WINGENDER 1990).

The influence of exogenous alginates on the expression of extracellular lipase activity was studied in growing and nongrowing cells of different nonmucoid wild-type strains of *P. aeruginosa*. When the bacteria were grown in liquid media for up to 24 h, extracellular lipase activities were strongly enhanced in cultures containing exogenous alginate compared to cultures without alginate (WINGENDER ET AL. 1987; WINGENDER 1990). However, the growth kinetics of the bacteria were not markedly affected by the presence of alginate in the media. Alginate had no significant influence on the activities of other extracellular enzymes, including protease, phospholipase C, and alkaline phosphatase, indicating that alginate-mediated increase in extracellular enzyme activity was specific for lipase.

Extracellular lipase stimulation by alginate was characterized in more detail under defined conditions in short-term experiments, using late-logarithmic cells of *P. aeruginosa* suspended in Tris buffer. When incubated in the presence of exogenous alginates (1 mg ml^{-1}) for 30 min, the extracellular lipase activity of the bacteria rapidly increased to maximally 23-fold compared to untreated cell suspensions. The stimulatory effect was the same, whether commercial algal alginates with different molecular masses and mannuronate to guluronate ratios or purified alginates from mucoid *P. aeruginosa* and *A. vinelandii*, respectively, were added to the bacteria (WINGENDER AND WINKLER 1984). Removal of O-acetyl groups from the bacterial alginates did not change their biological activity. The stimulation of extracellular lipase activity was dependent on de novo protein synthesis, since inhibitors of transcription (rifampicin) and translation (chloramphenicol) completely suppressed the alginate-mediated effect (WINGENDER 1990). In all experiments with growing and nongrowing bacteria, exogenous alginate was not degraded and utilized as a carbon and energy source by the bacteria nor did alginate increase extracellular lipase activity when added to cell-free culture supernatants. These observations precluded an indirect metabolic effect on the release of lipase or a direct stimulatory effect of the polysaccharides on the enzyme activity.

Originally, WINKLER AND STUCKMANN (1979) had formulated the detachment hypothesis to explain the elevated extracellular lipase activities of *Serratia marcescens* in the presence of exogenous polysaccharides other than alginate. It was assumed that the polysaccharides displaced lipase molecules from binding sites on the cell surface, thus mediating the release of cell-bound enzyme. Interestingly, a similar mechanism has also been described for mammalian lipases, which were released from endothelial cells, liver cells, and alveolar macrophages after addition of glycosaminoglycans such as heparin, heparan sulphate, and dermatan sulphate (OLIVECRONA ET AL. 1977; MAHONEY ET AL. 1982; OGATA AND HIRASAWA 1982). The release of lipase from polysaccharidic binding sites on the epithelial cell surface was mediated by the binding of the exogenous polysaccharides to lipoprotein lipase molecules (OLIVECRONA ET AL. 1977). In accordance with the detachment hypothesis (WINKLER AND STUCKMANN 1979), it was assumed that enhancement of extracellular lipase activity by alginate in *P. aeruginosa* was also due to the displacement of cell surface-bound lipase molecules through an interaction with exogenous alginate, triggering the synthesis and secretion of new enzyme molecules to occupy again the free lipase binding sites at the outer membrane (WINGENDER 1990). Lipase has been described to be associated with lipopolysaccharide (LPS) at the outer membrane of nonmucoid *P. aeruginosa*, before the enzyme is released as a lipase-LPS complex (JAEGER ET AL. 1991). Purified LPS was also able to enhance extracellular lipase activity, although its effect was less than that of exogenous alginate (WINGENDER 1990). Most of the lipase-stimulating activity of LPS resided in its polysaccharide portion, whereas the lipid A portion was only weakly active. Since alginate could be replaced by LPS as a lipase-stimulating substance, it was hypothesized that alginate acted either by displacing lipase molecules from LPS-containing binding sites on the cells surface, or by solubilizing lipase-LPS complexes from the outer membrane (WINGENDER 1990).

In mucoid strains of *P. aeruginosa*, the simultaneous production of alginate and lipase occurs (WINGENDER 1990). In view of the stimulating effect on lipase release by exogenous alginates observed in nonmucoid wild-type strains, it was assumed that alginate might also influence extracellular lipase activity in mucoid strains. When mucoid strains were compared with isogenic nonmucoid strains under conditions of alginate-slime production, the mucoid bacteria displayed significantly higher extracellular lipase activities. Several independently isolated mucoid mutants, obtained by selection for carbenicillin resistance, revealed 2- to 12-fold higher lipase activities than their nonmucoid parental wild-type strains when grown either on agar medium or in liquid culture; six out of seven clinical mucoid strains isolated from cystic fibrosis patients showed up to 9-fold higher extracellular lipase activities than their spontaneous nonmucoid revertant strains (WINGENDER 1990). Elevated extracellular lipase activities in mucoid strains were supposed to be based on the same mechanism as proposed to explain extracellular lipase stimulation by exogenous alginate in nonmucoid strains. This was supported by the observation that, under conditions of repressed endogenous alginate production, all mucoid strains were also able to respond to exogenous algal and purified bacterial alginates by a strong enhancement of extracellular lipase activities in the same way as originally discovered in nonmucoid wild-type bacteria.

Several observations indicated that direct or indirect interactions between alginate and extracellular lipase occurred that were the underlying cause for lipase release and accumulation of the enzyme in the slime matrix of mucoid *P. aeruginosa*.

From in vitro experiments using lipase preparations from *P. aeruginosa* and other microorganisms, it was concluded that alginate was able to associate with lipases with concomitant change of enzyme properties (WINGENDER ET AL. 1987):

– First, adsorption chromatography on a column of glass beads resulted in the almost complete binding of lipase to the glass surfaces; the enzyme activity could be completely eluted with the non-ionic detergent Triton X-100. However, pre-incubation of lipase in the presence of alginate prevented approximately 80% of the enzyme activity from binding to the glass matrix. It was assumed that the association of alginate with lipase prevented the enzyme from adopting a conformational state favorable for its hydrophobic interaction with the glass surface in the absence of the polysaccharide (WINGENDER ET AL. 1987). In contrast to alginate, the neutral polysaccharide dextran, which displayed no significant lipase-stimulating effect in the biological experiments, also had no influence on the elution behavior of lipase. Thus, dextran did not prevent lipase from adsorbing to the glass surface, suggesting a certain degree of specificity of the lipase-polysaccharide interactions.

- Second, lipase activity proved to be more resistant to heat inactivation at 70°C in the presence of alginate than in its absence (WINGENDER ET AL. 1987). Similarly, it was reported that the extracellular polysaccharide xanthan from *X. campestris* was also able to stabilize extracellular lipase activity from *P. aeruginosa* at 55°C, when xanthan was added to lipase-containing supernatants (LEZA ET AL. 1996). In contrast, dextran did not protect lipase from heat inactivation. These observations indicate that some polysaccharides mediate the protection of lipase from thermal denaturation, probably by forming stable polysaccharide-enzyme complexes.

- Third, an increase in lipase activity of up to 85% was observed in the presence of Triton X-100. This effect was strongly reduced by alginate, suggesting an inhibition of detergent binding to the enzyme by the polysaccharide.

- Finally, extracellular lipase activity was almost completely recovered from culture supernatants of *P. aeruginosa* by ethanolic co-precipitation with alginate, whereas ethanol treatment in the absence of alginate almost completely destroyed lipase activity. The interaction between extracellular lipase and alginate in bacterial cultures was supposed to be predominantly polar, since addition of NaCl impaired co-precipitation, whereas Triton X-100 did not. The interaction between alginate and extracellular lipase was reversible, since separation of enzyme and polysaccharide was achieved by ion-exchange chromatography on a column of DEAE-Sephadex A-25 (WINGENDER ET AL. 1987).

The in vitro binding studies demonstrated that the interactions between alginate and lipase seemed to be based on weak binding forces, which are typical of interactions between EPS in bacterial biofilms (FLEMMING 1996; MAYER ET AL. 1999). It was speculated that the extracellular lipase from *P. aeruginosa* displayed different binding properties due to conformational changes depending on whether the immediate environment was hydrophilic or hydrophobic (WINGENDER ET AL. 1987). In this respect, extracellular lipase may resemble mammalian lipoprotein lipase in possessing interdependent binding sites for substrate molecules, lipid-water interfaces, polyanions such as uronic acid-containing polysaccharides, and detergents (CRYER 1985). The ability of alginates to interact with enzymes and other proteins seems to be a general property of these polysaccharides. The formation of complexes between alginate and various proteins such as trypsin, α-chymotrypsin, albumins, and myoglobin has been demonstrated (BRAUDO ET AL. 1975; IMESON ET AL. 1977; SCHWENKE ET AL. 1977). Electrostatic interactions between alginate in the gel state and the cationic polypeptide poly-L-lysine have been described by THU ET AL. (1996). Poly-L-lysine was applied in the coating of calcium alginate beads for the stabilization and strengthening of the gel beads used for cell encapsulation. Temperature and pH stability of urease immobilized in alginate-xanthan gels were significantly higher than that of free enzyme (ELÇIN 1996), apparently due to a protective effect of the polysaccharides. Recently, the interaction of alginate from *P. aeruginosa* with human leukocyte elastase has been reported (YING ET AL. 1996). It was proposed that each elastase molecule interacted with 19 uronic acid residues of alginate, predominantly through electrostatic forces. In addition, alginate reduced the association rate between elastase and alpha 1-proteinase inhibitor, but increased its association rate with secretory leukoprotease inhibitor (YING ET AL. 1996). It was assumed that alginate in microcolonies of mucoid *P. aeruginosa* represented an important factor in determining the local concentration of leukocyte elastase and in influencing the protease-antiprotease balance in the infected lungs of cystic fibrosis patients.

These observations suggest that the interaction between certain polysaccharides, including alginate, and lipase with concomitant change of enzyme properties and location may be a common phenomenon in pro- and eukaryotic cells. Extracellular lipase stimulation by exogenous alginate originally discovered in vitro in nonmucoid strains of *P. aeruginosa* may reflect the in vivo processes of alginate slime formation of mucoid bacteria in biofilms. In mucoid bacteria, the interaction between lipase and alginate is expected to occur after secretion, since separate pathways of secretion have been described for the enzyme and the polysaccharide (Fig. 1).

Alginate-lipase complexes may temporarily remain associated with the cell surface, before being extruded into the intercellular space within the biofilm (Fig. 1). Detachment of lipase from surface binding sites by alginate from the same cell may trigger novel enzyme synthesis and secretion of the enzyme. Thus, in biofilms of mucoid bacteria, alginate production may result in an "auto-stimulatory" effect with respect to extracellular lipase formation. In the biofilm matrix, lipase may remain noncovalently associated with alginate, resulting in increased stability and accumulation of enzyme activity in the microenvironment of the slime-embedded cells. Similarly, stabilization of other extracellular enzymes due to their noncovalent interaction with extracellular polysaccharides produced by the same bacterium has been described for an agarase from a *Cytophaga* species (DUCKWORTH AND TURVEY 1969), a β-lactamase from *Bacillus cereus* (KUWABARA AND LLOYD 1971) and a protease activity within an extracellular protein-polysaccharide-lipid complex from slime-forming *Myxococcus virescens* (GNOSSPELIUS 1978). Thus, it may be a general property of extracellular polysaccharides to function as a matrix for enzymes in biofilms with concomitant changes of enzyme properties. Enhanced de novo synthesis and release of lipase as well as concentration of enzyme activity in the EPS matrix may be the reason for higher extracellular lipase activities in mucoid strains of *P. aeruginosa* than in nonmucoid strains.

In addition to the evidence for lipase-alginate interactions from in vitro experiments as described above, confirmation of lipase association with alginate surrounding the cells came from the observation that treatment of mucoid bacteria with a purified alginate lyase resulted in an additional release of lipase activity without killing the cells (WICKER-BÖCKELMANN ET AL. 1987). This effect was due to the alginate-degrading activity of the enzyme, since it was shown that (i) the enzyme effectively dissolved the slime of mucoid strains, (ii) heat inactivation of alginate lyase abolished the lipase-releasing effect, and (iii) the enzyme was significantly less effective in nonmucoid strains of *P. aeruginosa* (Fig. 2).

In summary, the transition of the nonmucoid planktonic state to the mucoid biofilm phenotype in *P. aeruginosa* seems to be accompanied by alterations in synthesis and/or secretion of extracellular enzymes. These processes are at least partially influenced by the interaction between alginate and enzyme. This interaction constitutes one of several factors determining the ratio between cell-bound and cell-free enzyme molecules and thus contributes to the location and level of enzyme activities within the biofilm matrix.

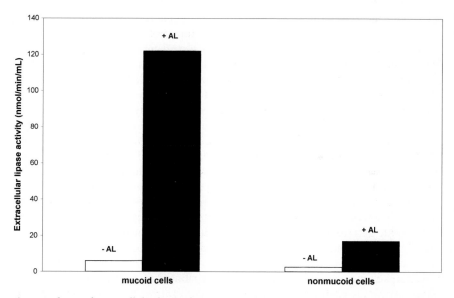

Fig. 2. Release of extracellular lipase from a mucoid strain of *P. aeruginosa* and an isogenic nonmucoid revertant strain in the absence or presence of an exogenous alginate lyase (WICKER-BÖCKELMANN ET AL. 1987). Unwashed bacteria from the stationary growth phase were suspended in 10 mmol l^{-1} phosphate buffer, pH 5.8, containing 10 mmol l^{-1} MgCl$_2$, and incubated at 37 °C for 3 h without (–AL) or with (+AL) the addition of an alginate lyase. Cell-free supernatants were assayed for lipase activity using emulsified *p*-nitrophenlypalmitate as substrate. The alginate lyase (endomannuronidase) was purified from *Bacillus circulans* (WICKER-BÖCKELMANN ET AL. 1987)

4
Consequences of Enzyme-Polysaccharide Interactions in Biofilms

It remains to be established if polysaccharide-mediated enhancement and stabilization of extracellular enzyme activities also apply to enzymes other than
those of *P. aeruginosa* and if this phenomenon is typical of the biofilm mode of
growth. Occasional reports in the literature indicate that growth in a biofilm
markedly influences the expression, stabilization, and accumulation of extracellular enzyme activities in the microenvironment (EPS matrix) of the immobilized cells. Extracellular protease activity of *Staphylococcus epidermidis* studied at varying specific growth rates was generally higher by biofilms than by
planktonic populations (EVANS ET AL. 1994). With increasing growth rates, soluble protease activity increased as production of extracellular polysaccharide
decreased, presumably due to a less effective entrapment of protease within the
polysaccharide matrix of the biofilm (EVANS ET AL. 1994). In biofilms of
P. aeruginosa, the bacteria revealed lower levels of β-lactamase induction by β-
lactam antibiotics than planktonic populations (GIWERCMAN ET AL. 1991). One
reason for this effect was supposed to be the trapping and accumulation of β-
lactamase enzyme within the biofilm to such levels that afforded protection of
the sessile bacteria from β-lactam antibiotics by inactivating them.

Extracellular enzymes can form stable complexes with humic compounds
(BURNS 1989), which are abundant in aquatic and terrestrial ecosystems and
may be present in significant amounts in soil and water biofilms. In soil biofilms, humic-enzyme complexes, but also enzyme complexes with clay particles,
were supposed to contribute significantly to total enzyme activity, giving the
soil environment a persistent extracellular catalytic activity independent of the
existing microflora (BURNS 1989). Complexes between enzymes and humic
matter from soil have been reported to be extremely resistant to thermal denaturation, dehydration, and proteolysis (BURNS 1989). The formation of poly-
phenolic-enzyme complexes can partially or entirely inactivate enzymes in a reversible manner (WETZEL 1991). It was proposed that, in these associations, enzymes are protected from the environment, are stored in a suppressed but
chemically active state, and can be reactivated at the same or displaced sites to
full enzyme activity (WETZEL 1991). Thus, it can be assumed that potential enzyme activities in humic acid-containing biofilms may be higher than normally
observed.

Perhaps some of the various functions attributed to polysaccharides in biofilms may be the retention, accumulation, stabilization, and modulation of enzyme activities within biofilms due to the nonenzymatic interaction between
enzyme proteins and polysaccharide polymers. This has occasionally been hypothesized in the literature, although without direct experimental evidence. In
a structural-functional model for river epilithon, LOCK ET AL. (1984) suggested
the polysaccharide matrix to be a site for attached enzymes analogous to en-
zyme-humus complexes in soil. The advantages to microorganisms in aquatic
biofilms would be the availability of low-molecular-weight reaction products in
the immediate vicinity of the immobilized cells, the long-term existence of different enzymes from the same and previous generations excluding the require-

ment of novel enzyme synthesis, and the initiation of enzyme induction in adjacent cells by high levels of products released by the immobilized enzyme (LOCK ET AL. 1984). The stabilization of extracellular enzymes due to their interaction with matrix polysaccharides may result in persistence of enzyme capacity within the biofilm, that is independent of the usual forms of regulation and control of enzyme synthesis and secretion. Thus, enzyme activities may be decoupled from their producer cells. This may be one reason why the activities of certain enzymes were found not to be correlated with parameters for cell or biomass densities in biofilms (JONES AND LOCK 1989; SINSABAUGH ET AL. 1991). The reservoir of extracellular degradative enzymes within the polysaccharide matrix was assumed to buffer the biofilm community to sudden changes in the composition and concentration of dissolved organic matter in the bulk water phase (JONES AND LOCK 1989). The accumulation and stabilization of polysaccharide-bound enzymes within the biofilm matrix seem to be essential processes that have to be considered in the understanding of structure and function of biofilms in different ecosystems.

Until now, secretion has mostly been studied separately for selected proteins and polysaccharides under conditions of laboratory cultures. It will be necessary to develop integrated experimental approaches allowing one to look at secretion of proteins and polysaccharides in parallel and under natural growth conditions in a biofilm environment. Future research should concentrate on answering the following questions:

1. Are those genes co-regulated which encode components of the protein and polysaccharide secretion systems?
2. Which components can influence such a co-regulation in a biofilm situation?
3. Would inhibitors of secretion also prevent biofilm formation?
4. What are the ecological consequences of the interactions of enzymes with other EPS components within biofilms composed of natural mixed populations?

References

Akatsuka H, Kawai E, Omori K, Shibatani T (1995) The three genes *lipB*, *lipC*, and *lipD* involved in the extracellular secretion of the *Serratia marcescens* lipase which lacks an N-terminal signal peptide. J Bacteriol 177:6381–6389

Akatsuka H, Binet R, Kawai E, Wandersman C, Omori K (1997) Lipase secretion by bacterial hybrid ATP-binding cassette exporters: molecular recognition of the lipBCD, PrtDEF, and HasDEF exporters. J Bacteriol 179:4754–4760

Allison DG, Ruiz B, SanJose C, Jaspe A, Gilbert P (1998) Extracellular products as mediators of the formation and detachment of *Pseudomonas fluorescens* biofilms. FEMS Microbiol Lett 167:179–184

Becker A, Katzen F, Puhler A, Ielpi L (1998) Xanthan gum biosynthesis and application: a biochemical/genetic perspective. Appl Microbiol Biotechnol 50:145–152

Beveridge TJ, Makin SA, Kadurugamuwa JL, Li Z (1997) Interactions between biofilms and the environment. FEMS Microbiol Rev 20:291–303

Binet R, Létoffé S, Ghigo JM, Delepelaire P, Wandersman C (1997) Protein secretion by Gram-negative bacterial ABC exporters – a review. Gene 192:7–11

Bitter W, Koster M, Latijnhouwers M, de Cock H, Tommassen J (1998) Formation of oligomeric rings by XcpQ and PilQ, which are involved in protein transport across the outer membrane of *Pseudomonas aeruginosa*. Mol Microbiol 27:209–219

Blenkinsopp SA, Lock MA (1990) The measurement of electron transport system activity in river biofilms. Wat Res 24:441–445

Boyd A, Chakrabarty AM (1994) Role of alginate lyase in cell detachment of *Pseudomonas aeruginosa*. J Bacteriol 60:2355–2359

Braatz JA, Heath EC (1974) The role of polysaccharide in the secretion of protein by *Micrococcus sodonensis*. J Biol Chem 249:2536–2547

Braudo EE, Strelzowa SA, Tolstogusow WB (1975) Einfluß von sauren Polysacchariden auf die Eigenschaften der Pankreasproteinasen. Nahrung 19:903–910

Braun P, Tommassen J, Filloux A (1996) Role of the propeptide in folding and secretion of elastase of *Pseudomonas aeruginosa*. Mol Microbiol 19:297–306

Burns RG (1989) Microbial and enzymic activities in soil biofilms. In: Characklis WG, Wilderer PA (eds) Structure and function of biofilms. Wiley, Chichester, pp 333–349

Chróst RJ (1991) Environmental control of the synthesis and activity of aquatic microbial ectoenzymes. In: Chróst RJ (ed) Microbial enzymes in aquatic environments. Springer, Berlin Heidelberg New York, pp 29–59

Confer DR, Logan BE (1997) Molecular weight distribution of hydrolysis products during biodegradation of model macromolecules in suspended and biofilm cultures. I. Bovine serum albumin. Wat Res 31:2127–2136

Confer DR, Logan BE (1998) Location of protein and polysaccharide hydrolytic activity in suspended and biofilm wastewater cultures. Wat Res 32:31–38

Cornelis GR, Wolf-Watz H (1997) The *Yersinia* Yop virulon: a bacterial system for subverting eukaryotic cells. Mol Microbiol 23:861–867

Cryer A (1985) Lipoprotein lipase: molecular interactions of the enzyme. Biochem Soc Trans 13:27–28

De Rosa S, Sconza F, Volterra L (1998) Biofilm amount estimation by fluorescein diacetate. Wat Res 32:2621–2626

Duckworth M, Turvey JR (1969) An extracellular agarase from a *Cytophaga* species. Biochem J 113:139–142

Duong F, Soscia A, Lazdunski A, Murgia M (1994) The *Pseudomonas fluorescens* lipase has a C-terminal secretion signal and is secreted by a three-component bacterial ABC-exporter system. Mol Microbiol 11:1117–1126

Duong F, Eichler J, Price A, Leonard RM, Wickner W (1997) Biogenesis of the Gram-negative bacterial envelope. Cell 91:567–573

Elçin YM (1996) Encapsulation of urease enzyme in xanthan-alginate spheres. Biomaterials 16:1157–1161

Ertesvåg H, Valla S (1998) Biosynthesis and applications of alginates. Polym Degrad Stabil 59:85–91

Evans E, Brown MRW, Gilbert P (1994) Iron chelator, exopolysaccharide and protease production in *Staphylococcus epidermidis*: a comparative study of the effects of specific growth rate in biofilm and planktonic culture. Microbiology 140:153–157

Filloux A, Michel G, Bally M (1998) GSP-dependent protein secretion in Gram-negative bacteria: the Xcp system of *Pseudomonas aeruginosa*. FEMS Microbiol Rev 22:177–198

Flemming H-C (1996) The forces that keep biofilms together. In: Sand W (ed) Biodeterioration and biodegradation. Dechema Monographs 133, VCH, Weinheim, pp 311–316

Fletcher J, Nair S, Poole S, Henderson B, Wilson M (1998) Cytokine degradation by biofilms of *Porphyromonas gingivalis*. Curr Microbiol 36:216–219

Frølund B, Griebe T, Nielsen PH (1995) Enzymatic activity in the activated-sludge floc matrix. Appl Microbiol Biotechnol 43:755–761

Gacesa P (1998) Bacterial alginate biosynthesis – recent progress and future prospects. Microbiol 144:1133–1143

Giwercman B, Jensen ET, Høiby N, Kharazmi A, Costerton JW (1991) Induction of β-lactamase production in *Pseudomonas aeruginosa* biofilm. Antimicrob Agents Chemother 35:1008–1010

Gnosspelius G (1978) Myxobacterial slime and proteolytic activity. Arch Microbiol 116:51–59

Govan JRW (1990) Characteristics of mucoid *Pseudomonas aeruginosa* in vitro and in vivo. In: Gacesa P, Russell NJ (eds) (1990) Pseudomonas infection and alginates. Biochemistry, genetics and pathology. Chapman and Hall, London, pp 50–75

Govan JRW, Deretic V (1996) Microbial pathogenesis in cystic fibrosis: mucoid *Pseudomonas aeruginosa* and *Burkholderia cepacia*. Microbiol Rev 60:539–574

Grabert E, Wingender J, Winkler, UK (1990) An outer membrane protein characteristic of mucoid strains of *Pseudomonas aeruginosa*. FEMS Microbiol Lett 68:83–88

Grobe S, Wingender J, Trüper HG (1995) Characterization of mucoid *Pseudomonas aeruginosa* strains isolated from technical water systems. J Appl Bacteriol 79:94–102

Hueck CJ (1998) Type III protein secretion systems in bacterial pathogens of animals and plants. Microbiol Mol Biol Rev 62:379–433

Imeson AP, Ledward DA, Mitchell JR (1977) On the nature of the interaction between some anionic polysaccharides and proteins. J Sci Food Agric 28:661–668

Jaeger K-E, Kharazmi A, Høiby N (1991) Extracellular lipase of *Pseudomonas aeruginosa*: biochemical characterization and effect on human neutrophil and monocyte function in vitro. Microb Pathog 10:173–182

Jaeger K-E, Ransac S, Dijkstra BW, Colson C, van Heuvel M, Misset O (1994) Bacterial lipases. FEMS Microbiol Rev 15:29–63

Jaeger K-E, Schneidinger B, Liebeton K, Haas D, Reetz MT, Philippou S, Gerritse G, Ransac S, Dijkstra BW (1996) Lipase of *Pseudomonas aeruginosa*. In: Nakazawa T, Furukawa K, Haas D, Silver S (eds) Molecular biology of pseudomonads. ASM Press, Washington, DC, pp 319–330

Jahn A, Nielsen PH (1996) Extraction of extracellular polymeric substances (EPS) from biofilms using a cation exchange resin. Wat Sci Tech 32:157–164

Jones SE, Lock MA (1989) Hydrolytic extracellular enzyme activity in heterotrophic biofilms from two contrasting streams. Freshwat Biol 22:289–296

Jones SE, Lock MA (1991) Peptidase activity in river biofilms by product analysis. In: Chróst RJ (ed) Microbial enzymes in aquatic environments. Springer, Berlin Heidelberg New York, pp 144–154

Kadurugamuwa JL, Beveridge TJ (1995) Virulence factors are released from *Pseudomonas aeruginosa* in association with membrane vesicles during normal growth and exposure to gentamic: a novel mechanism of enzyme secretion. J Bacteriol 177:3998–4008

Kawai E, Akatsuka H, Idei A, Shibatani T, Omori K (1998) *Serratia marcescens* S-layer protein is secreted extracellularly via ATP-binding cassette exporter, the Lip system. Mol Microbiol 27:941–952

Kuwabara S, Lloyd PH (1971) Protein and carbohydrate moieties of a preparation of β-lactamase II. Biochem J 124:215–220

Laurent P, Servais P (1995) Fixed bacterial biomass estimated by potential exoproteolytic activity. Can J Microbiol 41:749–752

Lemmer H, Roth D, Schade M (1994) Population density and enzyme activities of heterotrophic bacteria in sewer biofilms and activated sludge. Wat Res 28:1341–1346

Leza A, Palmeros B, García JO, Galindo E, Soberón-Chávez G (1996) *Xanthomonas campestris* as a host for the production of recombinant *Pseudomonas aeruginosa* lipase. J Ind Microbiol 16:22–28

Li X, Tetling S, Winkler UK, Jaeger K-E, Benedik MJ (1995) Gene cloning, sequence analysis, purification and secretion by *Escherichia coli* of an extracellular lipase from *Serratia marcescens*. Appl Environ Microbiol 61:2674–2680

Li Z, Clarke AJ, Beveridge TJ (1998) Gram-negative bacteria produce membrane vesicles which are capable of killing other bacteria. J Bacteriol 180:5478–5483

Lock MA, Wallace RR, Costerton JW, Ventullo RM, Charlton SE (1984) River epilithon: toward a structural-functional model. Oikos 42:10–22

Mahoney EM, Khoo JC, Steinberg D (1982) Lipoprotein lipase secretion by human monocytes and rabbit alveolar macrophages in culture. Proc Natl Acad Sci USA 79:1639–1642

May TB, Chakrabarty AM (1994) *Pseudomonas aeruginosa*: genes and enzymes of alginate synthesis. Trends Microbiol 2:151–157

Mayer C, Moritz R, Kirschner C, Borchard W, Maibaum R, Wingender J, Flemming H-C (1999) The role of intermolecular interactions: studies on model systems for bacterial biofilms. Int J Biol Macromol (in press)

Missiakis D, Raina S (1997) Protein folding in the bacterial periplasm. J Bacteriol 179: 2465–2471

Ogata F, Hirasawa Y (1982) Heparin-released triglyceride lipase from Chang liver cells. Biochem Biophys Res Commun 106:397–399

Olivecrona R, Bengtsson G, Marklund S-E, Lindahl U, Höök M (1977) Heparin-lipoprotein lipase interactions. Fed Proc 36:60–65

Priest FG (1992) Enzymes, extracellular. In: Lederberg J (ed) Encyclopedia of microbiology. Academic Press, San Diego, pp 81–93

Pugsley AP (1993) The complete secretory pathway in Gram-negative bacteria. Microbiol Rev 57:50–108

Raina S, Missiakis D (1997) Making and breaking disulfide bonds. Annu Rev Microbiol 51:179–202

Rehm BHA, Valla S (1997) Bacterial alginates: biosynthesis and applications. Appl Microbiol Biotechnol 48:281–288

Rehm BHA, Boheim G, Tommassen J, Winkler UK (1994) Overexpression of *algE* in *Escherichia coli*: subcellular localization, purification, and ion channel properties. J Bacteriol 176:5639–5647

Roberts IS (1995) Bacterial polysaccharides in sickness and in health. Microbiology 141: 2023–2031

Roberts IS (1996) The biochemistry and genetics of capsular polysaccharide production in bacteria. Annu Rev Microbiol 50:285–315

Salmond GPC, Reeves PJ (1993) Membrane traffic wardens and protein secretion in Gram-negative bacteria. Trends Biochem Sci 18:7–12

Schulte G, Bohne L, Winkler U (1982) Glycogen and various other polysaccharides stimulate the formation of exolipase by *Pseudomonas aeruginosa*. Can J Microbiol 28: 636–642

Schwenke KD, Kracht E, Mieth G, Freimuth U (1977) Proteingewinnung unter Einsatz von komplexbildenden Stoffen. 2. Mitt. Zur Bildung unlöslicher Komplexe zwischen Sonnenblumenalbuminen und Alginat bzw. Pektin. Nahrung 21:395–403

Servais P, Laurent P, Randon G (1995) Comparison of the bacterial dynamics in various French distribution systems. Aqua 44:10–17

Simonen M, Palva J (1993) Protein secretion in *Bacillus* species. Microbiol Rev 57:109–137

Sinsabaugh RL, Repert D, Weiland T, Golladay SW, Linkins AE (1991) Exoenzyme accumulation in epilithic biofilms. Hydrobiol 222:29–37

Sutherland IW (1990) Biotechnology of microbial exopolysaccharides. Cambridge University Press, Cambridge

Teuber M, Brodisch KEU (1977) Enzymatic activities of activated sludge. European J Appl Microbiol 4:185–194

Thu B, Bruheim P, Espevik T, Smidsrød O, Soon-Shiong P, Skjåk-Bræk G (1996) Alginate polycation microcapsules. I. Interaction between alginate and polycation. Biomaterials 17: 1031–1040

Tommassen J, Filloux A, Bally M, Murgier M, Lazdunski A (1992) Protein secretion in *Pseudomonas aeruginosa*. FEMS Microbiol Rev 103:73–90

Urbain V, Block JC, Manem J (1993) Bioflocculation in activated sludge: an analytical approach. Wat Res 27:829–838

Van Heijne G (1986) A new method for predicting signal sequence cleavage sites. Nucleic Acids Res 14:4683–4690

Vetter YA, Deming JW, Jumars PA, Krieger-Brockett BB (1998) A predictive model of bacterial foraging by means of freely released extracellular enzymes. Microb Ecol 36:75–92

Wetzel RG (1991) Extracellular enzymatic interactions: storage, redistribution, and interspecific communication. In: Chróst RJ (ed) Microbial enzymes in aquatic environments. Springer, Berlin Heidelberg New York, pp 6–28

Wicker-Böckelmann U, Wingender J, Winkler UK (1987) Alginate lyase releases cell-bound lipase from mucoid strains of *Pseudomonas aeruginosa*. Zentr Bakt Parasit Infekt Hyg A266:379–389

Wingender J (1990) Interactions of alginate with exoenzymes. In: Gacesa P, Russell NJ (eds) Pseudomonas infection and alginates. Chapman and Hall, London, pp 160–180

Wingender J, Winkler UK (1984) A novel biological function of alginate in *Pseudomonas aeruginosa* and its mucoid mutants: stimulation of exolipase. FEMS Microbiol Lett 21: 63–69

Wingender J, Volz S, Winkler UK (1987) Interaction of extracellular *Pseudomonas* lipase with alginate and its potential use in biotechnology. Appl Microbiol Biotechnol 27:139–145

Winkler UK, Stuckmann M (1979) Glycogen, hyaluronate, and some other polysaccharides greatly enhance the formation of exolipase by *Serratia marcescens*. J Bacteriol 138: 663–670

Wuertz S, Pfleiderer P, Kriebitzsch K, Späth R, Griebe T, Coello-Oviedo D, Wilderer PA, Flemming H-C (1998) Extracellular redox activity in activated sludge. Wat Sci Tech 37: 379X–384

Xun L, Mah RA, Boone DR (1990) Isolation and characterization of disaggregatase from *Methanosarcina mazei* LYC. Appl Environ Microbiol 56:3693–3698

Ying QL, Kemme M, Simon SR (1996) Alginate, the slime exopolysaccharide of *P. aeruginosa*, binds human leukocyte elastase, retards inhibition by alpha 1-proteinase inhibitor, and accelerates inhibition by secretory leukoprotease inhibitor. Am J Respir Cell Mol Biol 15: 283–291

Subject Index

Printing (Computer to Film): Saladruck, Berlin
Binding: Stürtz AG, Würzburg

DATE DUE

DEMCO INC 38-2971